Lecture Notes in Computer Science 15905

Founding Editors

Gerhard Goos
Juris Hartmanis

AF172537

Editorial Board Members

The series Lecture Notes in Computer Science (LNCS), including its subseries Lecture Notes in Artificial Intelligence (LNAI) and Lecture Notes in Bioinformatics (LNBI), has established itself as a medium for the publication of new developments in computer science and information technology research, teaching, and education.

LNCS enjoys close cooperation with the computer science R & D community, the series counts many renowned academics among its volume editors and paper authors, and collaborates with prestigious societies. Its mission is to serve this international community by providing an invaluable service, mainly focused on the publication of conference and workshop proceedings and postproceedings. LNCS commenced publication in 1973.

Michael H. Lees · Wentong Cai ·
Siew Ann Cheong · Yi Su · David Abramson ·
Jack J. Dongarra · Peter M. A. Sloot
Editors

Computational Science – ICCS 2025

25th International Conference
Singapore, Singapore, July 7–9, 2025
Proceedings, Part III

 Springer

Editors
Michael H. Lees ⓘ
University of Amsterdam
Amsterdam, The Netherlands

Siew Ann Cheong ⓘ
Nanyang Technological University
Singapore, Singapore

David Abramson ⓘ
The University of Queensland
Brisbane, QLD, Australia

Peter M. A. Sloot ⓘ
University of Amsterdam
Amsterdam, The Netherlands

Wentong Cai ⓘ
Nanyang Technological University
Singapore, Singapore

Yi Su
Institute for High Performance Computing
A*STAR
Singapore, Singapore

Jack J. Dongarra ⓘ
The University of Tennessee
Knoxville, TN, USA

ISSN 0302-9743 ISSN 1611-3349 (electronic)
Lecture Notes in Computer Science
ISBN 978-3-031-97631-5 ISBN 978-3-031-97632-2 (eBook)
https://doi.org/10.1007/978-3-031-97632-2

Preface

Welcome to the 25th International Conference on Computational Science (ICCS - https://www.iccs-meeting.org/iccs2025/), held on July 7–9, 2025 at Nanyang Technological University (NTU), Singapore.

This 25th edition in Singapore marked our return to a fully in-person event. Although the challenges of our present times are manifold, we have always tried our best to keep the ICCS community as dynamic, creative, and productive as possible. We are proud to present the proceedings you are reading as a result.

ICCS 2025 was jointly organized by Nanyang Technological University, the A*STAR Institute of High Performance Computing, the University of Amsterdam, and the University of Tennessee.

Considered one of the most developed countries in the world, the island country of Singapore is a major aviation, financial, and maritime shipping hub in Asia. Singapore is multilingual, multiethnic, and multicultural, and as such a very popular, safe tourism destination.

NTU Singapore is a public university ranked among the world's best, with 35,000 students, and home to the world-renowned autonomous National Institute of Education and S. Rajaratnam School of International Studies. In addition to many research institutes and centers at the university, college, and school levels, NTU also hosts two National Research Foundation (NRF) and Ministry of Education (MOE) Research Centers of Excellence, namely the Singapore Center for Environmental Life Sciences Engineering (SCELSE) and the Institute for Digital Molecular Analytics & Science (IDMxS), and 11 Corporate Labs in partnership with various industries. ICCS 2025 took place on the One-north campus.

The Institute of High Performance Computing (IHPC) is a national research institute under the Agency for Science, Technology and Research (A*STAR), dedicated to advancing science and technology through computational modeling, simulation, AI, and high-performance computing. With a multidisciplinary team of scientists and engineers, IHPC drives innovation across sectors such as advanced manufacturing, microelectronics, sustainability, maritime, and biomedical sciences. It leads Singapore's national efforts in hybrid quantum-classical computing and digital twin platforms, and partners extensively with industry and government agencies to translate deep tech into real-world impact.

The International Conference on Computational Science is an annual conference that brings together researchers and scientists from mathematics and computer science as basic computing disciplines, as well as researchers from various application areas who are pioneering computational methods in sciences such as physics, chemistry, life sciences, engineering, arts, and humanitarian fields, to discuss problems and solutions in the area, identify new issues, and shape future directions for research.

The ICCS proceedings series has become a primary intellectual resource for computational science researchers, defining and advancing the state of the art in this field.

We are proud to note that this 25th edition, with 23 workshops (the Workshops on Computational Science), one co-located event (the Asian Network of Complexity Scientists Workshop), and over 300 participants, kept to the tradition and high standards of previous editions.

The theme for 2025, "**Making Complex Systems tractable through Computational Science**", highlighted the role of Computational Science in tackling the complex problems of today and tomorrow. This conference was a unique event, focusing on recent developments in scalable scientific algorithms; advanced software tools; computational grids; advanced numerical methods; and novel application areas. These innovative novel models, algorithms, and tools drive new science through efficient application in physical systems, computational and systems biology, environmental systems, finance, and others.

ICCS is well known for its lineup of keynote speakers. The keynotes for 2025 were:

- **Johan Bollen**, Indiana University Bloomington, USA
- **Jack Dongarra**, University of Tennessee, USA
- **Mile Gu**, Nanyang Technological University, Singapore
- **Erika Fille Legara**, Center for AI Research|Asian Institute of Management, Philippines
- **Yong-Wei Zhang**, Institute of High Performance Computing, A*STAR, Singapore

This year, the main track of ICCS registered 162 submissions, of which 64 were accepted as full papers, and 52 as short papers. There were on average 2.4 single-blind reviews per submission.

We would like to thank all committee members from the main track and workshops for their contribution to ensuring a high standard for the accepted papers. We would also like to thank *Springer, Elsevier,* and *Intellegibilis* for their support. Finally, we appreciate all the local organizing committee members for their hard work in preparing this conference.

We hope you enjoyed the conference and the beautiful country of Singapore.

July 2025

Michael H. Lees
Wentong Cai
Siew Ann Cheong
Yi Su
David Abramson
Jack J. Dongarra
Peter M. A. Sloot

Organization

Program Committee Chairs

Peter M. A. Sloot	University of Amsterdam, The Netherlands
Jack J. Dongarra	University of Tennessee, USA
Michael H. Lees	University of Amsterdam, The Netherlands
David Abramson	University of Queensland, Australia
Wentong Cai	Nanyang Technological University, Singapore
Cheong Siew Ann	Nanyang Technological University, Singapore
Su Yi	Institute for High Performance Computing, A*Star, Singapore

Local Program Committee at NTU Singapore

Ee Hou Yong	Nanyang Technological University, Singapore
Kang Hao	Nanyang Technological University, Singapore

Publicity Chairs

Leonardo Franco	University of Málaga, Spain
Muhamad Azfar Ramli	Institute for High Performance Computing, A*Star, Singapore

Impact Chair

Valeria Krzhizhanovskaya	University of Amsterdam, The Netherlands

Outreach Chair

Alfons Hoekstra	University of Amsterdam, The Netherlands

Program Committee Chair – Workshops on Computational Science

Maciej Paszynski   AGH University of Krakow, Poland

Program Committee – Workshops on Computational Science

Amanda S. Barnard   Australian National University, Australia
Yongjie Jessica Zhang   Carnegie Mellon University, USA

Reviewers

Julen Alvarez-Aramberri   University of the Basque Country, Spain
Philipp Andelfinger   Nanyang Technological University, Singapore
Adrian Bekasiewicz   Gdańsk University of Technology, Poland
Nik Brouw   University of Amsterdam, Netherlands
Roland V. Bumbuc   University of Amsterdam, Netherlands
Wentong Cai   Nanyang Technological University, Singapore
Pedro J. S. Cardoso   Universidade do Algarve, Portugal
Eddy Caron   ENS-Lyon/Inria/LIP, France
Lock-Yue Chew   Nanyang Technological University, Singapore
Ana Cortes   Universitat Autònoma de Barcelona, Spain
Daan Crommelin   CWI Amsterdam, Netherlands
Carlo Cunha   Northern Arizona University, USA
Bartosz Czaplewski   Gdańsk University of Technology, Poland
Venkata Rupesh Kumar Dabbir   Google LLC, USA
Eric Dignum   University of Amsterdam, Netherlands
Vitor Duarte   Universidade NOVA de Lisboa, Portugal
Mariusz Dzwonkowski   Medical University of Gdańsk, Poland
Nahid Emad   Paris-Saclay University, France
Roberto R. Expósito   Universidade da Coruña, CITIC, Spain
Ruy Freitas Reis   Universidade Federal de Juiz de Fora, Brazil
Wlodzimierz Funika   AGH University of Krakow, Poland
Victoria Garibay   University of Amsterdam, Netherlands
Paweł Gepner   Warsaw University of Technology, Poland
Alex Gerbessiotis   New Jersey Institute of Technology, USA
Maziar Ghorbani   Brunel University London, UK
Konstantinos Giannoutakis   University of Macedonia, Greece
Jorge González-Domínguez   Universidade da Coruña, Spain
Yuriy Gorbachev   Soft-Impact LLC, Russia
Michael Gowanlock   Northern Arizona University, USA

George Gravvanis Democritus University of Thrace, Greece
Derek Groen Brunel University London, UK
Loïc Guégan UiT the Arctic University of Norway, France
Rafiazka Hilman University of Amsterdam, Netherlands
Cillian Hourican University of Amsterdam, Netherlands
Neil Huynh Institute of High Performance Computing,
　　　　　A*STAR, Singapore
Alireza Jahani Brunel University London, UK
Song Jie Institute of High Performance Computing,
　　　　　A*STAR, Singapore
Zhong Jin Computer Network Information Center, Chinese
　　　　　Academy of Sciences, China
David Johnson Uppsala University, Sweden
Takahiro Katagiri Nagoya University, Japan
Sotiris Kotsiantis University of Patras, Greece
Sergey Kovalchuk Huawei, Russia
Valeria Krzhizhanovskaya University of Amsterdam, Netherlands
Michael Kuhn Otto von Guericke University Magdeburg,
　　　　　Germany
Jaeyoung Kwak Nanyang Technological University, Singapore
Michael Lees University of Amsterdam, Netherlands
Malcolm Low Singapore Institute of Technology, Singapore
Lukasz Madej AGH University of Science and Technology,
　　　　　Poland
Tomas Margalef Universitat Autònoma de Barcelona, Spain
Paula Martins University of Algarve, Portugal
Pedro Medeiros Universidade Nova de Lisboa, Portugal
Isaak Mengesha University of Amsterdam, Netherlands
Marianna Milano Università Magna Græcia di Catanzaro, Italy
Dhruv Mittal University of Amsterdam, Netherlands
Francisco J. Moreno-Barea Universidad de Málaga, Spain
Marcin Paprzycki IBS PAN and WSM, Poland
Giulia Pederzani Universiteit van Amsterdam, Netherlands
Alberto Perez de Alba Ortiz University of Amsterdam, Netherlands
Dana Petcu West University of Timisoara, Romania
Jolan Philippe IMT Atlantique, France
Dirk Pleiter University of Groningen, Netherlands
Alexander Pyayt EPAM Systems, Russia
Rick Quax University of Amsterdam, Netherlands
Muhamad Azfar Ramli Institute of High Performance Computing,
　　　　　A*STAR, Singapore
Amir Raoofy Technical University of Munich, Germany

Contents – Part III

ICCS 2025 Main Track Full Papers

Control Synthesis of Homogeneous Approximations of Nonlinear Systems 3
Marcin Korzeń, Jarosław Woźniak, Grigory Sklyar,
and Mateusz Firkowski

A Robust Ensemble Malware Detector Against Powerful Adversaries 17
Shangyuan Zhuang, Wei Zhang, Fengqi Liu, Jiyan Sun, Yinlong Liu,
Liru Geng, and Wei Ma

Physics-Aware Compression of Plasma Distribution Functions
with GPU-Accelerated Gaussian Mixture Models 33
Andong Hu, Luca Pennati, Ivy Peng, and Stefano Markidis

Hierarchical Structural Information – Theory and Applications 48
Marzena Bielecka, Andrzej Bielecki, Aleksander Suchorab,
and Igor Wojnicki

A Connectionist Approach to Federated Digital Twins 60
Christian Vergara-Marcillo, Rami Bahsoon, Nikos Tziritas,
and Georgios Theodoropoulos

Low Latency Recoding CORDIC Algorithm for FPGA Implementation 75
Pawel Poczekajlo, Leonid Moroz, Ewa Deelman, Michela Taufer,
Pawel Gepner, and Jerzy Krawiec

Global Optimization of Microwave Circuits Using Dimensionality
Reduction and Multi-fidelity EM Simulations 90
Slawomir Koziel, Anna Pietrenko-Dabrowska, and Leifur Leifsson

Understanding the Limitations of Deep Transformer Models for Sea Ice
Forecasting 104
Julia Borisova, Andrey Kuznetsov, Gleb Solovev, and Nikolay O. Nikitin

Microscopic Binary Engagement Model 119
Marco Lemos, Pedro J. S. Cardoso, and João M. F. Rodrigues

Dead Gate Elimination 135
Yanbin Chen, Christian B. Mendl, and Helmut Seidl

Advances in Adapting Memory-Bound CFD Computations to RISC-V
Multicore Architecture 151
   Tomasz Olas, Lukasz Szustak, Roman Wyrzykowski, Mateusz Olas,
   and Marco Lapegna

Detecting Potential HIV Inhibitors Using the Cross Siamese Network 167
   Konrad Witkowski, Agnieszka Duraj, and Piotr S. Szczepaniak

Generation of Quality Green's Function Libraries in Complex
Three-Dimensional Crustal Structures by Adaptive Mesh Refinement 183
   Kai Nakao, Hideaki Ito, Tsuyoshi Ichimura, Kohei Fujita,
   Lalith Wijerathne, and Muneo Hori

An Iterative Scheme for the Solidification Benchmark Modeling 198
   Xiaoyu Feng, Huangxin Chen, Bo Yu, and Shuyu Sun

Robust, Efficient, and Long-Time Accurate Schemes to Simulate Gas
Storage in Geological Formation 213
   Huangxin Chen, Yuxiang Chen, Jisheng Kou, Shuyu Sun, Dunhui Xiao,
   Xuejun Xu, Haitao Yu, Tao Zhang, and Xiaoying Zhuang

A Hybrid Approach for Medical Deepfake Detection Using Depth-Wise
Convolutions in Vision Transformer and Frequency Domain Analysis 228
   R. Dhanyalakshmi, Alexander Zakharov, Natalia Romanchuk,
   J. Anitha, and Jude Hemanth

Static Load Balancing for Molecular-Continuum Flow Simulations
with Heterogeneous Particle Systems and on Heterogeneous Hardware 241
   Amartya Das Sharma, Louis Viot, Piet Jarmatz, Hauke Preuß,
   and Philipp Neumann

Joint Spatial-Temporal Representation for Host Intrusion Detection System 256
   Hao Li, Zehui Wang, Shang Shang, Zhengwei Jiang, Qiuyun Wang,
   Fangli Ren, and Baoxu Liu

Will It Blend? Mixing Numerical and Machine-Learned Physics Quantities
for Accurate on-the-Fly Surrogate Modeling 270
   Michael Tynes, Kyle Chard, Ian Foster, and Logan Ward

Biological Community Detection with Graph Neural Network
and Network Curvature Analysis on Gene Co-expression Networks 285
   Marianna Milano, Pietro Cinaglia, Mario Cannataro,
   and Pietro Hiram Guzzi

Incorporating Performance Ordering in MCDA: A Study of the Frobenius
SPOTIS Method   297
   Andrii Shekhovtsov, Jean Dezert, and Wojciech Sałabun

Author Index   311

ICCS 2025 Main Track Full Papers

Control Synthesis of Homogeneous Approximations of Nonlinear Systems

Marcin Korzeń⬛, Jarosław Woźniak$^{(\boxtimes)}$⬛, Grigory Sklyar⬛,
and Mateusz Firkowski⬛

Faculty of Computer Science and Information Technology, West Pomeranian
University of Technology in Szczecin, Żołnierska str. 49, 71-210 Szczecin, Poland
{mkorzen,jwozniak,gsklyar,mateusz.firkowski}@zut.edu.pl
http://www.wi.zut.edu.pl/

Abstract. The objective of the paper is to describe computational methods of control synthesis for a certain class of nonlinear driftless control systems. Such systems are previously found to be simplifications (called homogeneous approximations) of more complicated nonlinear systems that still preserve most crucial properties of the original ones like controllability. The class of systems in question have a special feed-forward form that is sufficiently easy to integrate and allows to solve concrete problems in control theory. Here we continue our research with describing the computational procedure for control synthesis as the extension of existing software libraries in Python language. We show that our approach leads to faster computation times compared to standard methods. The results are illustrated with some numerical experiments and simulations.

Keywords: control synthesis · nonlinear system · nonlinear approximation · homogeneous approximation · computational procedures

1 Introduction

In this paper we continue our investigation from [11], namely we consider control systems – nonlinear with respect to state and linear with respect to controls – namely (driftless) systems of the form

$$\dot{x} = \sum_{i=0}^{m} X_i(x)u_i, \tag{1}$$

where $X_i(x)$ are real analytic vector fields in the neighborhood of the origin in $\mathbb{R}^{N+1}$. It is one of class of systems widely considered in modern control theory.

Control theory is an interdisciplinary science that combines analysis and mathematical modeling of systems treated as dynamical systems with control. Most often, a dynamical system is described using different types of differential

M. H. Lees et al. (Eds.): ICCS 2025, LNCS 15905, pp. 3–16, 2025.
https://doi.org/10.1007/978-3-031-97632-2_1

equations, e.g. linear/nonlinear or ordinary/partial. One of the most important property of control theory is controllability, which describes possibility of finding the external input that carries out the system in a finite period of time to a given state under the fulfillment of initial conditions.

The intensive development of modern control theory led to the introduction of new methods that allowed to analyze the properties of systems described by nonlinear differential equations. The classical approaches based on linearization and the other one on concepts of differential geometry (Lie brackets theory) (cf. [18] on both approaches). Linearization, i.e. replacing a nonlinear system with a linear one, although being useful in many cases, may result in the loss of structural properties of the original system and therefore cannot be freely applied. It means that in order to approximate complex nonlinear systems we should use simpler nonlinear systems. One of possible ways of choosing those simpler systems is a homogeneous approximation method and was extensively developed over the last four decades [1, 2, 4, 6, 9, 14, 15]. This lead to the systems that have a special, feed-forward form, that allows to consider new approaches to their further analysis. In this paper we will present some new numerical methods of control synthesis for such systems. It is worth to notice that all homogeneous systems can be described in the feed-forward form, but not the other way around. Thus the methods presented below can be applied in general for a larger class of systems, that is all feed-forward systems.

2   Numerical Control Synthesis in Details

Control System. Here we pick up the investigations from [11] and consider homogeneous approximation of system (1) with two controls, u_0 and u_1, and we assume that it is already given in the feed-forward form,

$$
\begin{pmatrix} \dot{x}_0 \\ \dot{x}_1 \\ \dot{x}_2 \\ \vdots \\ \dot{x}_N \end{pmatrix} = \begin{pmatrix} f_0 \\ f_1(x_0) \\ f_1(x_0, x_1) \\ \vdots \\ f_N(x_0, \ldots, x_{N-1}) \end{pmatrix} u_0 + \begin{pmatrix} g_0 \\ g_1(x_0) \\ g_1(x_0, x_1) \\ \vdots \\ g_N(x_0, \ldots, x_{N-1}) \end{pmatrix} u_1 \qquad (2)
$$

that is f_k and g_k depend only on x_0, ..., x_{k-1} for all $k = 1, \ldots, N$, and f_0, g_0 are constants. For such a system, with given controls u_j $(j = 0, 1)$ we can describe its behavior using only a successive integration of consecutive differential equations. Direct feed-forward integration can be troublesome because of nested integrations and some kind of interpolation of intermediate results would be helpful. Thus, a proper numerical representation for control signal and state variables is an important issue in numerical practice.

Numerical Representation. For this tasks we can use representations based on Chebyshev polynomials. Chebyshev interpolation poses a very robust and convenient framework for a different types of computations with continuous functions [3,7,10,12]. We can use this framework for solving differential equations in a feed-forward form both for the integration and for storing intermediate results. If the functions f_k & g_k and controls u_j are polynomials, then using Chebyshev interpolation we can obtain fully numerically accurate solutions. To implement such solutions we only need operations of integration, multiplication and addition of Chebyshev interpolators (or equivalently Chebyshev polynomials).

We assume that the control signal is a polynomial of degree n written in the Chebyshev basis of the first kind T_j $(j = 0, \ldots, n)$, that is

$$f_n(t) = \frac{1}{2}a_0 + \sum_{j=1}^{n} a_j T_j(t).$$

We will not use this form directly, but the barycentric interpolation on the Chebyshev points of the second kind (or the Chebyshev extreme points) instead [3].

The difference between Chebyshev interpolator and Chebyshev polynomial approximation is that in the first case we represent functions by the sequence of function values in Chebyshev nodes and in the second case we represent functions via sequence of coefficients in Chebyshev polynomials base. Both cases are equivalent and one can easily change the representation from one form to another, if needed. For example, for evaluating functions represented in Chebyshev nodes we have a nice barycentric interpolation procedure [3], the analogue for a polynomial representation is Clenshaw algorithm [5].

Both representations are widely used in practice, e.g. packages Chebfun [7] and PaCal [10] are based on barycentric interpolation, while package polynomial.chebyshev from numpy provides common operations on Chebyshev polynomial expansion. The transformation between both representations (called Chebyshev transformation), i.e. transition from the interpolation using Chebyshev nodes to coefficients of expansion in the Chebyshev basis, can be effectively performed using Fast Fourier Transform [12, ch. 4].

Representation using interpolation is very convenient for most algebraic operations on such interpolators and function evaluation via barycentric procedure. However, the cumulative integration is more effective when performed using Chebyshev expansion [12, ch. 2], as we have a direct formula

$$\int_0^t f_n(\tau)d\tau = \frac{1}{2}c_0 + \sum_{j=1}^{n+1} c_j T_j(t),$$

where constant c_0 is determined from the initial condition, and

$$c_j = \frac{a_{j-1} - a_{j+1}}{2j}, \text{ for } j \geq 1,$$

with $a_{n+1} = a_{n+2} = 0$.

Representation by Chebyshev Interpolation. The interpolator Y is represented by values of functions y_j in Chebyshev nodes t_j:

$$Y = \begin{pmatrix} t_0, & t_1, & \ldots, & t_n \\ y_0, & y_1, & \ldots, & y_n \end{pmatrix}.$$

All operations used in system trajectory determining procedure are closed with respect to Chebyshev representation. For example, the sum of interpolators with the same set of Chebyshev nodes is given by

$$Y + Z = \begin{pmatrix} t_0, & t_1, & \ldots, & t_n \\ y_0 + z_0, & y_1 + z_1, & \ldots, & y_n + z_n \end{pmatrix},$$

and multiplication is given by

$$YZ = \begin{pmatrix} t_0, & t_1, & \ldots, & t_{n+m} \\ Y(t_0)Z(t_0), & Y(t_1)Z(t_1), & \ldots, & Y(t_{n+m})Z(t_{n+m}) \end{pmatrix},$$

where

$$Y(t) = \frac{\displaystyle\sum_{j=0}^{n} \frac{w_j}{t - t_j} y_j}{\displaystyle\sum_{j=0}^{n} \frac{w_j}{t - t_j}},$$

$t_j = \cos \dfrac{j\pi}{n}$ are Chebyshev nodes of the second kind, y_i are interpolated function values, and w_i are proper (barycentric) weights (cf. [3]) defined as the sequence

$$w = \left(\frac{1}{2}, -1, 1, -1, \ldots, (-1)^{n-2}, (-1)^{n-1}, \frac{(-1)^n}{2} \right).$$

Recalling that the class of analyzed systems has a special feed-forward form, we can easily approach it directly using a presented Chebyshev framework. Computations with this representation are fast and numerically very stable, accuracy is typically close to the machine precision. When we restrict control signals u_i and functions f_k & g_k to polynomials then the proposed procedure will be analytically exact.

Representation by Piecewise Functions. Chebyshev expansion is valid only for regular continuous functions on closed interval (see e.g. [12]), but we can extend such framework to continuous piecewise functions, where each segment is represented by a Chebyshev interpolator. This is a natural extension and to perform algebraic operations we should find a common nested set of intervals. This representation is a bit more complicated as we have to aggregate computations from each sub-interval.

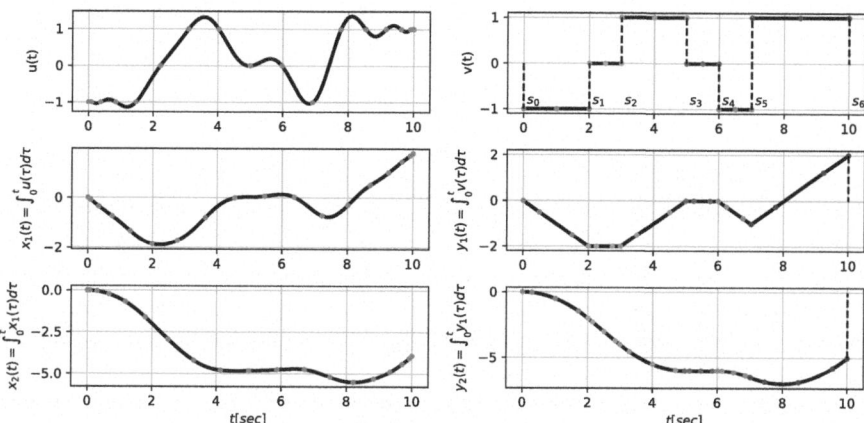

Fig. 1. Illustration of the proposed integration procedure using Chebyshev interpolation (left side) and piecewise Chebyshev interpolation (right side) for the system $\dot{x}_1 = u(t)$, $\dot{x}_2 = x_1$. Interpolation nodes are presented as dots, s_ℓ denote switching points for the case of the bang-bang control.

Note that piecewise constant functions play an important role in the control theory, and they are the main tool in the optimal control synthesis investigations [18]. In the case of bang-bang controls (piecewise constant functions) the resulting trajectories will consist also of piecewise, but not constant functions in general. Figure 1 presents comparison of two considered representations and results of simple preliminary computations. Notice that each integration step increases degrees of polynomials and, in consequence, adds some additional nodes to the interpolators.

Numerical Synthesis of the Controls. The control synthesis task is to find such control signals u_i, that would take the system for a given initial point $X(0)$ to the origin $\mathcal{O}$. Both considered types of representations provide two different ways to produce control signals using polynomial control and bang-bang control. We consider two optimization procedures for bang-bang control.

Details of all three considered approaches are as follows.

1. **Chebyshev or polynomial control**: In this case we represent control signals via Chebyshev barycentric interpolation, and system coordinates are represented also by Chebyshev barycentric interpolation on interval $[0, T_f]$, where time of system evolution, T_f, is given up front. To find the controls we use the optimization procedure with goal function Q_P of the form

$$Q_P(U_0, U_1) = \|x_f\|_2^2 + \alpha \sum_{k=0}^{n} \|x_k\|_2^2$$

where

$$U_0 = \begin{pmatrix} t_1, & t_2, & \ldots, & t_n \\ u_{01}, & u_{02}, & \ldots, & u_{0n} \end{pmatrix}, \quad U_1 = \begin{pmatrix} t_1, & t_2, & \ldots, & t_n \\ u_{11}, & u_{12}, & \ldots, & u_{1n} \end{pmatrix},$$

u_{ij} are values of control signals u_i ($i = 0,1$) in Chebyshev nodes t_j ($j = 0, 1, \ldots, n$) scaled to the interval $[0, T_f]$, $\alpha > 0$ is a (small) regularization parameter, $\|x_f\|_2^2 = \sum_{k=0}^{n} |x_k(T_f)|^2$, $\|x_k\|_2^2 = \int_0^{T_f} |x_k(t)|^2 dt$. As final time T_f is fixed, Chebyshev nodes t_j are also fixed, and the cost function depends on interpolated nodes values u_{ij} only. As the optimization procedure we use here the l-bfgs-b procedure from scipy [13], with additional bounds on maximal values of nodes of the control signals.

2. **Bang-bang control 0**: Here we represent the control signals using piecewise constant functions on interval $[0, T_f]$ (with fixed T_f) and switching points $0 \le s_1 < \ldots < s_p \le T_f$, while system coordinates are represented by piecewise Chebyshev interpolation. In this case the cost function Q_0 is similar to the previous case, namely

$$Q_0(U_0, U_1) = \|x_f\|_2^2 + \alpha \sum_{k=0}^{n} \|x_k\|_2^2$$

where

$$U_i(t) = \begin{cases} 1, & \text{for } t \in [s_{4\ell}, s_{4\ell+1}) \\ 0, & \text{for } t \in [s_{4\ell+1}, s_{4\ell+2}) \\ -1, & \text{for } t \in [s_{4\ell+2}, s_{4\ell+3}) \\ 0, & \text{for } t \in [s_{4\ell+3}, s_{4\ell+4}), \end{cases}$$

$s_0 = 0$, $s_{p+1} = T_f$, with constrains

$$s_1 \ge 0,$$
$$s_\ell \le s_{\ell+1} \, (\ell = 0, 1, \ldots, p-1),$$
$$s_p \le s_{p+1} = T_f,$$

but in this case the cost function depends on switching points s_j only (see Fig. 1). Both control functions U_i are optimized using independent sequences of switching points. As we have linear constraints here this time the optimization procedure we use is trust region method, namely scipy's trust-constr procedure.

3. **Bang-bang control 1**: After careful investigation of the above optimization problem we observed that we can simplify the procedure by using interval lengths instead of switching points as variables in the procedure. In this case we are able to drop fully linear constraints and use the bound-type constraints. Here again we represent the control signals using piecewise constant functions, but on unknown at first interval $[0, T_f]$, and by widths of interpolation intervals $\Delta s_\ell := s_\ell - s_{\ell-1} \ge 0$, $\ell = 1, 2 \ldots, p+1$ (instead switching

points), and system coordinates are again represented by piecewise Chebyshev interpolation. As the optimization procedure we use the same procedures – the l-bfgs-b procedure from scipy [13] – with bounds on maximal values of nodes of the control signals, but this time we are able to optimize the time T_f of control of the system, if possible, by including an additional term to the cost function. Thus, the new cost function Q_1 takes the form

$$Q_1(U_0, U_1) = \|x_f\|_2^2 + \alpha \sum_{k=0}^{n} \|x_k\|_2^2 + \beta \sum_{\ell=1}^{p} \Delta s_\ell,$$

where $\alpha > 0$ and $\beta > 0$ are (small) regularization parameters,

$$U_i(t) = \begin{cases} 1, & \text{for } t \in [s_{4\ell}, s_{4\ell+1}) \\ 0, & \text{for } t \in [s_{4\ell+1}, s_{4\ell+2}) \\ -1, & \text{for } t \in [s_{4\ell+2}, s_{4\ell+3}) \\ 0, & \text{for } t \in [s_{4\ell+3}, s_{4\ell+4}), \end{cases}$$

and $s_\ell = \sum_{i=1}^{\ell} \Delta s_i$, with constrains

$$\Delta s_\ell \geq 0 \, (\ell = 1, 2 \dots, p + 1).$$

As the reference procedure for all three cases we will use a general purpose ode solver odeint function from scipy. Experimental part was prepared using Python computational environment with numpy [8] libraries mainly for arrays, scipy [17] for optimization and ODE solvers.

3 Numerical Experiments

Systems. In the experimental part we consider three control systems: one based on real-life model, *sys0*, of the form

$$\begin{pmatrix} \dot{x}_0 \\ \dot{x}_1 \\ \dot{x}_2 \\ \dot{x}_3 \\ \dot{x}_4 \end{pmatrix} = \begin{pmatrix} 1 \\ 0 \\ 0 \\ 0 \\ 0 \end{pmatrix} u_0 + \begin{pmatrix} 0 \\ 1 \\ x_0 \\ -\frac{x_0^2}{2} \\ \frac{x_0^3}{6} \end{pmatrix} u_1, \tag{3}$$

which is the homogeneous approximation of the truck with trailers system (see [11]), and two artificially generated ones, namely *sys1*, that is

$$\begin{pmatrix} \dot{x}_0 \\ \dot{x}_1 \\ \dot{x}_2 \\ \dot{x}_3 \\ \dot{x}_4 \end{pmatrix} = \begin{pmatrix} 1 \\ 0 \\ x_1 \\ 0 \\ \frac{x_1 x_2}{6} \end{pmatrix} u_0 + \begin{pmatrix} 0 < \\ x_0 \\ 0 \\ -\frac{x_2^2}{4} \\ 0 \end{pmatrix} u_1 >, \tag{4}$$

and *sys2*, given by

$$
\begin{pmatrix} \dot{x}_0 \\ \dot{x}_1 \\ \dot{x}_2 \\ \dot{x}_3 \\ \dot{x}_4 \end{pmatrix} = \begin{pmatrix} 1 \\ 0 \\ 0 \\ 0 \\ 0 \end{pmatrix} u_1 + \begin{pmatrix} 0 \\ 1 \\ -x_0 x_1 < \\ -\frac{x_{0>} x_1^2}{2} \\ -\frac{x_0 x_1^3}{6} \end{pmatrix} u_2.
\tag{5}
$$

Results. We use two quality measures in all cases, that is ℓ^2-norm of differences between the obtained final position and the goal (here the origin $\mathcal{O}$),

$$
\|x_f\|_2 = \sqrt{\sum_{k=0}^{4} |x_k(T_f)|^2},
\tag{6}
$$

and ℓ^1-norm of those differences,

$$
\|x_f\|_1 = \sum_{k=0}^{4} |x_k(T_f)|,
\tag{7}
$$

but in the case of piece-wise algorithm 1 (bang-bang control 1) we also compare the time of systems' evolution, T_f, against other algorithms, where we used the same fixed value of $T_f = 8$. In all cases we check computation times, namely we find time of establishing the control signals (t_{cont}) and time of constructing the trajectories using our feed-forward methods (t_{f-f}) and compare it to the reference time (t_{ref}) of trajectory constructions with the general purpose ode solver odeint function from scipy. Detailed results are summarized in Table 1, Figs. 2, 3 and 4 present sample trajectories for different initial positions.

Concerning computational time, one can observe that piece-wise algorithm 1 (bang-bang 1 controls) was able to cut the systems' evolution time, in two cases halving it, which is very promising for the applications for controllability related problems for original, more complicated nonlinear systems (cf. [11,16]). In two cases ('*sys0*' and '*sys1*') control synthesis computation times of all algorithms were comparable, but for system '*sys2*' Chebyshev algorithm was significantly faster. Direct comparison of our methods (Chebyshev and piecewise functions) is not simple as the number of nodes influences very differently the computational complexity in both methods. For example, for five switching points we obtain five interpolators and 15 interpolating nodes (three per each point), and calculating values of the function requires more numerical steps than calculating the analogous value for Chebyshev polynomial of 15°C. Experiments show significant time advantage over piecewise algorithms for the same number nodes used, especially for trajectory computation times. Time needed for synthesis control

Table 1. Comparison of results of control synthesis experiments, where $|x_0|_2$ denotes ℓ^2 norm of initial state of the system, $|x_f|_2$ – ℓ^2 norm of final state of the system, $|x_f|_1$ – ℓ^1 norm of final state of the system, T_f – total time of system evolution, t_{cont} – computation time of control synthesis, t_{f-f} – computation time of finding the trajectory using our feed-forward approach, t_{ref} – computation time of finding the trajectory using reference procedures, i.e. function **odeint** from scipy.

system	method	nodes	$|x_0|_2$	$|x_f|_2$	$|x_f|_1$	T_f	t_{cont}	t_{f-f}	t_{ref}
sys0	cheb	5	1.11	**0.00155**	0.00220	8	1.46	0.00320	0.304
		10	1.11	0.00214	**0.00155**	8	6.51	0.00385	0.357
		20	1.11	0.00354	0.00217	8	13.2	0.00395	0.329
		50	1.11	0.00814	0.00399	8	41.6	0.00383	0.338
	piece0	5	1.11	0.01030	0.00769	8	23.8	0.0229	0.468
		7	1.11	0.01020	0.00595	8	39.2	0.0280	0.393
		9	1.11	0.00884	0.00379	8	59.1	0.0318	0.423
	piece1	5	1.11	0.00740	0.00594	4.14	14.2	0.0157	0.170
		7	1.11	0.00610	0.00384	3.77	39.4	0.0206	0.146
		9	1.11	0.00759	0.00330	**3.61**	50.1	0.0234	0.162
sys1	cheb	5	1.11	0.00364	0.01630	8	1.92	0.00350	0.318
		10	1.11	0.00368	0.01380	8	5.93	0.00402	0.343
		20	1.11	**0.00266**	0.00595	8	18.6	0.00400	0.284
		50	1.11	0.0103	0.01020	8	45.8	0.00409	0.273
	piece0	5	1.11	0.0314	0.01710	8	29.2	0.0282	0.467
		7	1.11	0.0244	0.01060	8	48.1	0.0330	0.436
		9	1.11	0.0131	0.01160	8	74.5	0.0405	0.413
	piece1	5	1.11	0.0135	0.00827	5	30.3	0.0256	0.276
		7	1.11	0.00738	**0.00274**	3.99	41.6	0.0270	0.174
		9	1.11	0.00724	0.00335	4.56	63.0	0.0308	0.198
sys2	cheb	5	2.19	0.0542	0.0191	8	2.25	0.00411	0.342
		10	2.19	0.0529	0.0232	8	5.83	0.00479	0.284
		20	2.19	0.0534	0.0188	8	15.4	0.00551	0.300
		50	2.19	0.0579	0.0217	8	68.8	0.00642	0.373
	piece0	5	2.19	0.0600	0.0318	8	33.4	0.0339	0.308
		7	2.19	0.0536	0.0286	8	72.0	0.0526	0.427
		9	2.19	0.0532	0.0243	8	135	0.0713	0.573
	piece1	5	2.19	0.0460	0.0222	**6.60**	31.9	0.0317	0.263
		7	2.19	0.0294	0.0210	7.91	105	0.0472	0.357
		9	2.19	**0.0269**	**0.0111**	7.36	168	0.0672	0.545

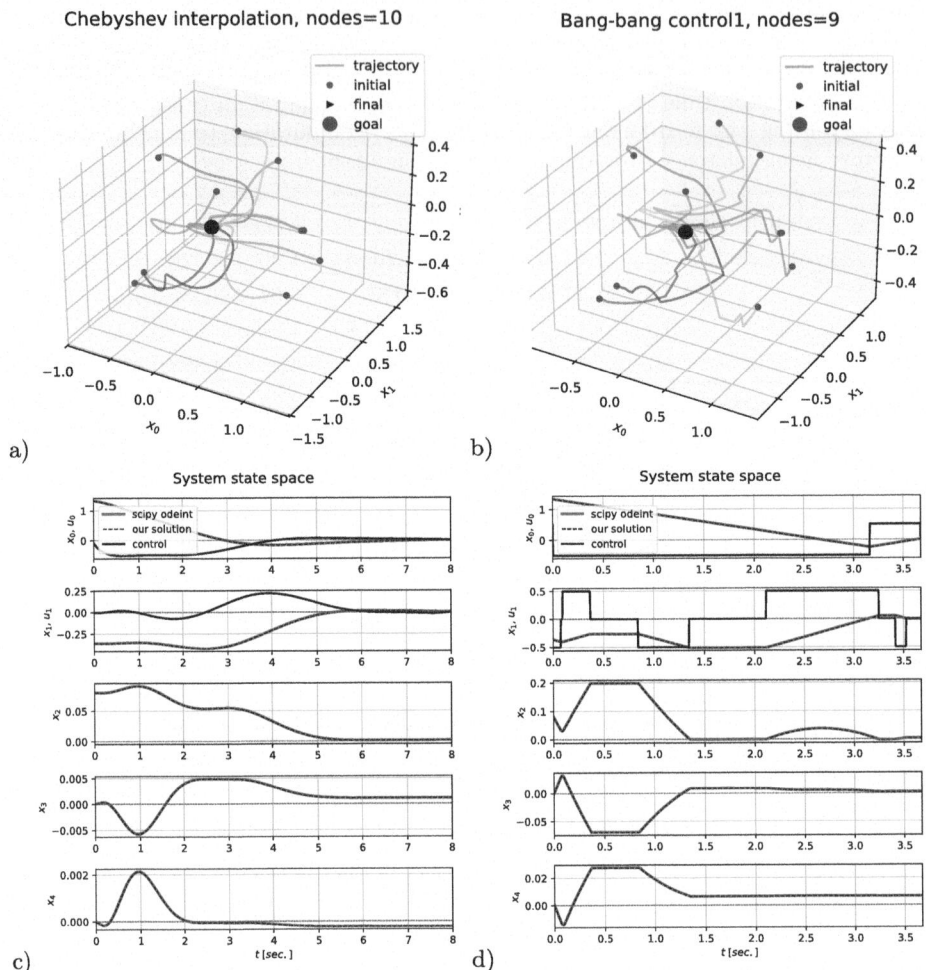

Fig. 2. Illustration of the proposed approach with the system 'sys0' (3); a,b) projection of system's trajectories to 3D subspaces starting from different initial points. c,d) exemplary solution: control signals u_0, u_1 (solid black), black dashed lines represent solution in our representation – system's state space in separate coordinates x_k, $i = 0, 1, 2, 3, 4$, red solid lines – the same solution given by scipi.odeint.

does not show such large differences, most probably due to the fact that the goal function Q_P has an easier form but has to be called more times in the optimization process. Experiments show that those synthesis times are comparable in the cases of 50 Chebyshev nodes and only 7 piecewise nodes (switching points). In all 30 experiments computational times of generating systems' trajectories were much faster using our new method than with the general reference procedures (function odeint from scipy), being from ten times to even one hundred times faster.

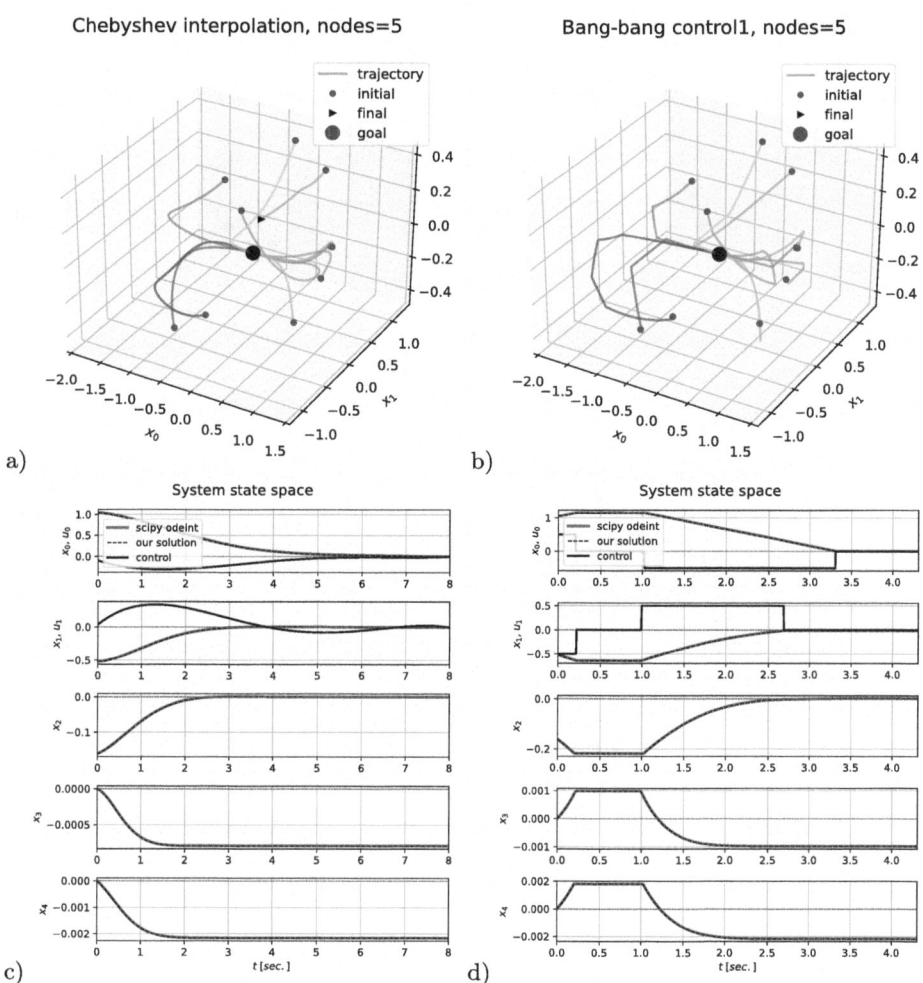

Fig. 3. Illustration of the proposed approach with the system '*sys1*' (4); a,b) projection of system's trajectories to 3D subspaces starting from different initial points. c,d) exemplary solution: control signals u_0, u_1 (solid black), black dashed lines represent solution in our representation – system's state space in separate coordinates x_k, $i = 0, 1, 2, 3, 4$, red solid lines – the same solution given by scipi.odeint.

Concerning quality of the goal objectives – reaching the origin $\mathcal{O}$ – one can observe that in all cases our trajectories tended in the direction of the goal. All goal reaching measures give comparable results between Chebyshev and piecewise methods, but in general piece-wise algorithm 1 (bang-bang 1 controls) seems to give slightly better results than piece-wise algorithm 0 (bang-bang 0 controls).

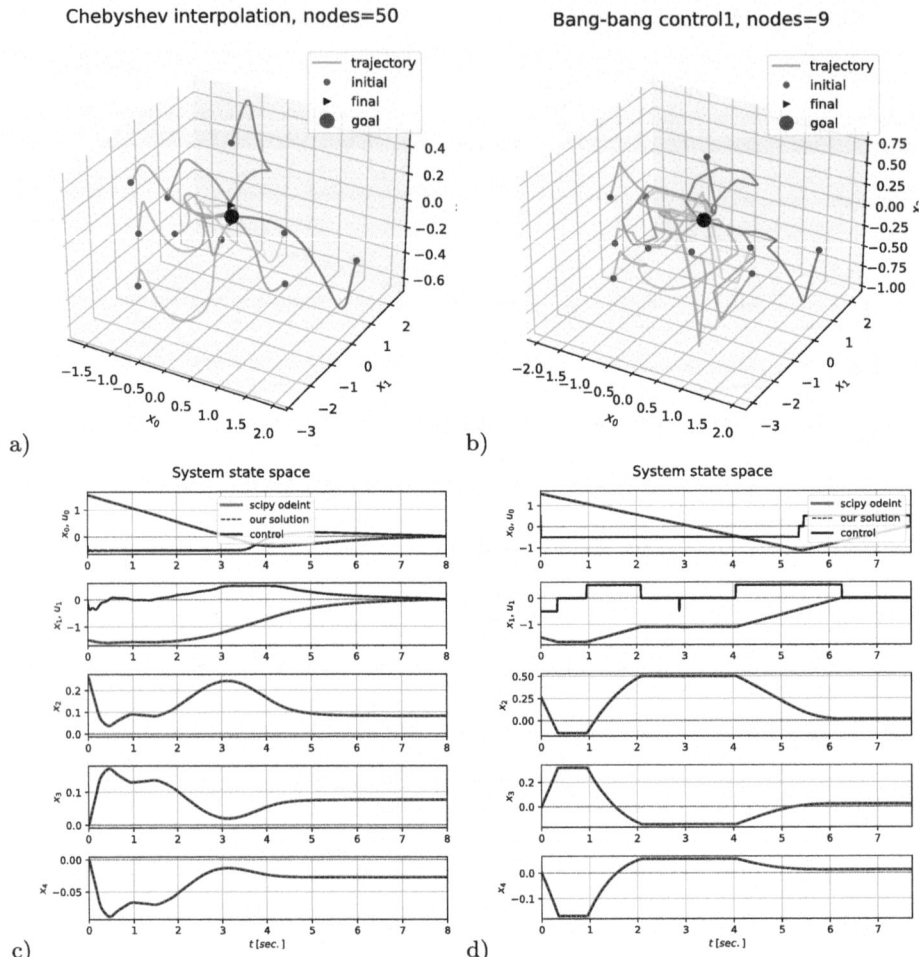

Fig. 4. Illustration of the proposed approach with the system 'sys2' (5); a,b) projection of system's trajectories to 3D subspaces starting from different initial points. c,d) exemplary solution: control signals u_0, u_1 (solid black), black dashed lines represent solution in our representation – system's state space in separate coordinates x_k, $i = 0, 1, 2, 3, 4$, red solid lines – the same solution given by scipi.odeint.

4   Summary

In this paper we presented the procedure of determining controls for homogeneous nonlinear systems from a computational point of view, and we provided the numerical experiments with some selected nonlinear control systems in the feed-forward form, that could arise from homogeneous approximations. After comparing the system trajectories, we briefly discussed the quality of such controls. The experiments confirmed that the theoretical results concerning homogeneous approximations of nonlinear systems can be used in practice, and the need to construct suitable software libraries to be used in possible applications – e.g. in practical control design – is evident.

The approach presented in this paper can be further investigated, for example trying to minimize not only time, but also energy norm of control signals.

Disclosure of Interests. The authors have no competing interests to declare that are relevant to the content of this article.

References

1. Agrachev, A., Marigo, A.: Nonholonomic tangent spaces: intrinsic construction and rigid dimensions. Electron. Res. Announ. Am. Math. Soc. **9**, 111–120 (2003)
2. Bellaïche, A.: The tangent space in sub-Riemannian geometry. Prog. Math. **144**, 4–78 (1996)
3. Berrut, J.P., Trefethen, L.N.: Barycentric lagrange interpolation. SIAM Rev. **46**(3), 501–517 (2004)
4. Bianchini, R., Stefani, G.: Graded approximation and controllability along a trajectory. SIAM J. Control Optimiz. **28**, 903–924 (1990)
5. Clenshaw, C.W.: A note on the summation of Chebyshev series. Math. Tables Other Aids Comput. **9**, 118–120 (1955)
6. Crouch, P.E.: Solvable approximations to control systems. SIAM J. Control Optim. **22**, 40–54 (1984)
7. Driscoll, T.A., Hale, N., Trefethen, L.N.: Chebfun Guide. Pafnuty Publications (2014). http://www.chebfun.org/docs/guide/
8. Harris, C.R., et al.: Array programming with NumPy. Nature **585**(7825), 357–362 (2020)
9. Hermes, H.: Nilpotent and high-order approximations of vector field systems. SIAM Rev. **33**, 238–264 (1991)
10. Jaroszewicz, S., Korzeń, M.: Arithmetic operations on independent random variables: a numerical approach. SIAM J. Sci. Comp. **34**(4), A1241–A1265 (2012)
11. Korzeń, M., Sklyar, G.M., Ignatovich, S.Y., Woźniak, J.: Computational aspects of homogeneous approximations of nonlinear systems. Proc. ICCS **2024**, 368–382 (2024)
12. Mason, J., Handscomb, D.: Chebyshev Polynomials. CRC Press, Boca Raton (2002)
13. Morales, J.L., Nocedal, J.: Remark on "algorithm 778: L-bfgs-b: fortran subroutines for large-scale bound constrained optimization". ACM Trans. Math. Softw. **38**(1) (2011)
14. Sklyar, G.M., Ignatovich, S.Y.: Approximation of time-optimal control problems via nonlinear power moment min-problems. SIAM J. Control Optim. **42**, 1325–1346 (2003)
15. Sklyar, G.M., Ignatovich, S.Y.: Description of all privileged coordinates in the homogeneous approximation problem for nonlinear control systems. C. R. Math. Acad. Sci. Paris **344**, 109–114 (2007)

16. Sklyar, G.M., Ignatovich, S.Y.: Free algebras and noncommutative power series in the analysis of nonlinear control systems: an application to approximation problems. Dissertationes Math. (Rozprawy Mat.) **2014**, 1–88 (2014)
17. Virtanen, P., et al.: SciPy 1.0: fundamental algorithms for scientific computing in python. Nat. Methods **17**, 261–272 (2020). https://doi.org/10.1038/s41592-019-0686-2
18. Zabczyk, J.: Mathematical Control Theory: An Introduction. Birkhäuser, Basel (2008)

A Robust Ensemble Malware Detector Against Powerful Adversaries

Shangyuan Zhuang[1,2], Wei Zhang[3], Fengqi Liu[3], Jiyan Sun[1(✉)],
Yinlong Liu[1,2], Liru Geng[1,2], and Wei Ma[1]

[1] Institute of Information Engineering, Chinese Academy of Sciences, Beijing, China
{zhuangshangyuan,sunjiyan,liuyinlong,gengliru,mawei}@iie.ac.cn
[2] School of Cyber Security, University of Chinese Academy of Sciences,
Beijing, China
[3] Henan Jiuyu EPRI Electric Power Technology Co., LTD., Henan, China

Abstract. Considering the vulnerability of machine learning to adversarial attacks, the state-of-the-art malware detectors are designed with appropriate ensembles. However, most ensemble detectors are developed based on unreasonable and weak threat assumptions, which do not match the characteristics of real-world adversaries with powerful adaptive mixed capabilities. Such detectors will unavoidably suffer severe failures in practical deployment. To this end, we build two realistic powerful adversary models and propose *NashAE* as a robust malware detector based on a novel Game-theory-enabled ensemble adversarial training approach against them. Specifically, *NashAE* establishes a Minimax Game where the adversary and detector compete on opposing targets. By solving the Nash equilibrium of the game, *NashAE* can obtain the optimal ensemble adversarial training strategy under adversaries' constantly adaptive attacks. Since the game has no closed-form solution, we further develop a simplified solution scheme based on Bayesian optimization to find the approximate Nash equilibrium of the game. We conduct comprehensive experiments with 10 baseline detection models on 2 malware datasets. Experimental results show that *NashAE* can achieve a stable detection rate of 58% higher than advanced methods after only 15 iterations against the most powerful adversaries.

Keywords: Malware detector · Adversarial defense · Nash equilibrium solution · Game theory · Ensemble learning · Network security

1 Introduction

The deep integration of 5G and satellite networks as communication carriers for critical infrastructure (e.g., new distribution grid systems) has expanded the propagation radius of malware from terrestrial systems to space networks. One of the mainstream solutions identifies malware based on network traffic since traffic can capture all behaviors in the network [15,17,21]. By analyzing whether

M. H. Lees et al. (Eds.): ICCS 2025, LNCS 15905, pp. 17–32, 2025.
https://doi.org/10.1007/978-3-031-97632-2_2

the traffic is benign or malicious based on machine learning techniques, malware can be detected. Although these detectors have achieved certain success in practice, they are vulnerable to adversarial attacks [10]. Through adding carefully designed adversarial perturbations to malware traffic, the adversarial attack can trick detectors to identify malware traffic as benign. Then these adversarial malware traffic will escape the detector's detection, so that the corresponding malware can commit malicious activities without any hindrance.

To robustly detect malware traffic even in the presence of adversarial perturbations, detectors based on ensemble learning are considered a promising solution, since adversaries have to generate adversarial malware traffic that can confront all base models [8,14]. Nonetheless, the transferability of adversarial samples renders simple ensemble detectors still unreliable [20]. To further protect against those transfer-based attacks, ensemble adversarial training approaches have been introduced [7,9,16,18], which allow the detector to pre-learn the features of the adversarial traffic in the ensemble training phase. However, when facing real-world adversaries, these advanced detectors are still indefensible because their ensemble strategies are usually developed based on unrealistic and weakness adversarial attack assumptions. More precisely, the characteristics of adversarial traffic learned by them do not match the real adversaries.

Firstly, existing ensemble adversarial training strategies are usually tailored to individual adversarial perturbation types, whereas real-world adversaries are much more sophisticated and powerful. For one thing, in reality, there are multiple adversaries from different organizations with different goals to generate various types of adversarial perturbations. For another, adversaries will have varied attack strategies thanks to open-source adversarial attack tool [2]. Aiming to achieve strong evasion performance, powerful adversaries tend to launch hybrid attacks that mix multiple attack strategies. Further, there are also more powerful adversaries even can adaptively adjust their hybrid attack strategies based on the detector's detection results. With this regard, in the ensemble adversarial training phase, practical detectors have to consider adversaries with mixed attack schemes or even adaptive mix attack schemes, rather than the current simple individual attack schemes.

Secondly, existing ensemble adversarial training strategies usually neglect the traffic-space constraints and knowledge limitation of practical adversarial attacks [3]. For one thing, after adding adversarial perturbations, adversarial malware traffic must still follow the specific semantics of communication protocols. Only such adversarial traffic is effective. However, existing detectors directly learn all adversarial traffic features, causing not only limited improvement in robustness but also leading to detection performance degradation and sometimes even creating more vulnerabilities for detectors. For another, adversaries have no access to their target detectors in reality, while most ensemble detectors learn features of the adversarial samples generated based on the white-box assumptions [13,19]. It is clear that such adversarial traffic features are not fully consistent with adversarial traffic generated in real-world settings. This further limits the robustness improvement by the ensemble adversarial training.

To address the above challenges, we build two realistic threat models and propose *NashAE* as a robust ensemble malware detector based on a novel Game-theory-enabled ensemble adversarial training. By building and solving a Minimax Game where the adversary and detector compete based on the opposite target in the ensemble adversarial training phase, *NashAE* automatically updates its ensemble strategy according to adversaries' constantly adaptive attack scheme. The main contributions of our work are summarized as follows:

- **NashAE Detector against Real-World Powerful Adversaries.** We design *NashAE* ensemble detector which is based on game theory to address the challenge of practical adversaries' adaptive mixed adversarial attacks. By establishing a Minimax Game, *NashAE* simulates the process that the adversaries continuously adjust their attack scheme and detector continuously selects its optimal ensemble strategy. Based on this more realistic dynamic process, *NashAE* has greater potential to defend against the admixture and changing nature of adversarial attackers in real world.
- **Practical Threat Model Building.** We build two practical threat models, M-TBA and AM-TBA, that more closely reflect real-world powerful attacks. Under the M-TBA model, adversaries launch black-box attacks that are mixed and conform to the traffic-space constraints. Under the AM-TBA model, adversaries launch black-box attacks that can adjust their mixed schemes with a certain adaptive ability and conform to the traffic-space constraints. In particular, via a well-designed remapping function, the adversarial traffic can strictly satisfy the traffic-space constraints.
- **Simplified Solution to Nash Equilibrium.** Since Minimax Game has no closed solution, typical gradient-based algorithms will fail to solve the Nash equilibrium that ends the game. To address this problem and achieve rapid response in security field, we provide a simple solution method with relatively less calculation times. Specifically, we convert the Minimax Game into two related Markov Decision Processes (MDPs) and solve them by Bayesian optimization. The solutions of MDPs can be considered as the approximate Nash equilibrium, which is used to obtain optimal ensemble strategy.
- **Comprehensive evaluations of NashAE.** We evaluate our realistic attack and defense methods with 2 encrypted malware traffic datasets using 10 detection strategies in both M-TBA model and AM-TBA model. Experimental results demonstrate that *NashAE* can defend against realistic attacks with stable high performance after only 15 iterations, outperforming other advanced methods by at least 40%. Besides, we verify the feasibility of black-box attacking methods based on the transferability of adversarial perturbations according to our experiments.

2   Related Works

2.1   Adversarial Attack to Malware Detectors

Machine learning techniques have been widely used in malware traffic detection [15,17,21]. Thanks to the powerful classification capability of machine learning, these detectors achieve certain success in practice. Unfortunately, Rigaki *et al.* [13] demonstrate that traffic detectors based on machine learning are vulnerable to adversarial attacks. By adding adversarial perturbations that maximize the loss function of target detectors, adversarial attacks can cause detector misclassification. Considering the widespread encryption protocols such as HTTPS, Novo *et al.* [10] implement adversarial attacks take account into encryption traffic. Since attackers hardly have knowledge of target detectors, transfer-based adversarial attacks [5,12,24] are proposed to perform black-box attacks. These attacks first create substitute models of their target detectors, and then generate adversarial malware traffic based on the fully accessible substitute models. Benefiting from the transferability of adversarial examples, this adversarial malware traffic has the potential to trick target malware detectors.

2.2   Ensemble Detectors Against Adversarial Attacks

Malware detectors based on ensemble learning are considered a potential way to defend against adversarial attacks [7]. Nevertheless, some works demonstrate that detectors based on ensemble learning are still insufficient to robustly defend against adversarial perturbations [22,23]. For example, Zhang *et al.* [23] show that discrete-valued tree ensemble models can be easily attacked by adversarial examples generated according to the models decision output.

To further improve the robustness of the ensemble model, ensemble adversarial training strategies [16] are proposed. Specifically, Tramer *et al.* [16] first augment training data with adversarial perturbations transferred from other models to increase the robustness of ensemble detectors. Following them, also some studies show robust defense ways for adversarial attacks via ensemble adversarial learning [9,18]. However, due to unrealistic assumptions of adversaries, these detector ensemble strategies are fixed and learn some features of adversarial traffic that do not satisfy the traffic-space constraints. This inevitably invalidates their robust defense strategies against realistically powerful adversaries. Further, Shu *et al.* [14] propose the Omni ensembles models, whose hyperparameters are controlled to make the attacker's target model far away. Based on their idea of hyper-parametric control, Omni has a certain ability to defend against powerful adversaries with mixed strategies (i.e., the M-TBA threat model). But it will be vulnerable to more powerful adversaries that can dynamically and adaptively adjust their mixed attack strategies (i.e., the AM-TBA threat model).

3   Overview of NashAE Ensemble Detector

In this section, we will briefly introduce the overall structure of our NashAE ensemble detector, which is capable of robustly defending against the admixture and adaptive nature of adversarial attackers in the real world.

As shown in Fig. 1, *NashAE* enhances detection robustness through a novel ensemble adversarial training approach. Considering that the most powerful adversaries have mixed attack schemes and adapt their attack schemes according

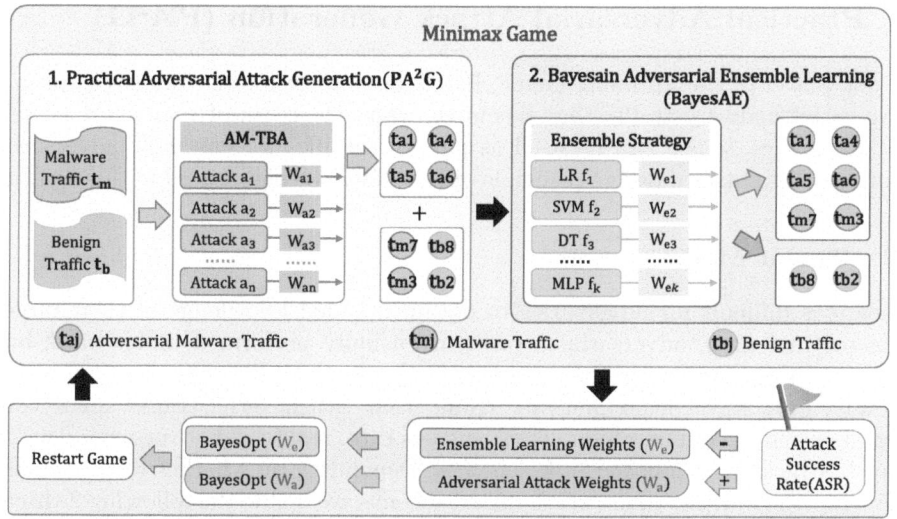

Fig. 1. The overall structure of *NashAE* ensemble adversarial training approach.

to attack performance and defense strategies, *NashAE* is adversarially trained with a dynamic ensemble strategy. To describe the dynamic process in which the detector adjusts its ensemble strategy in line with adversaries' constantly adaptive attack schemes, *NashAE* establishes a Minimax Game. The two players of the game are practical adversarial attack generation (PA^2G) and Bayesian adversarial ensemble learning (BayesAE).

Among them, PA^2G simulates a realistic adversary with mixed adaptive attack scheme, and BayesAE simulates an ensemble detector with dynamic ensemble strategies. Before the game starts, PA^2G has a range of base adversarial attack resources $a_1, a_2, ..., a_n$ that can generate a series of adversarial malware traffic t_a from the malicious traffic t_m. BayesAE has a range of base malware detection resources $f_1, f_2, ..., f_k$ that aim to detect all malware traffic. The two players align their strategies toward opposite targets during the game. PA^2G tries to find a series of base adversarial attack weights $\mathbf{w_a} = \{w_{a1}, w_{a2}, ..., w_{an}\}$ that maximum attack success rate (ASR), so that the realistic adversary can best fool the current ensemble detector. Meanwhile, BayesAE tries to adjust a series of ensemble weights $\mathbf{w_e} = \{w_{e1}, w_{e2}, ..., w_{en}\}$ that minimize the ASR, so that the ensemble detectors can best robust against the current attack. Let $g(\cdot)$ denote the objective ASR that the two players want to maximize/minimize, then the minimax game can be described as:

$$\min_{w_{ei} \in \mathbf{w}_e} \max_{w_{ai} \in \mathbf{w}_a} g(\mathbf{w}_e, \mathbf{w}_a) \tag{1}$$

Since they compete on opposite directions of ASR, the game will reach a Nash equilibrium, which is the solution of the most robust ensemble detector. Detailed strategies of the two players are described below.

4   Practical Adversarial Attack Generation (PA^2G)

As one player of the Minimax Game, PA^2G aims to generate a series of optimal adversarial malware traffic that meets the characteristics of real-world powerful adversaries, which can be used as part of samples for ensemble adversarial training. The detailed design principle of PA^2G will be elaborated in this section.

4.1   Threat Model

Since it is difficult for adversaries to obtain detailed knowledge of their target detector in reality, adversarial attacks are usually performed under black-box scenarios. Considering that the network will automatically drop non-compliant packets, only adversarial malware traffic that satisfies the traffic space constraints is effective. For a high attack success rate, real-world adversaries usually launch attacks that mix multiple schemes. Depending on whether the adversary can adaptively adjust their attack mixed scheme, we define the following 2 threat models, aiming to match the characteristics of a realistic and powerful adversary.

1. **M-TBA**: As a mixed traffic-space black-box attack, M-TBA has no adaptive capability. In this case, adversaries launch attacks based on the average mixed scheme, rather than mixed schemes with different weights. Thus it can be directly defended without establishing the Minimax Game.
2. **AM-TBA**: As an adaptive mixed traffic-space black-box attack, AM-TBA can constantly adapt its mixed scheme according to the current attack performance and defense strategies. In this dynamic case, we have to establish a game to solve the most robust detector ensemble strategy.

To obtain an optimal detection ensemble strategy with the upper robustness performance, PA^2G simulates the most powerful adversaries AM-TBA. It is clear that the detection strategy obtained from AM-TBA threat model also can robustly resist the M-TBA threat model. Detailed principles are as follows:

- **Principle of Mixed Attack** In the case of M-TBA, the adversary has multiple adversarial attack methods $a_1, a_2, ..., a_n$ with a series of fixed mixture weight factor $w_{a1}, w_{a2}, ..., w_{an}$, aiming to evade the detection of ensemble detector $\mathbf{f}$ that integrates multiple base detectors $f_1, f_2, ..., f_k$ with a certain strategy.
- **Principle of Adaptive Mixed Attack** In the case of AM-TBA, we solve adversary's optimal attack mixed scheme by establishing the Minimax Game. As shown in Fig. 1, there are multiple adversarial attack methods $a_1, a_2, ..., a_n$ in PA^2G. To effectively attack the ensemble detector $\mathbf{f}$ that integrates multiple base detectors $f_1, f_2, ..., f_k$ with a certain strategy, the adversary will mix $a_1, a_2, ..., a_n$ by a series of weight factor $w_{a1}, w_{a2}, ..., w_{an}$.
- **Principle of Black-Box Attack:** Since we consider black-box scenarios that adversary has no knowledge of $\mathbf{f}$, adversaries cannot directly find appropriate adversarial perturbations δ according to $\mathbf{f}$. To this end, they first locally

train an ensemble model f' as the substitute model of target detector $\mathbf{f}$. Then δ can be generated according to substitute model f', which is considered effective perturbation to $\mathbf{f}$. By adding δ to malware traffic t_m, PA^2G can obtain adversarial malware traffic t_a that has the potential to trick the current $\mathbf{f}$.

- **Problem Modeling:** For a successful attack in practice, t_a needs to be deceptive (i.e. misclassified as t_m by the detector based on its extracted feature $\phi(t_a)$), and satisfy traffic-space constraints Θ. This problem can be described as:

$$\text{argmin}_{t_a} \, f'\left(\phi'(t_a)\right) + \lambda\mathcal{L}(t_a, t_m) \qquad \text{s.t. } t_a \in \Theta(t_m) \tag{2}$$

where $\phi'(t_a)$ represents the extracted feature of adversarial malware traffic t_a, λ indicates a parameter that determines the feature transformation cost from f' to $\mathbf{f}$, $\mathcal{L}$ is the loss function of $\mathbf{f}$, and $\Theta(t_m)$ represents the traffic-space constraints with the same malicious function and communication protocol as t_m.

- **Adversarial Malware Traffic t_a Generation:** To generate adversarial malware traffic t_a satisfying the traffic-space constraints, we take the following two steps. First, we generate adversarial feature x_a from a clean input x_m in feature space. Second, we design a remapping function $\mathcal{M}$ to project x_a to obtain the ultimate traffic-space adversarial traffic t_a, which is satisfying the traffic-space constraints. Details will be introduced in Sect. 4.2 and Sect. 4.3, respectively.

4.2   Generating Adversarial Features x_a

To find the optimal adversarial perturbation δ that can trick the substitute model f' to identify adversarial malware feature $x_m + \delta$ as a benign feature x_b, we need to maximize the loss function $\mathcal{L}'$ of f'. Let y represents the output label by f' according to the input features, the optimization problem for optimal δ can be expressed as:

$$max \, \mathcal{L}'(f'(x_m + \delta), y) \tag{3}$$

To solve the above optimization problem, a common method is to find δ along the gradient ascent direction of loss function $\mathcal{L}'$. Unfortunately, there are non-differentiable tree models in the ensemble substitute model f' trained locally by the adversary. This will cause the classical gradient ascent method to not work. Inspired by [6], we consider evaluating the gradient of loss function based on Natural Evolutionary Strategies (NES). Then the gradient ascent algorithm can be performed according to the estimated gradients.

Specifically, loss function $\mathcal{L}'(\theta)$ can be approximated as $\mathcal{L}'(\theta + \xi)$ through adding a series of search noise ξ_i with distribution $\pi(\theta|x_m)$. Rather than directly maximizing $\mathcal{L}'(\theta)$, NES maximizes the expected value of the loss function under the search distribution. By sampling n points of ξ_i values that obey $\xi_i \sim \mathcal{N}(0, I)$, the gradient G can be estimated by:

$$G \approx \nabla\mathbb{E}[\mathcal{L}(\xi)] = \frac{1}{\sigma n}\sum_{i=1}^{n}\xi_i\mathcal{L}\left(\theta + \sigma\delta_i\right) \tag{4}$$

Based on the estimated gradient G of substitute model f' loss function, attackers can iteratively find optimal perturbation δ along the direction of gradient ascent. Then the adversarial features x_a can be generated by $x_a = x_m + \delta$. Thanks to the transferability of adversarial attacks, x_a can trick the target detector $\mathbf{f}$ with certain effectiveness.

4.3   Projecting Adversarial Features x_a to t_a in Traffic Space

Adversarial traffic generated in practice should strictly satisfy the traffic-space constraints, that is, follow the communication protocol and preserve the malicious functionality of malware. To enforce these domain constraints, we use the remapping function $\mathcal{M}$ to adjust the perturbed features. There are some inherent characteristics of network traffic, such as flow-based features cannot be changed by attackers, some features depend on other features. For a more reasonable feature remapping, we group traffic features into the following four categories:

1. **Features with unchangeable value:** To maintain the malicious function of the original malware, some features shouldn't be changed because they are extracted from backward packets (i.e., victim packets).
2. **Features with zero value:** To prevent modified adversarial traffic from being directly dropped by the network, the modified features have to maintain the correctness of the network protocol. This requires certain features to have eigenvalues of 0 (e.g. the value of 'TCP flag').
3. **Features with valid interval boundaries:** Based on our analysis of different types of malware traffic, we found that there are upper and lower bounds on the value of each feature for different malware traffic, i.e., these feature values have validity intervals. In order to maintain the original malicious function, it is necessary to ensure that the modified feature values remain within the valid interval boundaries.
4. **Features depending on other features:** Since existing feature extractors usually extract statistical features of encrypted traffic, there is a correspondence between some feature values computed based on the same attributes. That is, if one feature value is modified, the feature values of other related features need to be changed accordingly.

We design remapping function $\mathcal{M}$ to restrict the adversarial malware traffic generated by modifying the traffic features to satisfy the above four types of constraints. The remapping function $\mathcal{M}$ mainly takes three technical solutions, which are masking, clip, and equation constraints. For the features with unchangeable values, $\mathcal{M}$ uses the masking technique to refuse adversaries to perturb these features when generating adversarial samples. For the features with zero value and the features with valid interval boundaries, $\mathcal{M}$ uses the clip technique to ensure that the perturbed feature values are 0 or within bounds. For the features depending on other features, $\mathcal{M}$ uses equation constraints to maintain the correlation between the feature values. According to $\mathcal{M}$, we can map the adversarial malware features x_a to the adversarial malware traffic t_a that satisfies the traffic-space constraints. Obviously, t_a has the potential to evade detection by $\mathbf{f}$, and will not be dropped by the network since network can understand it.

5 Bayesian Adversarial Ensemble Learning (BayesAE)

As the other player of Minimax Game, BayesAE aims to obtain a more robust ensemble detector with a dynamic ensemble strategy. The ensemble strategy when the game reaches Nash equilibrium is capable to defend against the current real-world adaptive mixed adversarial attacks. To solve the Nash equilibrium with little computation times, BayesAE provides a simplified solution. The detailed design principle of BayesAE will be elaborated in this section.

5.1 Dynamic Ensemble Strategy of Detector

To deal with the adaptively mixed attack of real-world adversaries, BayesAE constantly adjusts the ensemble strategy of *NashAE* according to current adversaries' attack schemes. This process is described as a Minimax Game between PA^2G and BayesAE, which is shown in Fig. 1. Specifically, the adaptive adversary constantly updates its mixed attack scheme that is composed of multiple attacks $a_1, a_2, ..., a_n$ with different weights $\mathbf{w}_a = w_{a1}, w_{a2}, ..., w_{an}$. Each update is performed with the goal of maximizing its attack success rate (ASR). Aiming to defend against every mixed attack launched by the adversary, *NashAE* constantly reconfigures a series of ensemble weights $\mathbf{w}_e = w_{e1}, w_{e2}, ..., w_{en}$ of multiple detector models. Each reconfiguration is performed with the goal of minimizing the updated ASR.

In each round of the game, PA^2G and BayesAE evaluate ASR under the current strategies. To maximize and minimize ASR by the two players, we update the ensemble learning weights $\mathbf{w}_e$ of ensemble strategy and the adversarial attack weights $\mathbf{w}_a$ by Bayesian optimization. After multiple rounds of fictitious play, the ensemble strategy based on converged $\mathbf{w}_e$ will be robust to the adaptive mixed attack schemes. This game process can be formulated as:

$$\min_{w_{ei} \in \mathbf{w}_e} \max_{w_{ai} \in \mathbf{w}_a} g(\mathbf{w}_e, \mathbf{w}_a)$$
$$\text{s.t. } g(\mathbf{w}_e, \mathbf{w}_a) = \sum_{i=1}^{n} w_{ai} \mathcal{L}\left(a_i(x), y\right) \tag{5}$$

where $g = ASR$ is the objective function that represents what we want to minimize/maximize, $\mathcal{L}$ is the loss function of the ensemble detector. For $i \in [1, n]$, a_i is a series of base attacks and n is the number of base attack types.

5.2 Solving Approximate Nash Equilibria

In order to obtain the most robust ensemble strategy w_e^*, we need to find the Nash equilibrium solution of the above Minimax Game. Inspired by [3], we convert the Minimax Game into two Markov decision processes, aiming to simplify computation. The two Markov decision are the process of finding optimal w_e^* and the process of finding optimal w_a^*, respectively. Then the Nash equilibrium solution can be represented as (w_e^*, w_a^*).

With the goal of simplifying the solution process of (w_e^*, w_a^*), we consider solving an approximate Nash equilibrium rather than an exact Nash equilibrium. Since the Bayesian optimization algorithm can save time overhead by referring to previous evaluations when trying the next set of hyper-parameters, it can get better results with a small number of attempts. To this end, we use the Bayesian optimization (BayesOpt) to solve the approximate Nash equilibrium (w_e^*, w_a^*). Details are shown in Algorithm 1.

Algorithm 1. Solving Approximate Nash Equilibria

Input: f, X_a, X_e, S, w_e, w_a
Output: $w_e{}^*$, $w_a{}^*$
1: Initialize f, X_a, X_e, S, w_e, w_a
2: $w_a = f_s(f, X_a)$
3: $w_e = f_s(f, X_e)$
4: **while** : each $k = 1, 2, 3, ...$ **do**
5:     **while** : each $i = 1, 2, 3, ...$ **do**
6:         $p(y_a \mid D_a, w_a) = fit(M, D_a)$
7:         $w_{ai} = \arg\max_{w_{ai} \in X_a} S(w_a, p(y|w_a, D))$
8:         Calculate objective function $y_i = f(w_{ai})$
9:         $D_a = D_a \bigcup (w_{ai}, y_{ai})$
10:         Increment i
11:     **end while**
12:     Adversary update attack strategies w_{ai} according to $M(w_a)$
13:     **while** : each $j = 1, 2, 3, ...$ **do**
14:         $p(y \mid D_e, w_e) = fit(M, D_e)$
15:         $w_{ej} = \arg\min_{w_{ej} \in X_e} S(w_e, p(y|w_e, D_e))$
16:         Calculate objective function $y_{ej} = f(x_{ej})$
17:         $D_e = D_e \bigcup (w_{ej}, y_{ej})$
18:         Increment j
19:     **end while**
20:     Detector update ensemble strategies w_{ej} according to $M(w_e)$
21: **end while**

Specifically, there are two loops of training. The first loop is the entire game model, and the second loop is BayesOpt. In the first loops, the adversary and detector sample attack mixed strategy w_a and detector ensemble strategies w_e during each epsilon $c \in \{1, ..., k\}$. The current objective function g is solved and $\mathbf{w}_a^c$ is updated to $\mathbf{w}_a^{c+1}$. Simultaneously, the detector ensemble strategy $\mathbf{w}_e^c$ is also updated by g to obtain $\mathbf{w}_e^{c+1}$. In the second loop, the adversary first randomly based on sample function f_s selects several sets $w_{a1}, w_{a2}, ... w_{an}$ to train the prior function M that is fitted to datasets D_a, D_e, D_t, and get the target values y. Then M is used to fit w_a, y, and acquisition function S is used to select the best w_a^*, and get the new y. Similarly, the detector's parameters are updated in the same way as the adversary. Finally, at a Nash equilibrium, both the adversary and detector will not want to change their strategies (w_a^*, w_e^*) since they will not lead to further benefit.

6   Experiments and Results

In this section, we conduct comprehensive experiments to evaluate the performance of our proposed *NashAE*. The experimental setup and results analysis will be described in detail.

6.1   Experimental Settings

- **Datasets:** We use CTU13 [4] and Datacon2020-EMT [1] as encryption malware traffic datasets to evaluate our *NashAE*. Traffic in the two datasets is divided into training set, validation set, and test set in a ratio of 6:2:2.
- **Base Models for Ensemble Malware Detector:** We implement 4 classic machine-learning models as base models, which can be integrated with certain strategies to construct ensemble malware detectors. They are logistic regression (LR) model, support vector machines (SVM) model, decision trees (DT) model, and multi-layer perception (MLP) model, respectively.
- **Baselines of Ensemble Strategy:** To evaluate the robustness performance against adversarial attacks of our *NashAE*, we implement 4 typical ensemble strategies and 2 advanced robust defense strategy as baselines. The typical ensemble strategies include: **1)** Bagging, **2)** Adaboost, **3)** Gradient boosting decision trees (GBDT), and **4)** Stacking. The advanced ensemble strategies include **5)** Omni [14], and **6)** NashRL [3]. Besides, aiming to demonstrate the effectiveness of *NashAE*'s dynamic ensemble structure, we extra implement **7)** BayesAE, which removes adaptive adversaries from the Minimax Game.
- **Evaluation Metrics:** We use three typical machine learning metrics to evaluate the detection performance of malware detectors. To evaluate the attack performance of adversarial attacks, we extra design attack success rate (ASR) metric. To evaluate the robustness against adversarial attacks of malware detectors, we extra design detection rate (DR) metric. Details are as follows:
  1. **Typical metrics**: It includes precision (P), recall (R), and F1 score (F1).
  2. **ASR**: The attack success rate measures the proportion of traffic with successful attacks in all generated traffic.
  3. **DR**: The detection rate represents the proportion of the generated adversarial traffic that is detected by the detector within all adversarial traffic.
- **Experiment preparation:** For the convenience of subsequent experiments, we pre-trained a series of malware detectors on CTU13 dataset and Datacon-EMT dataset, respectively. As shown in Table 1, all of these detectors achieve satisfactory detection performance with high precision, recall, and F1 score.

6.2   Effectiveness Evaluation of Transfer-Based Adversarial Attacks

To evaluate the effectiveness of transfer-based adversarial attacks, we perform adversarial attacks on each of the above 11 malware detectors. All the optimal adversarial perturbations are found by NES algorithm. With the goal of transferability evaluation, we add adversarial perturbations generated from the current

Table 1. Detection performance without adversarial attacks

Detectors	CTU13			Datacon-EMT		
	R	P	F1	R	P	F1
LR	0.8961	0.9363	0.9380	**0.8589**	0.7248	0.7862
SVM	0.9151	0.9845	0.9485	**0.8643**	0.7208	0.7861
DT	**0.9958**	**0.9983**	**0.9970**	0.8381	0.8335	0.8358
MLP	0.9847	**0.9979**	**0.9912**	0.8326	0.8437	0.8381
Bagging	**0.9962**	**0.9979**	**0.9970**	**0.8873**	**0.8528**	**0.8697**
Adaboost	**0.9907**	**0.9953**	**0.9930**	0.8381	0.8464	0.8422
GBDT	**0.9958**	**0.9991**	**0.9975**	**0.9201**	**0.8599**	**0.8890**
Stacking	**0.9958**	**0.9983**	**0.9970**	**0.9136**	0.8486	**0.8799**
Omni	0.9571	**0.9969**	0.9766	**0.9639**	0.7569	0.8479
NashRL	0.9745	0.9899	**0.9901**	**0.9136**	0.8486	**0.8799**
BayesAE	**0.9943**	**0.9979**	**0.9912**	**0.8973**	**0.8638**	**0.8871**
NashAE	**0.9907**	**0.9953**	**0.9930**	**0.8873**	**0.8528**	**0.8697**

detector to the input traffic of the other 10 detectors. Based on this, we can obtain a cross-technology transferability matrix. As shown in Fig. 2, the number on each cell of the matrix represents the ASR based on the transfer between the row and column corresponding detectors. More intuitively, the redder the cell color, the higher the ASR of the corresponding transfer attack. Similarly, the bluer the color, the lower the ASR.

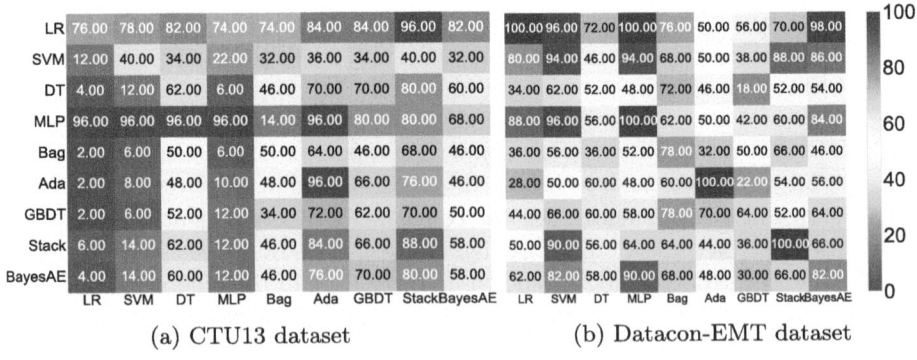

(a) CTU13 dataset   (b) Datacon-EMT dataset

Fig. 2. Transfer-based attack success rates on two datasets.

Specifically, we can find that dark red cells are usually transferred between the same or similar detector models. That is, the more similar the models are, the better the transfer will be. But there is a special case that the adversarial traffic generated for DT are almost useless for LR, which only achieve a 4% attack success rate on the CTU13 dataset. In addition, almost all ensemble models (Bagging, Adaboost, GBDT, Stack) have been attacked successfully with ASR up to 90%. Even though our BayesAE is relatively robust compared with other ensemble methods, it is still vulnerable to specific adversarial attacks. This means that a simple fixed ensemble strategy cannot achieve satisfactory robustness against even black-box adversarial attacks.

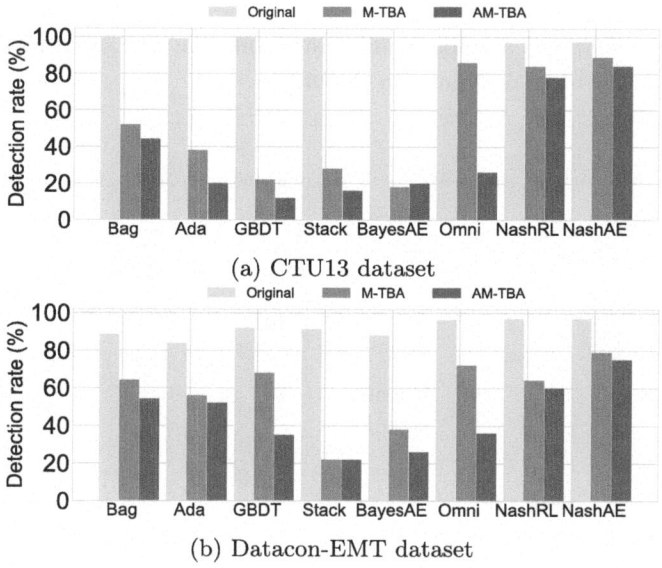

Fig. 3. Detection rate of 8 ensemble detectors under two threat models.

6.3   Robustness Evaluation Against M-TBA and AM-TBA

To further evaluate the robustness of ensemble detectors under realistic attack assumptions, we conduct adversarial attacks to five basic ensembles and three advanced ensembles under M-TBA and stronger AM-TBA attacks, respectively. As shown in Fig. 3, compared with the basic detectors, all of the three advanced ensemble detectors (Omni, NashRL and *NashAE*) exhibit strong robustness against an M-TBA attack. For example, on the CTU13 dataset, their detection rate is over 80%, while almost all of other ensemble detectors can only achieve a detection rate below 50%. Under AM-TBA, however, the robustness of Omni is greatly diminished, reaching a DR of only 22% on the CTU13 dataset, while NashRL and *NashAE* are significantly better and still achieve around 80% DR. Datacon-EMT performs similarly to CTU-13, and our *NashAE* can still attain a detection rate of over 75% despite adaptive attacks.

6.4   Convergence Performance Evaluation

As shown in Fig. 4, we further study the convergence performance of NashRL and *NashAE*. Although their iteration times of converge almost identical on both CTU13 and Datacon-EMT datasets, the detection rate of *NashAE* is steadily better than that of NashRL. Specifically, *DR* of NashRL is only 2.5% worse than NashAE on the CTU13 dataset. However, on the Datacon-EMT dataset, NashRL becomes extremely vulnerable to adversarial attacks after attack power has been doubled. The *DR* metric of NashRL is even less than 60%. In contrast, our NashAE can still achieve a maximum of 75% *DR*. This means that our method can converge to a stable equilibrium solution, and outperform NashRL by over 15% regardless of the attack strength.

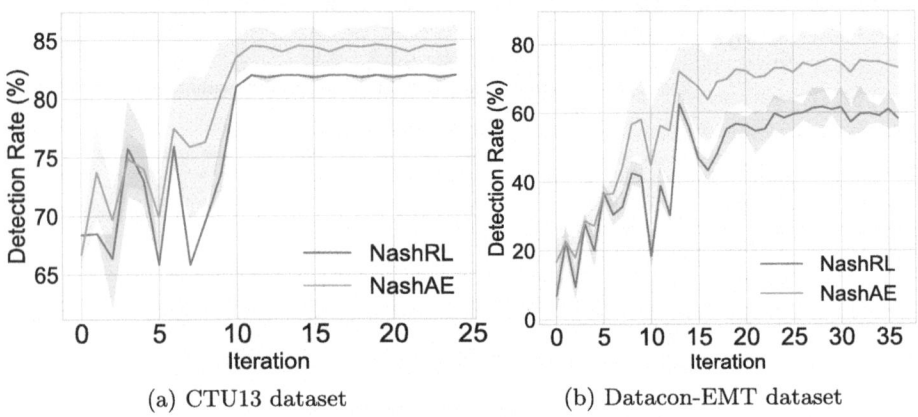

(a) CTU13 dataset   (b) Datacon-EMT dataset

Fig. 4. Iterations numbers of NashAE and NashRL on two datasets.

7   Conclusion

In this paper, we propose *NashAE* as an ensemble malware detector with dynamic ensemble strategies. As the first step towards training robust malware detectors by adversarial ensemble framework, *NashAE* provides security operation and maintenance personnel with a solution that can deal with powerful adversaries. Our work can be further improved, i.e., *NashAE* uses restricted feature-space attacks to mimic traffic-space attacks, while the traffic should be directly modified in a more realistic traffic-space adversarial attack. To solve this problem, the inverse feature-mapping [11] is a critical technique. However, converting a feature vector into a traffic-space object is difficult due to the feature mapping function is neither invertible nor differentiable. In the future, we will continue to explore solutions to this challenge for a more robust detector.

Acknowledgement. This study was supported in part by the Key Research Program of the Chinese Academy of Sciences under Grant NO. KGFZD-145-23-03, and the State Grid Henan Company's Industrial Unit Science and Technology Project under Grant 2024-KJ-01 (Research on High Security Access Technology for 5G New Distribution Grids).

References

1. https://datacon.qianxin.com/opendata/maliciousstream
2. https://github.com/trusted-ai/adversarial-robustness-toolbox
3. Dou, Y., Ma, G., Yu, P.S., Xie, S.: Robust spammer detection by nash reinforcement learning. In: Proceedings of the 26th ACM SIGKDD International Conference on Knowledge Discovery & Data Mining, pp. 924–933 (2020)
4. Garcia, S., Grill, M., Stiborek, J., Zunino, A.: An empirical comparison of botnet detection methods. Comput. Secur. **45**, 100–123 (2014)
5. Huang, Z., Zhang, T.: Black-box adversarial attack with transferable model-based embedding. arXiv preprint arXiv:1911.07140 (2019)
6. Ilyas, A., Engstrom, L., Athalye, A., Lin, J.: Black-box adversarial attacks with limited queries and information. In: International Conference on Machine Learning, pp. 2137–2146. PMLR (2018)
7. Kariyappa, S., Qureshi, M.K.: Improving adversarial robustness of ensembles with diversity training. arXiv preprint arXiv:1901.09981 (2019)
8. Li, D., Li, Q.: Adversarial deep ensemble: evasion attacks and defenses for malware detection. IEEE TIFS **15**, 3886–3900 (2020)
9. Li, Y., Song, Y., Jia, L., Gao, S., Li, Q., Qiu, M.: Intelligent fault diagnosis by fusing domain adversarial training and maximum mean discrepancy via ensemble learning. IEEE Trans. Ind. Inf. **17**(4), 2833–2841 (2020)
10. Novo, C., Morla, R.: Flow-based detection and proxy-based evasion of encrypted malware c2 traffic. In: Proceedings of the 13th ACM Workshop on Artificial Intelligence and Security, pp. 83–91 (2020)
11. Pierazzi, F., Pendlebury, F., Cortellazzi, J., Cavallaro, L.: Intriguing properties of adversarial ml attacks in the problem space. In: 2020 IEEE Symposium on Security and Privacy (SP'20), pp. 1332–1349. IEEE (2020)
12. Qiu, H., Dong, T., Zhang, T., et al.: Adversarial attacks against network intrusion detection in iot systems. IEEE Internet Things J. (2020)
13. Rigaki, M.: Adversarial deep learning against intrusion detection classifiers (2017)
14. Shu, R., Xia, T., Williams, L., Menzies, T.: Omni: automated ensemble with unexpected models against adversarial evasion attack. Empir. Softw. Eng. **27**(1), 1–32 (2022)
15. Tayyab, U., Khan, F.B., Durad, M.H., Khan, A., Lee, Y.S.: A survey of the recent trends in deep learning based malware detection. J. Cybersecur. Priv. **2**(4), 800–829 (2022)
16. Tramèr, F., Kurakin, A., Papernot, N., et al.: Ensemble adversarial training: attacks and defenses. arXiv preprint arXiv:1705.07204 (2017)
17. Urooj, B., Shah, M.A., Maple, C., Abbasi, M.K., Riasat, S.: Malware detection: a framework for reverse engineered android applications through machine learning algorithms. IEEE Access **10**, 89031–89050 (2022)
18. Wang, H., Wang, Y.: Self-ensemble adversarial training for improved robustness. In: International Conference on Learning Representations (2021)

19. Wu, D., Fang, B., Wang, J., Liu, Q., Cui, X.: Evading machine learning botnet detection models via deep reinforcement learning. In: ICC 2019-2019 IEEE International Conference on Communications (ICC), pp. 1–6. IEEE (2019)

20. Xie, C., Zhang, Z., Zhou, Y., et al.: Improving transferability of adversarial examples with input diversity. In: Proceedings of the IEEE/CVF Conference on Computer Vision and Pattern Recognition, pp. 2730–2739 (2019)

21. Xing, X., Jin, X., Elahi, H., Jiang, H., Wang, G.: A malware detection approach using autoencoder in deep learning. IEEE Access 10, 25696–25706 (2022)

22. Zhang, F., Wang, Y., Liu, S., Wang, H.: Decision-based evasion attacks on tree ensemble classifiers. World Wide Web 23(5), 2957–2977 (2020). https://doi.org/10.1007/s11280-020-00813-y

23. Zhang, F., Wang, Y., Wang, H.: Gradient correlation: are ensemble classifiers more robust against evasion attacks in practical settings? In: Hacid, H., Cellary, W., Wang, H., Paik, H.-Y., Zhou, R. (eds.) WISE 2018. LNCS, vol. 11233, pp. 96–110. Springer, Cham (2018). https://doi.org/10.1007/978-3-030-02922-7_7

24. Zhou, M., Wu, J., Liu, Y., Liu, S., Zhu, C.: Dast: data-free substitute training for adversarial attacks. In: Proceedings of the IEEE/CVF Conference on Computer Vision and Pattern Recognition, pp. 234–243 (2020)

Physics-Aware Compression of Plasma Distribution Functions with GPU-Accelerated Gaussian Mixture Models

Andong Hu[✉], Luca Pennati, Ivy Peng, and Stefano Markidis

KTH Royal Institute of Technology, Stockholm, Sweden
andonghu@kth.se

Abstract. Data compression is a critical technology for large-scale plasma simulations. Storing complete particle information requires Terabyte-scale data storage, and analysis requires ad-hoc scalable post-processing tools. We propose a physics-aware in-situ compression method using Gaussian Mixture Models (GMMs) to approximate electron and ion velocity distribution functions with a number of Gaussian components. This GMM-based method allows us to capture plasma features such as mean velocity and temperature, and it enables us to identify heating processes and generate beams. We first construct a histogram to reduce computational overhead and apply GPU-accelerated, in-situ GMM fitting within `iPIC3D`, a large-scale implicit Particle-in-Cell simulator, ensuring real-time compression. The compressed representation is stored using the `ADIOS 2` library, thus optimizing the I/O process. The GPU and histogramming implementation provides a significant speed-up with respect to GMM on particles (both in time and required memory at run-time), enabling real-time compression. Compared to algorithms like SZ, MGARD, and BLOSC2, our GMM-based method has a physics-based approach, retaining the physical interpretation of plasma phenomena such as beam formation, acceleration, and heating mechanisms. Our GMM algorithm achieves a compression ratio of up to 10^4, requiring a processing time comparable to, or even lower than, standard compression engines.

Keywords: Gaussian-Mixture-Model Compression · Compression Particle-in-Cell · Distribution Functions

1 Introduction

Plasma simulations are essential computational tools for understanding the dynamics of space, astrophysical systems, and fusion devices like toka-maks. Among the various computational techniques, the Particle-in-Cell (PIC) method [3,9] is one of the most advanced and widely used plasma simulation tools for studying the evolution of electrons and protons (called ions in plasma

© The Author(s), under exclusive license to Springer Nature Switzerland AG 2025
M. H. Lees et al. (Eds.): ICCS 2025, LNCS 15905, pp. 33–47, 2025.
https://doi.org/10.1007/978-3-031-97632-2_3

physics) under the influence of electromagnetic fields. The distribution function evolves by sampling computational particles and following their trajectories. The key information from PIC simulations is the data about particle positions and velocities, as it enables us to identify acceleration and heating mechanisms in the plasmas. Nonetheless, the storage requirements and post-processing costs significantly limit the volume of simulation data that can be saved and analyzed. In this work, we develop a compression method based on the Gaussian Mixture Model (GMM) [1] to tackle this challenge. Rather than saving all particle information, we perform a GMM operation, obtaining the weight, center, and covariance for a small number of Gaussians used to represent all the data, thus saving the storage by several factors.

In particular, we develop a compression technique to compress particle distribution functions. Electron and ion distribution functions are related to the probability of finding an electron or an ion at a certain position in the phase-space, namely a position (x, y, z) in Cartesian coordinates with certain velocity components (u, v, w). In a 3D3V simulation, a distribution function $f(\mathbf{x}, \mathbf{v})$ has six dimensions. Electron and ion particle distributions tend to be Gaussian in nature as a consequence of the H-Theorem and relaxation processes present in plasmas [11]. If particle distribution functions are different from Gaussian distribution functions, it hints at the presence of certain resonance phenomena that act only on a small part of the plasma populations, such as Landau damping or bump-on-tail. A Gaussian centered on a given velocity, different from zero, shows the presence of a beam with that specific bulk velocity. Similarly, an increase in the standard deviation of the Gaussian reflects a rise in the local plasma temperature. For these reasons, a GMM-based compression method is called *physics-aware*. This work provides the following contributions:

- We develop a methodology to reduce the memory and computational demands of GMM, applying it to an intermediate histogram instead of particle data directly.
- We develop a GPU implementation of the histogram-GMM pipeline and execute it during spare GPU cycles in the PIC simulation to compress plasma distribution functions.
- We show physics-aware compression using Gaussian parameters, and compare the information loss, compression rate and performance with several established compression algorithms.

2   Background and Related Work

Large-scale PIC simulations compute the trajectories of billions of computational particles, generating Terabyte-size data, including distribution functions. Figure 1 shows different possibilities of storing particle information for particle distribution functions. On the leftmost panel, there is the raw information of particles velocity: in this plot, each dot corresponds to a particle, requiring the storage of $2N_p$ floating point data, where N_p is the number of particles. One compression scheme for particle distribution function, is to introduce a 2D grid in the space u, v to discretize the distribution and function and count how many particles belong to each bin. This data reduction technique allows us to encode our distribution function in a $N_g \times N_g$ grid. The approach we pursue in this work is to use the GMM technique to fit the particle distribution function with a relatively small number of Gaussians (1–20) and store only the Gaussian parameters for the reconstruction. In the example of Fig. 1, we only need the parameters of two Gaussians (12 floating point values for this two-dimensional example) to capture the complexity of the given distribution function.

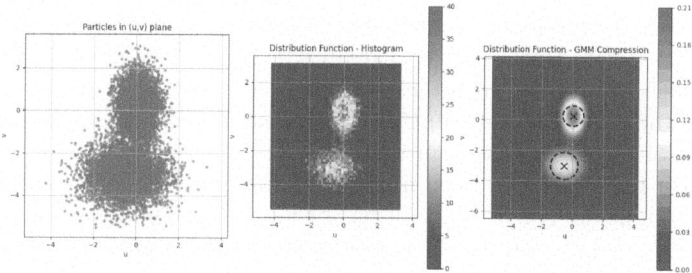

Fig. 1. The distribution function of 10,000 particles can be approximated with an 64×64 histogram (central panel) or a GMM approach storing the weights, the center and covariance matrices of two Gaussians (rightmost panel).

The basic idea of the GMM technique is to represent a probability distribution as a weighted sum of multiple Gaussian components, each defined by a mean vector and a covariance matrix [1,7]. Mathematically, a GMM models a probability density function (pdf) $p(\mathbf{x})$ as a weighted sum of M Gaussian components:

$$p(\mathbf{x}) = \sum_{i=1}^{M} \alpha_i \mathcal{N}(\mathbf{x}|\boldsymbol{\mu}_i, \boldsymbol{\Sigma}_i), \tag{1}$$

where α_i are the mixture weights, $\boldsymbol{\mu}_i$ are the means, and $\boldsymbol{\Sigma}_i$ are the covariance matrices. Given N observed data, each one denoted as $\mathbf{x}_n$, the GMM parameters $\{\alpha_i, \boldsymbol{\mu}_i, \boldsymbol{\Sigma}_i\}$ are estimated using the Expectation-Maximization (EM) algorithm [20], which maximizes the likelihood function $\mathcal{L}(\boldsymbol{\theta}|\mathbf{x})$, defined as: $\mathcal{L}(\boldsymbol{\theta}|\mathbf{x}) = \sum_{i=1}^{N} \left[\sum_{j=1}^{M} \alpha_j \mathcal{N}(\mathbf{x}_n|\boldsymbol{\mu}_j, \boldsymbol{\Sigma}_j) \right]$. The EM algorithm consists of a two-step iteration:

– **E-step:** Compute responsibilities for each Gaussian component for each observed data:

$$\gamma_i(\mathbf{x}_n) = \frac{\alpha_i \mathcal{N}(\mathbf{x}_n|\boldsymbol{\mu}_i, \boldsymbol{\Sigma}_i)}{\sum_{j=1}^{M} \alpha_j \mathcal{N}(\mathbf{x}_n|\boldsymbol{\mu}_j, \boldsymbol{\Sigma}_j)}. \tag{2}$$

– **M-step:** Update the parameters:

$$\alpha_i^{\text{new}} = \frac{1}{N} \sum_{n=1}^{N} \gamma_i(\mathbf{x}_n) \tag{3}$$

$$\boldsymbol{\mu}_i^{\text{new}} = \frac{\sum_{n=1}^{N} \gamma_i(\mathbf{x}_n)\mathbf{x}_n}{\sum_{n=1}^{N} \gamma_i(\mathbf{x}_n)} \tag{4}$$

$$\boldsymbol{\Sigma}_i^{\text{new}} = \frac{\sum_{n=1}^{N} \gamma_i(\mathbf{x}_n)(\mathbf{x}_n - \boldsymbol{\mu}_i)(\mathbf{x}_n - \boldsymbol{\mu}_i)^\top}{\sum_{n=1}^{N} \gamma_i(\mathbf{x}_n)} \tag{5}$$

The EM-GM technique is particularly suitable for describing the probability distribution associated with physical quantities since this algorithm conserves, exactly at each iteration – regardless of whether it has converged or not – the moments of the observed data up to the second moment [2]. This ensures the conservation of physical bulk quantities.

The GMM technique has been used before for compressing scientific data in several research fields. For instance, it has been used to compress image data [25] and medical ECG data [24]. In the area of computational plasma physics, GMM has been used for checkpointing and restart [2] and as a post-processing tool to characterize magnetic reconnection regions [6] in PIC simulations.

Several compression techniques have been developed for scientific computing data [4,13], consisting of floating-point number. Among the most used and successful compression methods, there is MGARD (MultiGrid Adaptive Reduction of Data) [14], developed at Oak Ridge National Lab for multi-level (inspired by Multigrid methods) data error-bounded lossy compression. The basic compression mechanism is based on decomposing into multiple levels (or grids). At each level, the coarse grained components are separated from the finer details, leading to an hierarchy of coarser and finer representations of the data. Another widely used compression scheme, targeting floating-point data and HPC simulation result, is ZFP [16], developed at Lawrence Livermore National Laboratory. The basics mechanisms are based on dividing the input into fixed-size blocks, reordering into a Z-order curve, and applying a customized Discrete Cosine Transform-like Transform, similarly to JPEG. In addition, another popular compression scheme is SZ (Squeeze in short) [5], developed at the Argonne National Laboratory. This method is based on a prediction step, based on a Lorenzo predictor [10] or linear regression, predict the value of each data point based on its neighboring values, error quantization and Huffmann encoding. A comparison of our proposed GMM-based compression with other compression schemes for scientific computing is presented in Table 1.

In this work, we design and implement a compression method based on GMM, and demonstrate its effectiveness in the iPIC3D code, a massively parallel PIC code [17]. The code is designed for space plasma simulations, focusing on studying phenomena such as plasma-wave interactions [26], magnetic reconnection [21], collisionless shocks [23], and global planetary magnetosphere dynamics [22]. iPIC3D uses the implicit moment PIC method [19], distinguishing it from standard explicit PIC approaches. This method relaxes numerical stability constraints related to time step and grid resolution, enabling simulations of large-scale systems such as planetary magnetospheres. iPIC3D is implemented in C++ with support for MPI and OpenMP, and it has demonstrated strong scalability, running on up to one million MPI processes [18]. Additionally, it supports heterogeneous architectures, including supercomputers with AMD and NVIDIA GPUs. To compare the results of our GMM-based compression scheme and other standard compression schemes, we developed an ADIOS 2 [8] I/O iPIC3D module to leverage different compression schemes.

Table 1. Comparison of GMM-Based Compression with other scientific data compression schemes.

Feature	GMM-Based Compression	Other Compression Schemes (ZFP, SZ, MGARD, ...)
Compression Type	Lossy, physics-aware Gaussian approximation	Lossy or lossless, numerical-based approximations
Data Representation	Stores Gaussian parameters (mean, covariance, weight) instead of full velocity distributions	Encodes data using fixed-size blocks, wavelets, or error-controlled quantization
Physics Preservation	Retains key plasma physics properties (bulk velocity, temperature, beams, heating)	May distort physical properties, especially in structured plasma distributions
Compression Ratio	Adaptive, based on the number of Gaussians; high reduction rates for Gaussian-like distributions	Fixed compression ratio; depends on error tolerance settings
Computational Cost	Iterative fitting; GPU-accelerated for in-situ execution	Faster for standard lossless/lossy compression; They can be computationally expensive
Error Control	Adjusts Gaussian component number dynamically to control error or prune Gaussian with relatively small weight	Provide absolute or relative error bounds
Interpretability	Directly relates to plasma physics; Gaussian parameters can be analyzed for physical insights	Mathematical compression

3 Methodology

3.1 Weighted Gaussian Mixture Model

We employ the GMM on particle velocity distribution functions to compress the data and perform in-situ analysis. Since we are mainly interested in full-scale three-dimensional simulations, for the feasibility of real-time in-situ compression, we treat each simulation subdomain independently, and we pre-process the raw particle data by binning the velocities into a histogram. Specifically, from the full 6D distribution function, we derive the 3D velocity distribution function integrating over each simulation subdomain: $f(u, v, w) = \int f(\mathbf{x}, \mathbf{v})d\mathbf{x}$.

Then, from the 3D $f(u, v, w)$, we create three 2D histograms that represent the reduced-dimensional distribution functions $f_1(u, v)$, $f_2(v, w)$ and $f_3(u, w)$. These histogram pdfs are eventually given as input data to GMM, thus, for each pdf we have a constant $N_b \times N_b$ number of observed data, where N_b is the number of histogram bins.

To retain the correct velocity distributions, we utilize a modified (weighted) GMM algorithm that assigns a weight to each observed data point based on the corresponding bin counts. The GMM fitting process relies on the EM algorithm, as detailed in Sect. 2, and the weight of each input data (w_n) is taken into account in the E step. Thus, Eqs. 3, 4, 5 are modified to Eqs. 6, 7, 8, respectively:

$$\alpha_i^{\text{new}} = \frac{\sum_{n=1}^{N} w_n \gamma_i(\mathbf{x}_n)}{\sum_{n=1}^{N} w_n}, \tag{6}$$

$$\mu_i^{\text{new}} = \frac{\sum n = 1^N w_n \gamma_i(\mathbf{x}_n)\mathbf{x}_n}{\sum_{n=1}^{N} w_n \gamma_i(\mathbf{x}_n)}, \tag{7}$$

$$\Sigma_i^{\text{new}} = \frac{\sum_{n=1}^{N} w_n \gamma_i(\mathbf{x}_n)(\mathbf{x}_n - \boldsymbol{\mu}_i^{\text{new}})(\mathbf{x}_n - \boldsymbol{\mu}_i^{\text{new}})^\top}{\sum_{n=1}^{N} w_n \gamma_i(\mathbf{x}_n)}. \tag{8}$$

Our GMM implementation guarantees the generality and robustness of the algorithm. Specifically, we implement an automatic tuning of the number of Gaussians in the mixture based on the mixing weights. Every ten EM iterations, a kernel checks whether any component has a mixing weight below a certain threshold; if so, the component is pruned before proceeding with the next iteration, and the weights of the other components are rescaled. To prevent excessively reducing the number of components, Gaussians are pruned one at a time. Thus, we start the GMM with M active components, and we end it with $1 \leq \hat{M} \leq M$ Gaussians. Additionally, we ensure the algorithm's numerical stability by implementing safety checks on the covariance matrix of each component. These checks guarantee that the property of being symmetric positive definite is maintained at every iteration. If, due to floating-point arithmetic, the matrix has a negative determinant, we increase the values of the main diagonal elements.

Normalization of the observed data is another beneficial step that enhances the algorithm's numerical stability. By rescaling the data to the $[-1, 1]$ range, we can work with values several orders of magnitude larger than the original data, making them less sensitive to rounding errors and floating-point arithmetic limitations.

The GMM algorithm is highly sensitive to the initial parameter estimates, and a well-chosen initialization is crucial for ensuring proper convergence. When in-situ data analysis (DA) is performed for the first time in the simulation, GMM is initialized with uniform mixing coefficients, variances equal to the species temperatures, and random means to ensure that the Gaussians cover the entire data range. In the subsequent DA steps, GMM is initialized using the parameters estimated in the previous DA cycle. The more frequently DA is executed, the less the data changes between consecutive GMM runs, leading to increasingly accurate initial parameter estimates.

In our integrated histogram-GMM DA pipeline, all calculations, beginning with velocity binning, are executed on the GPU, thus minimizing host-device data transfers. Specifically, raw particle data and histogram pdfs are not copied to the host, allowing real-time integration within the simulation workflow. The results of weighted GMM include parameters like mean vectors, covariance matrix, and weights for all clusters, along with some auxiliary values such as the number of EM steps and the log-likelihood. After GMM has converged, the parameters can be copied back to the host and then written to disk in different formats, such as JSON and BP5.

Regarding performance concerns, the buffers in the data analysis pipeline are reused in the whole life cycle of simulations, thereby minimizing latency caused by configurations. Customized reduction kernels are employed in the weighted GMM implementation for cross-platform compatibility. Tailored pre- and post-processing techniques are applied to reduce kernel launches in GMM regression.

3.2 ADIOS 2 in iPIC3D

In this work, we use ADIOS 2 to implement I/O of GMM-based compression and compare the performance of our GMM-based compressor with other established compressors. ADIOS 2 enables asynchronous (deferred) data output and provides operators for compression. Each particle attribute (position, velocity, charge, ID), in Structure-of-Array buffer, is mapped to an ADIOS 2 variable, then is output to Binary-Pack 5 (BP5) files.

The iPIC3D `initOutputFiles` function initializes the ADIOS 2 environment, defines output file structures, and sets up data writing parameters. Particle data is written in a structured format using the `appendParticleOutput` function, detailed in Listing 1.1. This function captures the simulation cycle and appends particle properties, such as position, velocity, charge, and ID, to the output file. It also employs ADIOS 2's `BeginStep` and `EndStep` to utilize deferred writes.

```
1  void ADIOS2Manager::appendParticleOutput(int cycle) {
2      ...
3      engineParticle.BeginStep();
4      auto cycleVar = _variableHelper<int>(ioParticle, "cycle");
5      engineParticle.Put<int>(cycleVar, cycle);
6      ...
7      engineParticle.EndStep();
8  }
```

Listing 1.1. Appending Particle Output.

A key benefit of ADIOS 2 is its built-in data compression operation function, which enables the specification of a compressor engine and the following compression of simulation data prior to storage. For example, Listing 1.2 details the instruction needed to apply compression to the variable `var` with the SZ compressor. The ADIOS 2 interface decouples the data and the main application code from the actual compression algorithm, this allows great flexibility since we can easily change the compressor back-end without the need to restructure the I/O.

```
1  var.AddOperation(adios2::ops::LossySZ, {{"accuracy", "0.00001"}});
```

Listing 1.2. Apply compression to data with a given engine.

In this work, we evaluate the performance and effectiveness of several non-physics-aware compression methods and compare them to our physics-aware GMM method. Specifically, we evaluate lossy methods, including SZ, ZFP, and MGARD, and lossless methods like BLOSC2 and BZIP2. All compressors are executed on the CPU and take host memory buffers as input; no GPU is involved in the process.

3.3   Compression Performances and Information Loss Evaluation

In this work, we analyse the computational performances and the compression rate of each compression method. Additionally, we assess the information loss in the compression stage comparing the compressed data with the ground truth.

Since in this work we focus on the particle velocity distribution, we evaluate the accuracy of the compressed data by employing the Jensen-Shannon Divergence (JSD) [15]. The JSD is an information-theoretic metric for comparing probability distributions, so it can be used to assess how well the compressed data retains the features of the original distribution. The JSD for two probability distributions P and Q is evaluated as reported in Eq. 9:

$$JSD(P||Q) = \frac{1}{2}D\left(P||M\right) + \frac{1}{2}D\left(Q||M\right),\tag{9}$$

where M is the mixture of the distribution $M = (P + Q)/2$ and $D(\cdot)$ denotes the Kullback-Leibler divergence [12], defined as:

$$D(P||Q) = \sum_{n=1}^{N} p(\mathbf{x}_n) \log\left(\frac{p(\mathbf{x}_n)}{q(\mathbf{x}_n)}\right).\tag{10}$$

The result of JSD falls between 0 and $\ln 2$. The lower the result, the closer the two distributions are to each other.

Additionally, we assess the quality of GMM, with Bayesian Information Criterion (BIC), reported in Eq. 11:

$$\mathrm{BIC} = -2\ln(\mathcal{L}) + k\ln(N),\tag{11}$$

where $\mathcal{L}$ denotes the maximized likelihood function at the end of the EM algorithm, k is the number of parameters estimated by the model and N the number of observed data. In the case of multivariate GMM applied to d dimensional data $k = M(1 + d(d + 3)/2)$, where M is the number of Gaussians in the mixture. The BIC metric penalizes mixtures with a high number of components.

We highlight here some crucial aspects related to ground truth identification and information loss evaluation for the GMM compression technique. In PIC simulations, due to the finite number of computational particles, the numerical distribution function is a noisy approximation of the true, smooth physical pdf, which is the correct ground truth, but is never available. Additionally, in our implementation, we utilize pre-processed binned data as input for the GMM algorithm. Therefore, we can consider the pdf extrapolated from the histogram (which is still noisy) as the ground truth. It is important to note that this pdf does not exactly match the numerical pdf since binning acts as a smoother. On the other hand, the GMM method inherently produces a smooth pdf that may ultimately offer a more accurate approximation of the true physical pdf than the noisy one provided by the computational particles. For this reason, evaluating the accuracy of the GMM pdf solely with a metric based on point-wise comparison with a noisy pdf, such as the JSD metric, may yield incomplete information.

In our analysis, we calculate the JSD metric to show the differences between the histogram and GMM pdf. We also compute the JSD between the raw particle data pdf and the histogram by binning the particle velocities into a highly refined 500×500 histogram (labeled as *original* data in the following) and comparing it with the linearly interpolated histogram used as the first stage of the GMM compression. Since the EM algorithm inherently ensures the conservation of quantities like the mean and variance of the data, we extend the evaluation of the GMM compression accuracy with a qualitative comparison between the GMM pdf and the histogram pdf shape.

In order to assess the performance, namely processing time and compression ratio, of the five standard compression engines and our GMM-based algorithm, we deploy all of them in the same GEM simulation run to ensure the utilization of the same raw data.

4 Experimental Setup

As a main demonstration for the GPU accelerated GMM compression, we run a 3D version of the standard plasma physics benchmark problem, called GEM challenge, with the iPIC3D code. This test mimics the plasma condition in the Earth magnetotail, initiates magnetic reconnection phenomena perturbing the magnetic field topology along a central line in anti-parallel configuration of the magnetic field, without the presence of a background magnetic field. The size of the simulation box, in units of ion skin depth (d_i), is $10 \times 10 \times 10$, the domain is discretized with a uniform Cartesian grid of $133 \times 133 \times 135$ cells, and we run the simulation for 3,000 steps. The domain is divided into $7 \times 7 \times 5$ subdomains, each one assigned to one MPI rank. Figure 2 shows an example of magnetic field configuration during magnetic reconnection. The area in the orange box represents the region from which we obtain the particle data, corresponding to the subdomain at the center. This region has dimensions of $L_x = L_y = 1.42\ d_i$ and $L_z = 2\ d_i$.

We run GMM every 50 simulation cycles, with an initial number of 12 Gaussians, for a maximum of 100 EM cycles. The automatic pruning method removes components with a mixing coefficient lower than 0.005. Raw particle data is pre-processed by binning the velocities into fixed-range histograms of 200×200 bins.

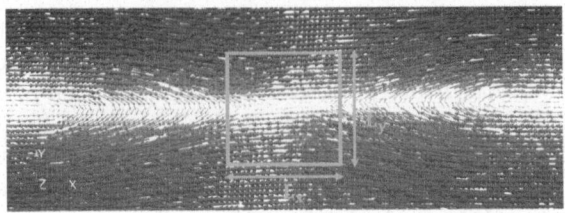

Fig. 2. Magnetic field line at the x-point during magnetic reconnection. The region where we take the data, with dimensions $L_x = L_y = 1.42\ d_i$, $L_z = 2\ d_i$ is highlighted in orange. (Color figure online)

We evaluate the performance of the different compression engines with a smaller GEM simulation, with a grid of $64 \times 64 \times 64$ cells divided into $2 \times 2 \times 2$ subdomains. We run the simulation for 4000 cycles. The output frequency for particle data is set to every 500 cycles, and particle velocity and charge data are written to BP5 files using ADIOS 2. For evaluation, we fix the number of GMM components to eight and disable features like pruning, with a 100×100 histogram as its pre-processor. The accuracy parameter of the lossy compressors is set to 0.00001, 0.01, and 0.01 for SZ, ZFP, and MGARD, respectively, to keep the output size as similar as possible. As for the lossless methods, BZIP2 uses $blockSize100k = 5$ and BLOSC2 uses $blosclz$ and $clevel = 9$.

We perform our tests on two main systems, involving both Nvidia and AMD GPUs. The first system is one node of the Sleipner cluster at KTH equipped with a Grace-Hopper 200 480 GB accelerator. We use this machine to compare the GMM algorithm with different compressors in terms of achievable compression ratio and execution time. The iPIC3D production runs involving in-situ GMM-based data compression are executed on the LUMI-G supercomputer. Each LUMI-G computing node is equipped with four AMD MI250x GPUs (each one with two GCDs and 128 GB HBM memory) and a 64-cores AMD EPYC Trento with a total of 512 GB CPU memory.

5   Results

5.1   GMM Compression Accuracy

In this section, we evaluate the accuracy of the GMM compression technique by comparing the reconstructed velocity pdf with the one obtained from the histogramming process, for both electron and ion species at the end of the large magnetic reconnection simulation. We primarily evaluate the overall GMM compression accuracy by qualitatively comparing the GMM pdf and the histogram pdf shape. Additionally, we report the point-wise difference between the two pdfs to spot regions where they might be greatly different.

The three panels in Fig. 3 show, respectively, from left to right, the histogram 2D pdf, the GMM reconstructed 2D pdf, and the point-wise absolute difference between the two pdfs for the electron species uv velocity. The JSD metric between the histogram and GMM pdfs is 0.0157. Data are retrieved from the subdomain containing the x-point.

To provide a better comparison between the histogram and the GMM pdfs, we report in Fig. 4 1D slices of the 2D pdf shown in Fig. 3, along both the two velocities u and v. Specifically, the left panel shows the 1D profile of the pdf along the u direction for five different v velocities, while the right panel reports the 1D profile of the pdf along the v direction for five u velocities. The GMM reconstructed pdf accurately reproduces the shape of the original pdf.

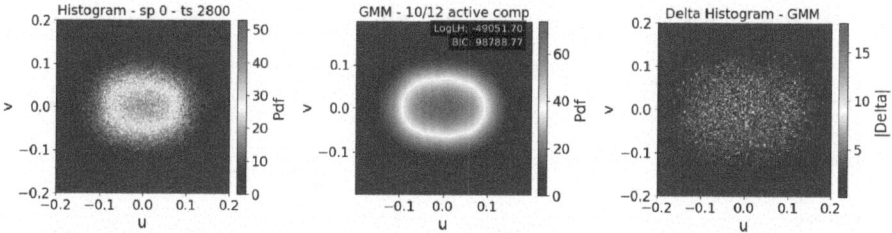

Fig. 3. Electron uv velocity pdf at the x-point location. Left panel: 2D pdf obtained from the histogram; central panel: 2D pdf reconstructed by GMM; right panel: absolute difference between the true and reconstructed pdf.

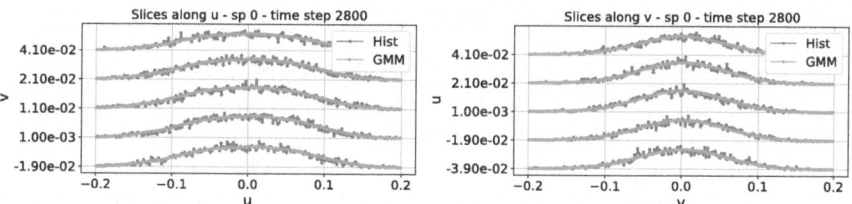

Fig. 4. Electron uv velocity pdf at the x-point location. 1D plots obtained as slices of the 2D pdf. In blue the true pdf obtained from the histograms, in orange the GMM reconstructed pdf. (Color figure online)

The three panels in Fig. 5 show respectively, from the left to the right, the histogram 2D pdf, the GMM reconstructed 2D pdf, and the point-wise absolute difference between the two pdfs for the ion species uw velocity. The JSD metric between the histogram and GMM pdfs is 0.0049. Data are retrieved from the subdomain containing the x-point.

In Fig. 6 1D slices of the 2D pdf shown in Fig. 5, along both the two velocities u and w, are reported. Specifically, the left panel shows the 1D profile of the pdf along the u direction for five different w velocities, while the right panel reports the 1D profile of the pdf along the w direction for five different u velocities. As for the electron pdf, GMM is capable of accurately reproducing the shape of the original pdf.

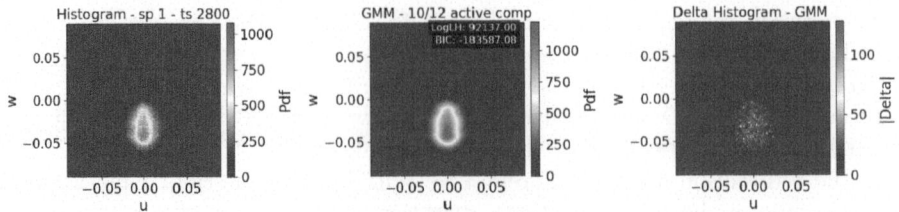

Fig. 5. Ion *uw* velocity pdf at the x-point location. Left panel: 2D pdf obtained from the histogram; central panel: 2D pdf reconstructed by GMM; right panel: absolute difference between the true and reconstructed pdf.

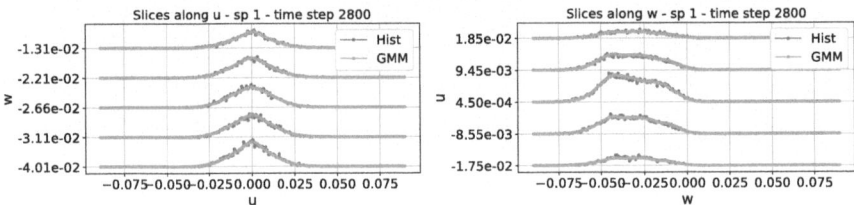

Fig. 6. Ion *uw* velocity pdf at the x-point location. 1D plots obtained as slices of the 2D pdf. In blue the true pdf obtained from the histograms, in orange the GMM reconstructed pdf. (Color figure online)

5.2   Compression Algorithms Evaluation

We detail in this section the performance of the different compression methods tested in this work on the small GEM simulation. Figure 7 shows the required time for compression and the compression ratio achieved by different algorithms. The histogram-GMM compressor requires a processing time in the order of 100 s ms, which is comparable to other engines like BLOSC2, SZ, ZFP and lower than engines like MGARD and BZIP2. The GMM with eight components reaches a compression ratio of 14 when compared to histogram data, and up to 10,000 when compared to raw particle data. GMM compression time is growing as the simulation goes, due to the increment in the number of EM iterations required to converge. This behavior is expected since the velocity pdf complexity increases as the simulation evolves. However, the increment in time can be limited by tuning the maximum number of EM iterations.

In our specific test case, the velocity pdfs do not significantly differ from Gaussians, thus both the histogram and GMM achieve a much higher compression ratio than non-physics-aware methods, with lower information loss (JSD < 0.1), as shown in Fig. 8. Clearly, the JSD metric is zero for lossless compressors like BZIP2 and BLOSC2.

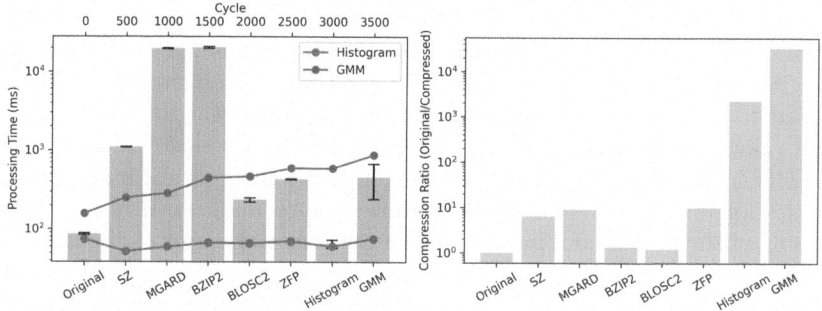

Fig. 7. Left panel: processing time (compression and I/O); right panel: compression ratio for different compression methods. The GMM time includes the time required to pre-process data with histogram. The red and green lines are the histogram and GMM processing time in different simulation cycles. (Color figure online)

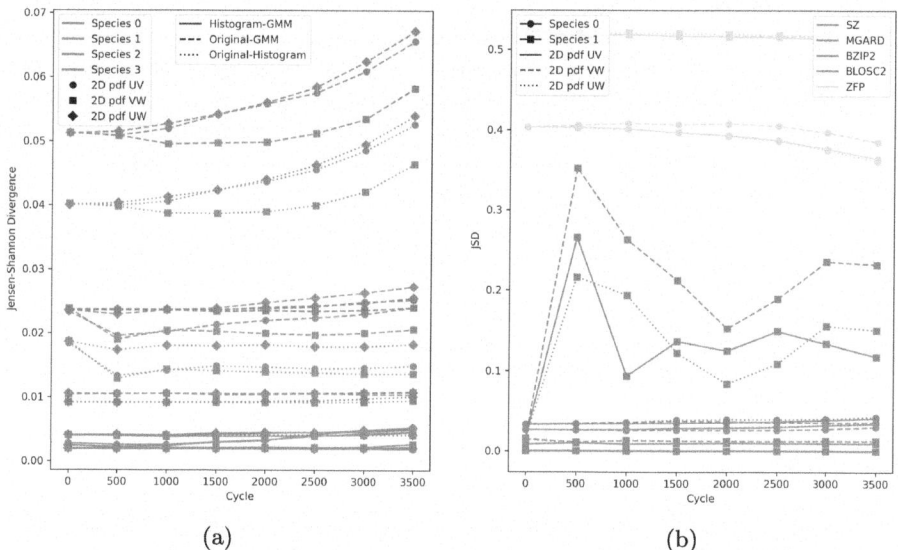

Fig. 8. (a): JSD between original data, histogram and GMM for different species and different 2D velocities pdf in subdomain 0. (b): JSD between compressors and the original particle data of subdomain 0; BZIP2 and BLOSC2 has zero JSD since they are lossless compressors. In both panels, species 0 (1) are main electrons (ions), while species 2 (3) are background electrons (ions).

6 Discussion and Conclusion

This work presented a GPU-accelerated Weighted Gaussian Mixture Model approach for compressing and analyzing high-dimensional data generated by PIC simulations. Weighted GMM was integrated into the iPIC3D code, and a new I/O module using the ADIOS 2 framework was implemented to compress particle velocity distributions. We showed only a small performance overhead while still achieving significant storage reduction and maintaining key statistical properties of the data. Our results demonstrated a compression rate exceeding 100x compared to established compression methods and a lower information loss in particle velocity distribution. We overlapped the iPIC3D electromagnetic solver on the CPU with GMM calculations on the GPU, showing utilization of GPU idle cycles for real-time GMM with minimized performance overhead.

Tuning of the GMM compressor parameters was the major challenge. Specifically, the choice of the number of GMM components critically influenced the accuracy and convergence steps since not all the distributions in the GEM simulation required the same number of clusters for adequate representation. To address this, we developed an adaptive algorithm that prunes Gaussians with negligible weights. Parameter initialization also proved difficult, particularly for complex particle distributions. Tackling this issue requires enhanced parameter initialization methods and feedback mechanisms between GMM results and subsequent analysis.

Future work will involve developing an efficient detection algorithm that, by exploiting the data compression achieved with GMM, can highlight deviations in the particle distributions from the initial pdfs in real-time while the simulation is evolving. This tool would be particularly beneficial for scientists, as it would allow them to identify relevant physical phenomena, such as heating or beam formation, with minimal computational cost.

Acknowledgment. The authors wish to thank Måns I. Andersson for the precious discussions. This work is funded by the European Union. This work has received funding from the European High Performance Computing Joint Undertaking (JU) and Sweden, Finland, Germany, Greece, France, Slovenia, Spain, and the Czech Republic under grant agreement No. 101093261, Plasma-PEPSC, https://plasma-pepsc.eu/.

References

1. Bishop, C.M., Nasrabadi, N.M.: Pattern Recognition and Machine Learning, vol. 4. Springer, Heidelberg (2006)
2. Chen, G., et al.: An unsupervised machine-learning checkpoint-restart algorithm using gaussian mixtures for particle-in-cell simulations. J. Comput. Phys. **436**, 110185 (2021)
3. Dawson, J.M.: Particle simulation of plasmas. Rev. Mod. Phys. **55**(2), 403 (1983)
4. Di, S., et al.: A survey on error-bounded lossy compression for scientific datasets. arXiv preprint arXiv:2404.02840 (2024)
5. Di, S., Cappello, F.: Fast error-bounded lossy hpc data compression with sz. In: 2016 IEEE International Parallel and Distributed Processing Symposium (IPDPS), pp. 730–739. IEEE (2016)
6. Dupuis, R., Goldman, M.V., Newman, D.L., Amaya, J., Lapenta, G.: Characterizing magnetic reconnection regions using gaussian mixture models on particle velocity distributions. Astrophys. J. **889**(1), 22 (2020)
7. Ghojogh, B., et al.: Fitting a mixture distribution to data: tutorial. arXiv preprint arXiv:1901.06708 (2019)

8. Godoy, W.F., et al.: Adios 2: the adaptable input output system. A framework for high-performance data management. SoftwareX **12**, 100561 (2020)
9. Hockney, R.W., Eastwood, J.W.: Computer Simulation Using Particles. CRC, Boca Raton (2021)
10. Ibarria, L., et al.: Out-of-core compression and decompression of large n-dimensional scalar fields. In: Computer Graphics Forum, vol. 22, pp. 343–348. Wiley (2003)
11. Jackson, J.D.: Classical Electrodynamics. Wiley, Hoboken (1999)
12. Kullback, S., Leibler, R.A.: On information and sufficiency. Ann. Math. Stat. **22**(1), 79–86 (1951)
13. Li, S., et al.: Data reduction techniques for simulation, visualization and data analysis. In: Computer Graphics Forum, vol. 37, pp. 422–447. Wiley (2018)
14. Liang, X., et al.: Mgard+: optimizing multilevel methods for error-bounded scientific data reduction. IEEE Trans. Comput. **71**(7), 1522–1536 (2021)
15. Lin, J.: Divergence measures based on the shannon entropy. IEEE Trans. Inf. Theory **37**(1), 145–151 (1991)
16. Lindstrom, P., et al.: ZFP: a compressed array representation for numerical computations. Int. J. High Perf. Comput. Appl. **39**(1), 104–122 (2025)
17. Markidis, S., et al.: Multi-scale simulations of plasma with ipic3d. Math. Comput. Simul. **80**(7), 1509–1519 (2010)
18. Markidis, S., et al.: The EPiGRAM project: preparing parallel programming models for exascale. In: Taufer, M., Mohr, B., Kunkel, J.M. (eds.) ISC High Performance 2016. LNCS, vol. 9945, pp. 56–68. Springer, Cham (2016). https://doi.org/10.1007/978-3-319-46079-6_5
19. Mason, R.J.: Implicit moment particle simulation of plasmas. J. Comput. Phys. **41**(2), 233–244 (1981)
20. McLachlan, G., Krishnan, T.: The EM Algorithm and Extensions. Wiley, Hoboken (1997)
21. Peng, I.B., et al.: Energetic particles in magnetotail reconnection. J. Plasma Phys. **81**(2), 325810202 (2015)
22. Peng, I.B., et al.: The formation of a magnetosphere with implicit particle-in-cell simulations. Procedia Comput. Sci. **51**, 1178–1187 (2015)
23. Peng, I.B., et al.: Kinetic structures of quasi-perpendicular shocks in global particle-in-cell simulations. Phys. Plasmas **22**(9) (2015)
24. Sahoo, R.R., Bhowmick, S., Mandal, D., Kumar Kundu, P.: A novel approach of gaussian mixture model-based data compression of ecg and ppg signals for various cardiovascular diseases. Biomed. Signal Process. Control **96**, 106581 (2024)
25. Sun, J., Zhao, Y., Wang, S., Wei, J.: Image compression based on gaussian mixture model constrained using markov random field. Signal Process. **183**, 107990 (2021)
26. Yu, Y., et al.: Pic simulations of wave-particle interactions with an initial electron velocity distribution from a kinetic ring current model. J. Atmos. Solar Terr. Phys. **177**, 169–178 (2018)

Hierarchical Structural Information – Theory and Applications

Marzena Bielecka⬤, Andrzej Bielecki⬤, Aleksander Suchorab⬤,
and Igor Wojnicki[✉]⬤

AGH University of Krakow, al.Mickiewicza 30, 30-059 Kraków, Poland
{bielecka,bielecki,asuchorab,wojnicki}@agh.edu.pl

Abstract. This paper introduces a novel measure to quantify structural information in hierarchical graphs. It addresses the limitation of current methods that do not adequately account for hierarchical structures. By considering inner structural information and distinguishability of higher-level vertices, the proposed measure captures the additional information generated by the hierarchy. The hypothesis that hierarchical graphs contain more structural information is validated using the "Countries" dataset. The results demonstrate a measurable increase in the information content when the hierarchical structure is considered, compared to a simple graph representation. This highlights the importance of recognizing and utilizing hierarchy to enhance the informational richness of graphs, potentially improving the performance of graph-based machine learning models.

Keywords: Information quantity · Structural information · Hierarchical information · Hierarchical data · Amount of information · Network analysis

1 Introduction and Motivation

The question of how much information is gathered in a graph, considering its topology and hence the structure, has been researched since the mid-20th century. The subject is brought to light by recent advances in machine learning in cases where there is a need to learn, classify, or regress information expressed with graphs. The reason for this is that information quantity addresses the quality of the data set. The more information there is in the graph, the more reliable Graph Neural Networks (GNNs) or graph embeddings would be.

Although the graph structure information quantity measure for general graphs is known, there is little research regarding hierarchical graphs. And this is the subject we want to address. We will focus on structural information in a case for which a hierarchical graph is available or a hierarchy can be identified unambiguously. In such a case, which is supported by real-world examples, additional information that is provided as edge or node labels, can express a hierarchy within the graph, transforming it into a hierarchical graph. As a result, a

© The Author(s), under exclusive license to Springer Nature Switzerland AG 2025
M. H. Lees et al. (Eds.): ICCS 2025, LNCS 15905, pp. 48–59, 2025.
https://doi.org/10.1007/978-3-031-97632-2_4

part of the graph semantics is turned into its structure. Our hypothesis is that the amount of structural information in such a structurally extended graph is greater than the initial one. To support this hypothesis, we would like to present a measure and apply it to a real-world example.

It should be stressed that this paper addresses the problem of the structure of relational data in the context of machine learning. The proposed methodology, however, goes beyond pure computer science and refers to cybernetics. Namely, the method can be used to analyse the complexity of multi-levelled social networks [2] as well as biological systems that are hierarchical by their nature [31]. The biological application are crucial because, in contemporary biology, the role of information and its processing is regarded as fundamental [11,19]. The presented approach can also be applied to theoretical foundations of embodied autonomous agents, that act in their environment and create knowledge about it [4]. Such knowledge is usually modelled as hierarchical formal structures.

In the following sections a brief introduction to complex data structures for machine learning and a proposal of an information quantity measure for hierarchical graphs are given. Furthermore, a real-world example dataset in the form of a graph with well established hierarchy is also presented along with the calculations of information quantity.

1.1   Graph Data Structures in Machine Learning

Graph data structures are fundamental for machine learning applications that require relational modeling. A graph consists of nodes (also called vertices) and edges, representing entities and their relationships. This structure allows machine learning models to effectively handle non-Euclidean data, making them well suited for social network analysis, recommendation systems, and biological modeling [24]. A major advantage of using graphs in machine learning is their ability to capture complex dependencies between data points. Unlike traditional vector-based models, graph-based approaches incorporate relational information, enabling better performance in structured data environments [21]. This is particularly valuable in semi-supervised learning, where labeled data is limited but unlabeled data can still contribute to learning by propagating information across connected nodes [28].

One of the most important techniques in graph-based learning is graph embedding, which transforms graph data into lower-dimensional vector representations while preserving structural properties.

As graph-based machine learning continues to evolve, more and more research is focused on enhancing scalability, interpretability, and robustness to expand its applicability across various domains.

1.2   Hierarchical Graph Data Structures and Their Applications in Machine Learning

Hierarchical graph data structures extend beyond traditional graphs by introducing multiple levels of abstraction, where each level represents a transformation of the previous one. These structures effectively model complex relationships in large-scale datasets, enabling machine learning algorithms to leverage multi-scale information efficiently. Unlike simple graphs, hierarchical graphs allow nested relationships while maintaining graph-based flexibility, making them ideal for applications in NLP, computer vision, and bioinformatics [23].

A significant use case of hierarchical graphs is the Hierarchical Graph Neural Networks (HGNNs), where graph representations are refined across multiple levels to enhance feature extraction. In NLP, they help capture sentence structures and contextual dependencies, improving text classification and sentiment analysis [32]. In bioinformatics, hierarchical graphs model molecular interactions, improving drug discovery and protein structure prediction [20]. Hierarchical structures also benefit computer vision, where multiresolution graph representations improve image segmentation, object recognition, and scene understanding. In autonomous systems, hierarchical graph-based models facilitate efficient decision making by structuring sensor data for multiscale reasoning [38]. As machine learning tasks grow in complexity, dynamic hierarchical graph models, that adapt to changing relationships, are developed. This is particularly useful in fraud detection, social network analysis, and recommendation systems, where interactions between entities evolve. Recent advances in Graph Attention Networks (GATs) with hierarchical structures have improved model interpretability and efficiency in handling large-scale data [27].

With continued advancements, hierarchical graph-based learning is expected to further enhance performance across AI-driven applications.

1.3   Graph Embeddings

A graph embedding technique transforms relationships in a graph into lower dimensional space, namely a vector with fixed length, while preserving as much as possible of the structural information and properties. Having a set of fixed-length vectors that represent nodes makes them easier to analyze and more suitable for machine learning. It enables tasks such as node classification, recommendation, link prediction, graph completion, or clustering, to name a few, on graph data with vector-based machine learning concepts and technologies.

There are multiple graph embedding techniques that can be used with varying applicability depending on the actual characteristics of the graphs [34] with some benchmarks [13] and taxonomy [9], available. One of the challenges is to verify if the embeddings preserve semantics [15]. As it has been observed, the sparsity of graphs degrades embedding performance greatly [29]. Performance gains reported for one graph type may not translate to other types of graphs. It is also challenging to compare different embedding methods and predict which one will be best for a given graph [13]. There is no standard way to quantify the advantages of one approach over another. Many methods are evaluated on dense, curated datasets that do not reflect the complexities of real-world graphs.

1.4   Graph Neural Networks

Graph Neural Networks are a class of deep learning models designed to process graph-structured data. Unlike traditional neural networks that operate on Euclidean data (e.g., images and text represented as vectors), GNNs can model relationships in non-Euclidean spaces. They achieve this by iteratively aggregating information from neighboring nodes, capturing both local and global dependencies within the graph [39]. A key property of GNNs is their ability to leverage relational information rather than treating data points as independent entities [18]. This makes them particularly effective in domains where interconnected data is essential. Their inductive learning capability enables them to generalize unseen graphs [35], and their flexibility allows them to process various types of graphs, including directed, undirected, weighted, and heterogeneous structures [37].

Compared to traditional machine learning models that rely on handcrafted features, GNNs automatically learn hierarchical and context-aware representations, improving performance in node classification, link prediction, and graph clustering. They are widely applied in social network analysis (e.g., community detection and friend recommendation) [18], biomedical research (e.g., protein structure prediction and drug discovery) [37], and fraud detection by analyzing transaction networks for anomalies. In natural language processing (NLP), they enhance tasks such as semantic parsing and document classification by capturing contextual relationships [8,35,36]. Furthermore, GNNs contribute to traffic prediction and optimization in autonomous systems [22]. By exploiting graph structures, they enable more accurate and context-aware predictions, making them a crucial tool in modern deep learning.

2   Entropy-Based Information Quantity for Graph-Based Hierarchical Structures

Before starting any machine learning process, the dataset has to be assessed if it is up to providing proper input for the model. Especially, it has to be confirmed if the data quantity and quality are sufficient. In case of vectors, there are several heuristics and methods that work well. For graphs, one should begin with assessment of how much structural information there is in it, to make sure that there is enough information to feed the subsequent learning process.

To measure amount of structural information in a graph, there are two starting points to consider. One point is to start with the graph entropy [26], which is based on information theory. Another one begins with an ontology, which results in the Hellerman's approach [14]. Even though their starting points are different, the end results are quite similar, being a single measure of information quantity. In this paper, we choose to apply the Hellerman's approach.

However, in case of graphs where a hierarchy can be identified, the proposed Hellerman's approach is lacking. This includes hierarchical graphs or graphs in which such hierarchy could be identified or inferred, based on semantic information such as edge or node labeling.

In this section, we put forward a proposal of such a measure that is suitable for hierarchical graphs or graphs with identifiable hierarchy.

2.1   Hierarchical Information

Hierarchical information appears when substructures of a given structure or a group of elements of a given set constitute a single element in a new structure or set at the higher level of hierarchy. Then, among such new elements novel relations can appear. Let us formalize this idea on the basis of the concept of structural information signalized in [3], worked out in detail in [5] and applied to cognitive maps in [6]. Thus, let X be an k-element set with a relation $\mathcal{R}$ on it that generates the directed graph (digraph) $G(X, \mathcal{R})$. Let us recall that the nodes of $G(X, \mathcal{R})$ correspond to elements of X, whereas the directed edges connect the nodes that correspond to elements in relation to each other [5]. To put it briefly, the form of the graph $G(X, \mathcal{R})$ is the structural information on the set X, that is generated by relation $\mathcal{R}$. The amount of such information can be calculated numerically by applying the concept worked out by Hellerman [14], as follows.

$$H = -N \sum_{k=1}^{K} \frac{n_k}{N} log \frac{n_k}{N} \tag{1}$$

where N is the number of vertices in the graph, K is the number of equivalence classes and n_k is the number of vertices belonging to class n_k.

It should be mentioned that the way in which the information in graphs can be calculated was studied since the fifties of the 20th century, but it was referred to entropy in graphs and not to ontological aspects of structural information as such [12, 25, 26, 30, 33]. The formulae used to calculate the amount of information in graphs were consistent in Shanon information theory, Hellerman's concept, and the idea presented in [5].

Let us consider the way in which the amount of information can be calculated in hierarchical structures. The mentioned set X with relation $\mathcal{R}$ constitutes the first level of the hierarchy, on which both the information and the way the amount of it is calculated, are given in [5] – see Fig. 1 as an example, where $X = \{x_1, \dots, x_{11}\}$ and relation $\mathcal{R}$ generate the digraph $G(X)$. Let $P(G)$ denote the set of all subgraphs of $G(X, \mathcal{R})$. Let N be the cardinality of $P(G)$. From the set $P(G)$ there are selected elements that will constitute components of the set, let us say $\mathscr{X}$, on the next level of hierarchy. Let us assume that $card\ \mathscr{X} = n \leq N$. Then, the amount of information H_c, generated by the choice of the number of the elements of $\mathscr{X}$, is equal to

$$H_c = -N \left(\frac{n}{N} \log \frac{n}{N} + \frac{N-n}{N} \log \frac{N-n}{N} \right). \tag{2}$$

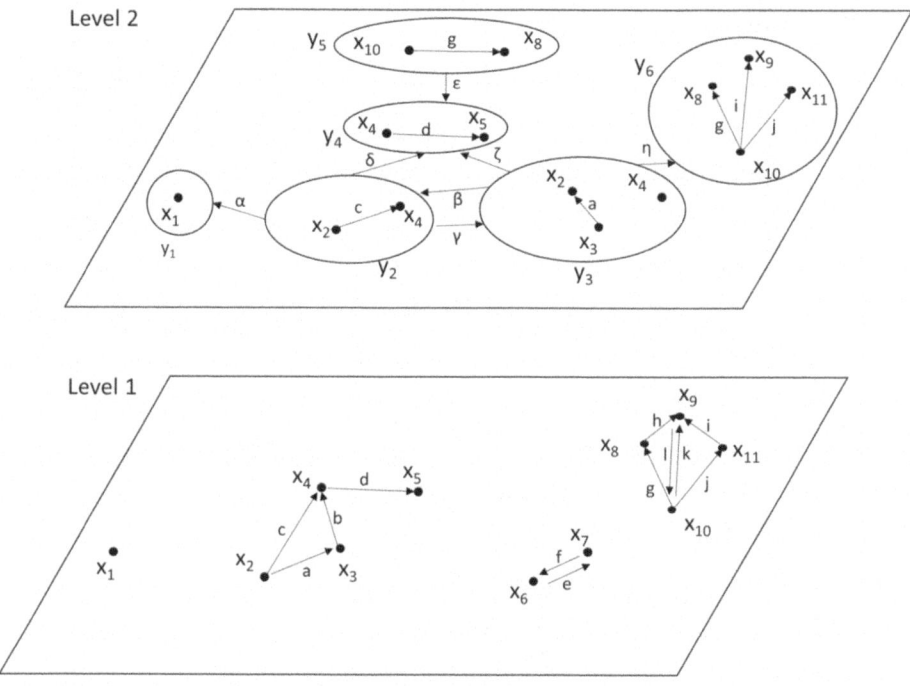

Fig. 1. Example of hierarchical information

The elements of $\mathscr{X}$ are graphs themselves, so each element has an inner structure and, as a consequence, inner structural information. Thus, for each element of $\mathscr{X}$, its amount of structural information can be calculated in the same way as for graph $G(X, \mathcal{R})$. Let a relation $\mathscr{R}$ be defined on $\mathscr{X}$. It generates the digraph $\mathscr{G}(\mathscr{X}, \mathscr{R})$ at the second level of the hierarchy – see Fig. 1. For this graph, the amount of information can be calculated in the same way as for graph $G(X, \mathcal{R})$.

In summary, selecting elements that constitute higher-level components in a hierarchical structure generates three types of information:

(a) information generated by number of the chosen elements – amount of this type of information is given by formula (2);
(b) inner structural information of every element on the higher level;
(c) information generated by the distinguishability of higher-level vertices resulting from their internal structure.

3 Computing the Hierarchical Information of an Example Dataset

Let us introduce an example, a real-world dataset, and calculate the amount of structural information for it. Then, let us identify its hierarchy and compare the amount of structural information when the hierarchy is taken into consideration.

3.1 The Dataset

The "Countries" dataset [7], provided by PyKEEN (Python KnowlEdge Embed-diNgs) [1], will be used to demonstrate the computation of hierarchical infor-mation. The motivation for using this dataset is its relatively small size, ease of interpretation of actual data, and the fact that it can be turned into a hierar-chical structure easily and consistently.

The dataset consists of countries, regions that group countries and continents that group regions. These three kinds of nodes are not explicitly differentiated, but the type of node can be deduced based on its connections, namely edge labels. The dataset contains two kinds of edges differentiated by labeling: "neighbor", a symmetric relationship denoting that two countries are next to each other, and "locatedin", which denotes a lower-level entity belonging to a higher-level entity. As a result, the semantic information encoded with the "locatedin" labels can be used to build the hierarchy providing additional structural information.

An excerpt from the dataset considering two countries, namely Czechia and Poland is given in Fig. 2. The hierarchical decomposition of the excerpt

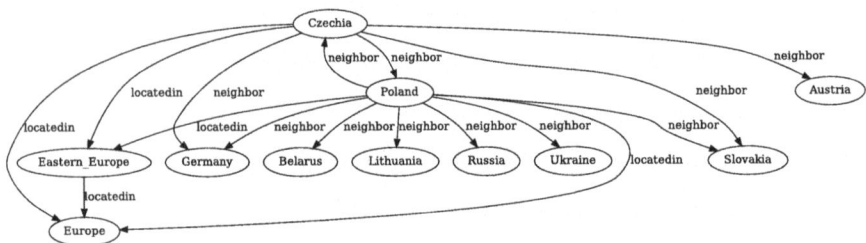

Fig. 2. An excerpt from the "Countries" dataset.

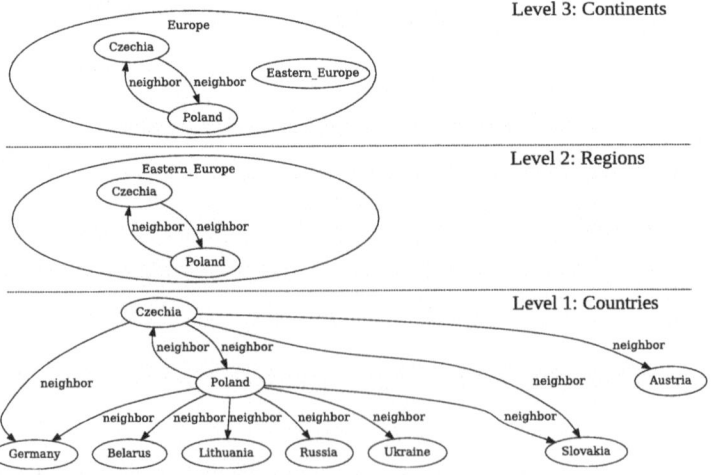

Fig. 3. An excerpt from the "Countries" dataset decomposed into a 3-level hierarchy, the decomposition is based on the "locatedin" label.

is given in Fig. 3. There are edges labeled "locatedin" between countries and regions, namely "Czechia"–"Eastern_Europe" and "Poland"–"Eastern_Europe", and between regions and continents, which is "Eastern_Europe"–"Europe". They create a three-level hierarchy with "Poland", "Czechia" and other neighboring countries at level 1, "Eastern_Europe" at the level 2, and "Europe" at the level 3.

3.2 Data Preparation

The dataset must be slightly adjusted, as there is a name conflict: *Micronesia* is both a country and a region, and in the original dataset there is a loop edge "Micronesia is located in Micronesia" which causes ontological ambiguity.It is just a mistake in the dataset.In order to fix it, the country of *Micronesia* will be assigned a non-conflicting name.

After converting the graph to a hierarchical form, there are 244 countries, 23 regions, and 5 continents, forming consecutive levels of hierarchy. The set of countries will constitute the initial set X from Sect. 2.1. The "neighbor" relation remains as connections between countries, like the relation $\mathcal{R}$ from Sect. 2.1, and the "locatedin" relation describes the composition of higher-level nodes from lower-level nodes. The initial graph $G(X)$ is generated by the relation $\mathcal{R}$.

3.3 Results

The calculations were performed using the Python programming language, with the *igraph* network analysis library [10] to handle the graphs, using the bliss isomorphism algorithm [16,17].

As described in Sect. 2.1, introducing hierarchy creates additional information which is dependent on two consecutive levels of hierarchy and includes information (a) generated by the number of chosen elements, (b) inner structural information and (c) generated by distinguishability.

Information (a) for the second layer (regions over countries) is 109.9, given by formula 2, choosing a set of 23 elements of higher level over a set of 244. Similarly, for the third layer (continents over regions), it is 17.4, choosing 5 elements over a set of 23.

Information (b) for the second layer is 570.8, as a sum of information given by formula 1 for each subgraph belonging to a higher-level vertex, where the equivalence classes are defined as in [5], equivalent to automorphism groups. Information (b) for the third layer is zero because there are no connections between regions in the dataset. It could be possible to infer missing neighbor relations between regions and continents based on connections of countries that belong to them, but it was decided against it, in order to avoid any unnecessary data transformations and also to prevent introduction of any inferred information which could affect final results.

Information (c) is 102.0 for the second layer and 4.9 for the third layer. It is calculated using formula 1 with vertices grouped into equivalence classes by inner structure distinguishability: two higher-level vertices are indistinguishable if the lower-level subgraphs belonging to them are isomorphic.

The amount of information in the entire graph in the Hellerman sense [14] as defined in [5] is 1956.1, without taking into account hierarchical structures. When interpreting the data set hierarchically, the amount of information in the initial graph is 1424.8, as the graph is now smaller.

The sum of information due to all the levels of hierarchy is 805.0, which when added to information generated by the initial graph constituting the first layer, is 2229.8. This value is greater than the amount of information in the graph in the non-hierarchical interpretation, which shows that the introduction of the hierarchy creates additional information, resulting in an increase of almost 14%. The comparison of the two interpretations can be seen in Fig. 4.

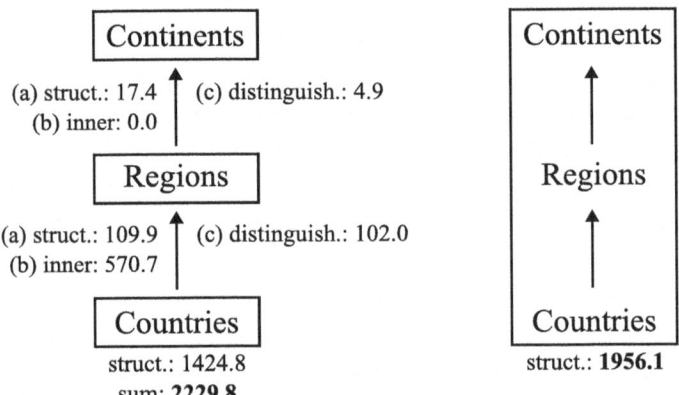

Fig. 4. Comparison of computing the information in the "Countries" dataset with a hierarchical interpretation (on the left) versus interpreting it as a single graph (on the right).

4   Summary

We propose a novel measure for quantifying information in hierarchical graphs which is influenced by existing concepts of graph entropy and especially Hellerman's ontological approach. It accounts for the additional information generated by the hierarchy itself, considering factors such as the number of chosen elements, inner structural information, and distinguishability of higher-level vertices.

To validate the hypothesis that hierarchical graphs contain more structural information, the proposed measure is applied to the "Countries" dataset. When comparing the information content, if treated as a simple graph versus a hierarchical one, the results demonstrate a significant increase in information if the hierarchical structure is taken into account. Identifying a single edge label that forms the hierarchy increases the amount of information by almost 14%. As a result, it supports the argument that recognizing and utilizing the hierarchy enhances the informational richness of graphs. It might provide better metrics for the quantity of information in graphs, indirectly improving the performance of graph-based machine learning models.

Further research focuses on investigating the relationship between the proposed measure and the quality of graph embeddings and the performance of graph neural networks. In particular, the following problems can be put as examples.

1. Will taking into account hierarchical information improve or optimize graph embedding algorithms?
2. There is a GNN with its input being a graph and output a vector. Can the hierarchical information within the input graph be taken into account and does it impact prediction quality?
3. Will including the hierarchical information in graph embedding algorithms pose any opportunities or risks.
4. How to encode the hierarchical information as a GNN input. How does it influence GNN's architecture?
5. Does the hierarchical information concept proposed in this paper optimize hierarchical graph embedding problem?

These are the most immediate goals, but the research can go further, beyond this scope, taking into account social networks, biology, biological systems or autonomous agents, as it was mentioned in the introduction.

References

1. Ali, M., et al.: PyKEEN 1.0: a python library for training and evaluating knowledge graph embeddings. J. Mach. Learn. Res. **22**(82), 1–6 (2021). http://jmlr.org/papers/v22/20-825.html
2. Amini, M.H., Imteaj, A., Pardalos, P.M.: Interdependent networks: a data science perspective. Patterns **1**(1), 9 (2020)
3. Bielecki, A.: The general entity of life: a cybernetic approach. Biol. Cybern. **109**(3), 401–419 (2015). https://doi.org/10.1007/s00422-015-0652-8
4. Bielecki, A.: The systemic concept of contextual truth. Found. Sci. **26**, 807–824 (2020)
5. Bielecki, A., Schmittel, M.: The information encoded in structures: theory and application to molecular cybernetics. Found. Sci. **27**, 1327–1345 (2022)
6. Bielecki, A., Stocki, R.: The concept of structural information and possible applications. Zagadnienia Filozoficzne w Nauce (Phil. Prob. Sci.) **75**, 157–183 (2023)
7. Bouchard, G., Singh, S., Trouillon, T.: On approximate reasoning capabilities of low-rank vector spaces. In: AAAI Spring Symposia (2015)
8. Bunke, H.: Graph-Based Tools for Data Mining and Machine Learning. Springer, Heidelberg (2003)
9. Cai, H., Zheng, V.W., Chang, K.C.C.: A Comprehensive Survey of Graph Embedding: Problems, Techniques and Applications (Feb 2018)
10. Csardi, G., Nepusz, T.: The igraph software package for complex network research. InterJ. Complex Syst. **1695** (2006). https://igraph.org
11. Davies, P.: The Demon in the Machine. How Hidden Webs of Information Are Solving the Mystery of Life, University of Chicago Press, Chicago (2019)

12. Dehmer, M., Mowshowitz, A.: A history of graph entropy measures. Inf. Sci. **181**, 57–78 (2011)
13. Goyal, P., Huang, D., Goswami, A., Chhetri, S.R., Canedo, A., Ferrara, E.: Benchmarks for Graph Embedding Evaluation (2019). https://doi.org/10.48550/arXiv.1908.06543
14. Hellerman, L.: Representation of living forms. Biol. Phil. **21**, 537–552 (2006)
15. Jain, N., Kalo, J.-C., Balke, W.-T., Krestel, R.: Do embeddings actually capture knowledge graph semantics? In: Verborgh, R., et al. (eds.) ESWC 2021. LNCS, vol. 12731, pp. 143–159. Springer, Cham (2021). https://doi.org/10.1007/978-3-030-77385-4_9
16. Junttila, T., Kaski, P.: Engineering an efficient canonical labeling tool for large and sparse graphs. In: Applegate, D., Brodal, G.S., Panario, D., Sedgewick, R. (eds.) Proceedings of the Ninth Workshop on Algorithm Engineering and Experiments and the Fourth Workshop on Analytic Algorithms and Combinatorics, pp. 135–149. SIAM (2007). https://doi.org/10.1137/1.9781611972870.13
17. Junttila, T., Kaski, P.: Conflict propagation and component recursion for canonical labeling. In: Marchetti-Spaccamela, A., Segal, M. (eds.) TAPAS 2011. LNCS, vol. 6595, pp. 151–162. Springer, Heidelberg (2011). https://doi.org/10.1007/978-3-642-19754-3_16
18. Khemani, B., Patil, S., Kotecha, K., Tanwar, S.: A review of graph neural networks: concepts, architectures, techniques, challenges, datasets, applications, and future directions. J. Big Data (2024)
19. Konieczny, L., Roterman, I., Spólnik, P.: Systems Biology. Functional Strategies of Living Organisms, Springer, Heidelberg (2023)
20. Li, L., Xiang, J., Chen, B., Heaney, C., Dargaville, S., Pain, C.: Implementing the discontinuous-galerkin finite element method using graph neural networks with application to diffusion equations. Neural Netw. **185** (2025)
21. Liquiere, M., Sallantin, J.: Structural machine learning with galois lattice and graphs (1998)
22. Liu, Z., Zhou, J.: Introduction to Graph Neural Networks. Cambridge University Press, Cambridge (2022)
23. Luo, B., Zhang, Z., Wang, Q., He, B.: Multi-chain graphs of graphs: a new approach to analyzing blockchain datasets. In: The Thirty-Eighth Conference on Neural Information Processing Systems (2024)
24. Makaro, I., Kiselev, D., Nikitinsky, N., Subelj, L.: Survey on graph embeddings and their applications to machine learning problems on graphs. PeerJ Comput. Sci. (2021)
25. Mowshowitz, A.: Entropy and the complexity of graphs: I. An index of the relative complexity of the graph. Bull. Math. Biophys. **30**, 175–204 (1968)
26. Mowshowitz, A.: Entropy and the complexity of graphs: II. The information content of digraphs and infinite graphs. Bull. Math. Biophys. **30**, 225–240 (1968)
27. Park, J., Han, M., Lee, K., Park, S.: Hierarchical graph attention network with positive and negative attentions for improved interpretability: ISA-PN. J. Chem. Inf. Model. (2024)
28. Perraudin, N.: Graph-based structures in data science: fundamental limits and applications to machine learning. Ph.D. thesis, EPFL (2017)
29. Pujara, J., Augustine, E., Getoor, L.: Sparsity and noise: where knowledge graph embeddings fall short. In: Proceedings of the 2017 Conference on Empirical Methods in Natural Language Processing, pp. 1751–1756. Association for Computational Linguistics, Copenhagen (2017). https://doi.org/10.18653/v1/D17-1184

30. Rashevsky, N.: Life, information theory, and topology. Bull. Math. Biophys. **17**, 229–235 (1955)
31. Rosslenbroich, B.: Endothermy. In: On the Origin of Autonomy. HPTLS, vol. 5, pp. 149–159. Springer, Cham (2014). https://doi.org/10.1007/978-3-319-04141-4_9
32. Shen, Y., Lin, Z.: Patentgrapher: a plm-gnns hybrid model for comprehensive patent plagiarism detection across full claim texts. IEEE Access (2024)
33. Trucco, E.: A note on the information content of graphs. Bull. Math. Biophys. **18**, 129–135 (1956)
34. Wang, Q., Mao, Z., Wang, B., Guo, L.: Knowledge graph embedding: a survey of approaches and Applications. IEEE Trans. Knowl. Data Eng. **29**(12), 2724–2743 (2017). https://doi.org/10.1109/TKDE.2017.2754499
35. Wu, Z., Pan, S., Long, G., Chen, F., Zhang, C.: A comprehensive survey on graph neural networks. IEEE Trans. Neural Netw. Learn. Syst. **32**, 4–24 (2020)
36. Xia, F., Sun, K., Yu, S., Aziz, A., Wan, L., Pan, S.: Graph learning: a survey. IEEE Trans. Artif. Intell. **2**, 109–127 (2021)
37. Zhang, X., Liang, L., Liu, L., Tang, M.: Graph neural networks and their current applications in bioinformatics. Front. Genet. **12** (2021)
38. Zhang, Y.: Constructing Knowledge Graphs with Language Models and Learning Hierarchies from Graphs Using Probabilistic Topic Modeling. Ph.D. thesis, University of Alberta (2024)
39. Zhou, J., Cui, G., Hu, S., Zhang, Z., Yang, C., Liu, Z.: Graph neural networks: A review of methods and applications. AI Open, Elsevier, Amsterdam (2020)

A Connectionist Approach to Federated Digital Twins

Christian Vergara-Marcillo[1,2](✉) (iD), Rami Bahsoon[2] (iD), Nikos Tziritas[3] (iD),
and Georgios Theodoropoulos[1,4](✉) (iD)

[1] Department of Computer Science and Engineering, Southern University of Science
and Technology (SUSTech), Shenzhen, China
theogeorgios@gmail.com
[2] University of Birmingham, Birmingham, UK
[3] University of Thessaly, Lamia, Greece
[4] Research Institute for Trustworthy Autonomous Systems, Southern University
of Science and Technology (SUSTech), Shenzhen, China

Abstract. Digital Twins (DTs) have driven significant innovation across
industries, creating virtual replicas of physical assets that enable continuous learning, optimization, and informed decision-making. Digital Twins
Systems of Systems (SoS) pose open challenges that relate to representation, orchestration, and management at scale and call for innovative
approaches for collaborative modelling of their ecosystem. Federated Digital Twins (FDTs) have emerged as a solution, enabling integration and
resource sharing between independent DTs, fostering collaboration, and
unlocking the full potential of interconnected systems. This work proposes a framework inspired by connectionism theory to model FDTs as
a system of systems, drawing on federated systems and cognitive neuroscience to facilitate collaboration and emergent communication patterns.
A Smart Connected Farming case study is used as a proof of concept for
the proposed framework.

Keywords: Digital Twins · Federated Systems · Connectionist
Theory · Smart Agriculture

1 Introduction

As dynamic virtual representations of physical assets or entities, ranging from
simple to complex systems [14], Digital Twins (DTs) facilitate insights, understanding, and informed decision-making, leveraging continuous real-time data
monitoring, what-if analysis, and predictions through simulation and learning
models. These advanced DT capabilities have been deployed due to multiple
cutting-edge enabling technologies, including the Internet of Things (IoT), Big
Data analytics, Artificial Intelligence (AI), edge and cloud computing [11]. While
individual DTs have demonstrated significant advantages in diverse domains, a
strong need has emerged for developing interconnected Digital Twin ecosystems

© The Author(s), under exclusive license to Springer Nature Switzerland AG 2025
M. H. Lees et al. (Eds.): ICCS 2025, LNCS 15905, pp. 60–74, 2025.
https://doi.org/10.1007/978-3-031-97632-2_5

to deal with large-scale and complex systems-of-systems [17]. As a result, the concept of *Federated Digital Twin (FDT)* has emerged as an approach for composing and coordinating DTs, each with its functional characteristics. Such composition may facilitate seamless coordination and collaborative decision-making [16,32,33,36]. Secure inter-twin communications (virtual-to-virtual) may enable cooperation and resource exchange within federated ecosystems while maintaining the operational autonomy of individual DTs [35]. Shared resources range from operational data reflecting physical system state to simulation model outputs, knowledge, learning models [23], and decisions made by agent-based DTs [32,33]. While FDTs hold great promise, their inherent complexity and dynamism give rise to substantial methodological and technological challenges spanning multiple domains, including simulation and analytics, software engineering methodologies, theoretical design frameworks, standards, and interoperability.

Aspiring to contribute to the design and development of FDTs, this paper investigates the concept of connectionism and the utilisation of *Connecitonism theory* principles as a suitable paradigm to study interrelationships between DTs within a federated environment. An FDT is conceptualized as a network of interconnected DTs wherein interrelationships and potential synergies are enabled across the federated network, regulating information flow based on local and global objectives. The paper proposes a connectionist-inspired methodology for capturing communication, interaction, and synergy among interdependent, potentially autonomous DTs. While connectionism is often associated with deep learning models, this paper draws inspiration from the fundamental principles and mechanisms studied in this theory and cognitive neuroscience to study interrelationships in FDTs. As a proof of concept, a precision agriculture use case within Smart Connected Farming (SCF) systems [25] is considered, wherein a network of cooperative Farm DTs are deployed as an FDT, virtually representing a community of real-world smart farms in an agri-food production region.

The contributions of the paper are as follows:

1. Conceptualization of FDTs as Networks of Interconnected DTs: The paper proposes a new perspective of Federated Digital Twins (FDTs) as a network of interconnected, cooperative DTs facilitating dynamic information flow across a federated ecosystem. This framework enables coordination and resource exchange while maintaining the autonomy of individual DTs.
2. Development of a Connectionist-Inspired Methodology for Federated Digital Twins (FDTs): This paper introduces a novel methodology inspired by connectionism theory to study and model the interrelationships between Digital Twins (DTs) within a federated system. The approach captures communication, interaction, and synergy among autonomous, interdependent DTs, enhancing their collaboration and decision-making.
3. Application to Smart Connected Farming (SCF): The paper demonstrates the practical applicability of the proposed FDT framework through a Smart Connected Farming (SCF) use case. This case study explores how a network of Farm DTs can be deployed in an agri-food production region, serving as

a proof of concept for the potential of FDTs in large-scale, complex systems like precision agriculture.

The remainder of the paper is organized as follows: Sect. 2 presents an overview of an FDT and current frameworks addressing federated ecosystems for DTs. Section 3 provides a detailed exploration of the connectionism theory, focusing on its application to FDTs. Section 4 outlines a proof-of-concept use case within the context of Smart Connected Farm, while Sect. 5 presents an experimental evaluation of the system. Section 6 concludes the paper by outlining potential areas for future work.

2   Federated Digital Twins

A network of connected DTs representing multiple interconnected physical assets, encompassing a wide range of complexities, has been analyzed by the Alan Turing Institute within the UK's national DT programme [13]. An ecosystem of connected DTs representing the highest level of integration for homogeneous and heterogeneous DTs, capable of capturing the intricacies of interconnected real-world systems across various spatiotemporal scales, leveraging shared resources and combined insights within federated environments [3]. Considering four potential architectural styles, FDTs have been conceptualized as a virtual medium for connecting autonomous DTs [32,33]. Security, standardization, and interoperability are crucial aspects that enable collaboration, informed decision-making, and systems optimization through advanced data analytics and simulations.

FDTs have also been proposed as part of an evolutionary development model for DTs, consisting of five layers to address their inherent complexity [19]. This model incorporates replication, intra-twin synchronization, modeling, and simulation with FDTs, enabling advanced services in the fourth stage via an intelligent platform. In urban and smart cities, the Internet of FDTs framework [36] is being developed as a unified platform for coordinating DT networks within the context of Society 5.0. A hierarchical architecture facilitates both horizontal and vertical interactions between local DTs, accounting for crucial factors such as inter-twin synchronization, dynamic resource allocation, and scalability. Similarly, the MATISSE project [6] envisions an FDT framework to enhance operations in industrial systems, reduce costs, and accelerate time-to-market by promoting the adoption of DTs in this domain. Model-driven engineering (MDE) techniques, general-purpose and domain-specific languages, the Functional Mockup Interface (FMI), and model transformations support federation across various large-scale industrial systems.

3   Connectionist Federated Digital Twins

3.1   Connectionism Theory

Connectionism, a concept from cognitive science, is a theory of information processing that emphasizes the parallel nature of cognition. It draws inspiration from the brain's neurophysiology to explain human cognitive abilities through mathematical and statistical principles in Artificial Neural Networks (ANNs) [27]. In contrast to the classical symbolic theory, connectionism posits that complex cognitive processes and mental phenomena *emerge* from the dynamic interactions of simpler processing units, similar to biological neurons. Neurocognition refers to modelling cognitive neuroarchitectures based on modern neuroscientific evidence [20]. The connectionist paradigm, also known as the Parallel Distributed Processing (PDP) framework, was extensively developed by the PDP Research Group, drawing on earlier connectionist ideas and neuroarchitectures [24]. This framework has profoundly influenced modern AI development, especially in Deep Learning [4]. However, its origins trace back to Aristotle's ideas on mental associations, later expanded by psychologists and neuropsychologists to explain complex cognitive functions associated with brain processes [21].

Connectionist models focus on the communication between presynaptic and postsynaptic neurons through activation states, with information flowing through synaptic links modulated by Hebbian learning principles [15]. Nonetheless, most computational models in ANNs are time-agnostic, overlooking the essential dynamism inherent in neural systems and cognitive processes, which prioritise time over order [28]. Contemporary neuroscience emphasises that biological intelligence involves cognition as an internal physical process unfolding through time [22]. From the perspective of dynamical systems in cognitive science, philosophy of mind, and neuroscience, cognition is viewed as the simultaneous, mutually influencing unfolding of complex temporal structures. Thus, *temporal dynamics* are crucial for understanding how neural connections organize and interrelate in dynamical spiking neural network models [31].

3.2   Connectionist-Inspired FDT Model

This paper leverages principles from connecitonism theory, synaptic communication (including synaptic weights), and Hebbian learning rules to conceptualize an FDT as a connectionist network of interconnected processing units (DTs) where synaptic weights modulate interrelationships and potential synergies between them based on *neural activity*. Relevant similarities between FDTs and connectionism (Table 1) motivate this study according to the fundamental properties outlined in the PDP framework [24]. Moreover, biologically plausible Spiking neuron models over ANNs are used to model dynamic DT state activation and information propagation across the federated environment.

In this paradigm, each DT is treated as a processing unit within a connectionist network (federation) utilizing Leaky-Integrate and Fire (LIF) neuron models [12]. Each processing unit accounts for two types of LIF models: (1) *Intra-twin LIF*, which produces neural spikes to propagate events and stimulate interconnected DTs, and (2) *Inter-twin LIF* which trigger operations and interactions in the receiving DT upon continuous input stimulation, as illustrated in Fig. 1. LIF models are used due to their biological plausibility, simplicity, low computational cost, and capability to compute temporal dynamics.

Table 1. Connectionist PDP properties applied to Federated Digital Twins

Connectionist (PDP) property	Federated Digital Twins
Set of processing units	Each processing unit in a connectionist network models activation states propagating through the network. Thus, in this study, each DT acts as a node within the federation, utilizing LIF spiking neuron models (intra and inter-twin) to capture its dynamics
States of activation	Activation states reflect the incoming stimuli. In an FDT, the internal state of each DT is determined by real-time inputs from its physical counterpart. When active, a DT emits an event (spike) to signal its state. Inter-twin activation triggers DT operations and interactions, modulating synaptic weights across interconnected DTs in the federation
Patterns of connectivity	Connectivity matrices represent the relationships between units. In the FDT, these matrices determine inter-twin connectivity, capturing aspects such as geographical distance, similarity, or communication relevance
Propagation and activation rules	Determines how DT outputs are transmitted across the network, modulated by the connection weights within the connectivity matrices (synaptic weights). This study considers two modes: (1) When DTs are directly connected, stimulation is based on the weight and output from external DT, emitting spikes; (2) when DTs are not directly connected, a probabilistic sigmoid function in the receiver is used to notify about potential connections on its connected nodes, enabling the emergence of new synaptic links
Algorithm for modifying patterns of connectivity	The algorithm that adjusts synaptic weights based on experience is crucial in connectionism. This study applies the *Spike Timing-Dependent Plasticity (STDP)* rule, a form of Hebbian learning, which modifies connectivity due to interactions between DTs
Representation of the environment	DTs are situated within a virtual environment that replicates the real world. For instance, in agricultural applications, the environment is represented by data on weather and soil conditions, which influence the behaviour of DTs

The threshold for each LIF model depends on the specific application, typically indicating a level of interest or saturation. For instance, intra-twin thresholds might denote the productivity levels of a smart farm in agriculture, the operational state of a machine on a manufacturing shop floor, or the traffic flow

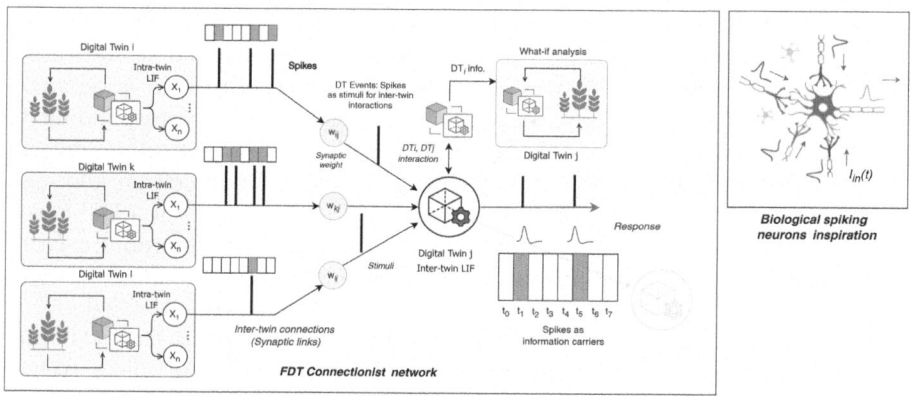

Fig. 1. FDT conceptualized as a connectionist network

limit in a cross intersection in transportation systems. Inter-twin LIF activation depends on the level of connectivity between two DTs. Connections between them are modeled as synapses, with the relevance of communication encoded in the synaptic weights that flow through the synaptic links. Over time, connectivity can adapt based on interaction and the significance of information (plasticity in neuroscience), influenced by parameters such as similarity, frequency, or trust, using a Hebbian Learning rule.

The Leaky-Integrate and Fire neuron [10] describes the evolution of a neuron's membrane potential in response to input stimuli, and is modeled as:

$$\tau \frac{dU(t)}{dt} = -U(t) + I_{in}(t)R \tag{1}$$

Where $U(t)$ is the membrane potential at time t, in volts (V) and τ is the membrane time constant (s), which determines how quickly $U(t)$ decays without input current. τ is calculated as $\tau = RC$, with R being the membrane resistance (ohms, Ω) and C the membrane capacitance (Farads, F). The term $dU(t)/dt$ represents the rate of change of membrane potential, while $I_{in}(t)$ is the input current in amperes (A), influencing the neuron's charge. Larger values of I_{in} promote significant changes in the membrane potential, towards its firing threshold ϑ. Equation 2 describes the discrete version of LIF used in this paper [10].

$$U(t + \Delta t) = U(t) + \frac{\Delta t}{\tau} \left(-U(t) + RI_{in}(t) \right) \tag{2}$$

The membrane potential $U(t)$ is interpreted as the evolving internal state of a DT at time t, where observations are treated as input currents $I_{in}(t)$, encoding information as charge in the LIF model. Two types of LIF neuron models, based on the two communication modes in DT networks [35], have been considered in this study:

- *Intra-twin LIF:* Models the internal state of a DT using the LIF neuron model, where the input current $I_{in}(t)$ represents local data sourced from sensors or generated by simulation models. This data is structured as a vector of characteristics $X = [x_1, x_2, ..., x_n]$, with each element x_i corresponding to a specific attribute of the DT. For example, in a smart farming context, attributes related to biomass productivity, influenced by irrigation and crop nutritional strategies, contribute to the internal activity level of a DT-Farm. The DT may emit an outgoing event (spike) if the accumulated input exceeds a threshold, indicating significant internal change.
- *Inter-twin LIF:* Captures inter-twin communication within the federation using the Leaky Integrate-and-Fire (LIF) neuron model, where each DT integrates incoming spikes modulated by synaptic weights. These spikes trigger internal processes in the receiving DT, which may include: (1) initiating interactions to request additional information from connected DTs, or (2) directly processing the content encoded in the event (spike), which includes performance metrics, model parameters, or predictive outputs. This model facilitates interoperability between DTs, which is a fundamental aspect of FDT systems.

The rationale for using two separate LIF models in each DT (processing unit) lies in the distinct nature of the information they handle. The proposed approach integrates intra-twin LIF (outgoing) spikes into external inter-twin LIF from connected DTs, influencing their behaviour (See Fig. 1).

3.3 Spike Generation and DT Events

When an intra-twin LIF fires due to accumulated input current (observations within the DT), it emits an event, represented as a spike, which is propagated to connected units in the network. Modeled as a $\delta - Dirac$ function, these "instant pulses" occur at a specific time t when a threshold ϑ is reached [12]. The temporal nature of spikes is suitable for encoding information and generating temporal dynamics across the network.

In this paper, each DT event carries information about the emitting DT in the form of $< P, Sp, t(s) >$, where P contains the current properties of the DT (static and dynamic attributes), Sp is the spiking state ($True$ or $False$) and t is the time when the spike is generated. Inter-twin spikes enable interactions between connected DTs across the federation. Successive spikes and the time between them may encode information, forming *spike trains* that represent the intensity of *neural activity*. The stimulation between two DTs is influenced by the temporal correlation and the synaptic weight in their connectivity, reflecting principles of neural plasticity.

3.4 Spike Timing-Dependent Plasticity

Spike Timing-Dependent Plasticity (STDP) is an unsupervised Hebbian learning mechanism that adjusts synaptic weights (w) based on the temporal relationships between pre- and postsynaptic spikes [26]. This mechanism embodies a core property in the PDP connectionist framework [24]. The weight between two processing units is reinforced if a presynaptic spike precedes a postsynaptic spike ($\Delta t > 0$), a process known as Long-Term Potentiation (LTP), indicating a potential causal relationship. In this paper, such stimulation initiates a process of interaction between DTs. If this interaction is beneficial, the synaptic weight between them is strengthened, fostering synergy. Conversely, synaptic weights are weakened if a presynaptic spike follows postsynaptic activity ($\Delta t < 0$) or if no meaningful interaction among DTs occurs. This process is known as Long-Term Depression (LTD). Following the STDP rule, the synaptic weight w at time t between DTs is updated as:

$$w_t = w_{t-1} + \Delta w_t. \tag{3}$$

$$\Delta w_t = \begin{cases} A_+ e^{-\frac{\Delta t}{\tau_+}}, & \text{if } \Delta t > 0 \quad \text{(post after pre)} \\ -A_- e^{\frac{\Delta t}{\tau_-}}, & \text{if } \Delta t < 0 \quad \text{(pre after post)} \end{cases} \tag{4}$$

Here, A_+ and A_- are the learning rate amplitudes for potentiation and depression, respectively, which determine the magnitude of the synaptic change. Parameters τ_+ and τ_- are the time constants that control the decay rate of potentiation and depression. $\Delta t = t_{\text{post}} - t_{\text{pre}}$ captures the time difference between postsynaptic and presynaptic spike events.

4   A Pilot Case

To demonstrate the proposed approach, this section applies the concepts presented in Sect. 3 to scenarios involving sustainable irrigation practices and cooperation in smart farming systems. In smart agriculture, DTs are increasingly being considered to enhance agri-food production through cyber-physical systems, thereby improving food security, sustainability, and waste management. By visualizing farming systems, DTs optimize resource efficiency and biodiversity conservation [2]. Reinforcement learning agents, utilizing synthetic data, minimize resource consumption while maximizing crop yields. *What-if* simulations can explore strategies to enhance carbon sequestration in croplands and pastures, including agroforestry [29].

On a larger scale, Smart Connected Farms (SCF) are transforming agriculture by leveraging high-precision sensors, Big Data, and AI for crop and climate monitoring. This promotes precision agriculture while providing socio-economic benefits and a framework for studying the impacts of climate change on agri-food systems [25]. Therefore, SCFs operating within federated environments utilizing DTs have the potential to optimize natural resources across vast agri-food production regions.

4.1   The Connectionist Model

A federation of DTs, conceptualized as a connectionist FDT, operates within a network of independent farms in the context of a Smart Connected Farming system, as illustrated in Fig. 2.

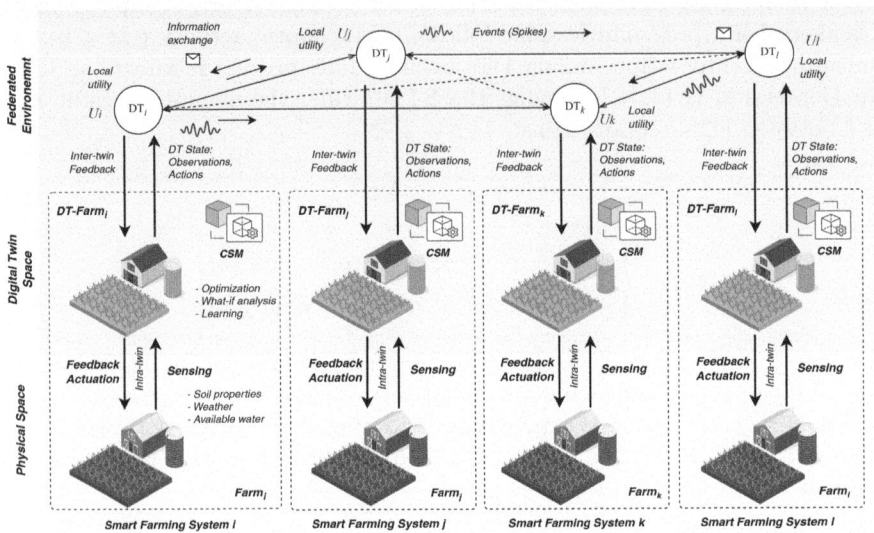

Fig. 2. Smart Connected DT-Farms as a Connectionist FDT

Given the complexities in agriculture and the daily challenges farmers face, interconnected DTs share information to optimize water resources in a decentralized, online manner within the federated ecosystem. An Agent-Based Modelling (ABM) approach enables autonomous decision-making based on local observations and utilities while integrating external influences from connected peers. Information is transmitted through events (spikes) that reflect productivity levels, facilitating communication within the federation and exchanging efficient strategies.

4.2   The Digital Twins Federation

At the core of each DT is a Crop Simulation Model (CSM) that simulates crop growth and abstracts biological processes. Developed by agronomists and biologists, CSMs enhance understanding of crop responses to varying conditions [8]. This study considers the WOrld FOod STudies (WOFOST) model [30], a dynamic, mechanistic CSM that simulates crop growth based on environmental factors (e.g., weather, soil, water, etc.). WOFOST operates on a daily time basis, modelling various crops and enabling annual production analysis. The simulated crop is a potato, with physical soil parameters provided by WOFOST in the Python Crop Simulation Environment (PCSE) [34].

In this scenario, each DT-Farm acts as a processing unit in a connectionist network, where intra-twin states and inter-twin communications are modeled using LIF neuron models. WOFOST defines various crop performance metrics, such as Total Above-Ground Production (TAGP), Leaf Area Index (LAI), and Total Weight of Storage Organs (TWSO), which are computed and retrieved through its implementation in the PCSE.

At the end of the growing season, TWSO $\left(\text{kg ha}^{-1}\right)$ is a key performance metric used to evaluate crop productivity and estimate potential profits [34]. Throughout the simulation, each DT-Farm monitors the daily biomass evolution, quantified by the changes in TWSO calculated as:

$$\Delta TWSO_t = TWSO_{t-1} - TWSO_t \tag{5}$$

Daily changes $\Delta TWSO$ are normalized relative to their yearly average (simulated), indicating productivity levels. This value is then transformed into the current inputs I_{in} to charge the intra-twin LIF model. When the neuron's membrane potential reaches a threshold (indicating an increase in productivity), a *spike* event is fired and transmitted to connected DTs to stimulate their inter-twin LIF model, triggering a what-if analysis. Both intra- and inter-twin LIF model parameters were calibrated using WOFOST outputs, with the following configuration: threshold $\vartheta = 0.2$, time step $\Delta t = 2\,\text{ms}$, resistance $R = 1\varOmega$, and capacitance $C = 5F$.

Six irrigation regimes (1–6), adapted from [18], were used to assess crop response to varying irrigation strategies. These regimes were classified into two groups based on frequency: low-frequency (every 4–6 days) and high-frequency (every 1–3 days). Initially, all sites operated under low-frequency schedules. However, 30% of the sites had a 50% weekly probability of assessing implications to use high-frequency irrigation, potentially offering benefits for local and similar agroecosystems. Daily irrigation (depending on the strategy) is determined based on crop evapotranspiration ET_c (cm), computed as:

$$ET_c = K_c \times ET_0 \tag{6}$$

Where ET_0 is the daily reference evapotranspiration derived using the Penman-Monteith model [1], which incorporates weather effects. K_c, a crop coefficient specific to crop developmental stages, was set for potatoes at 0.5 (initial), 1.15 (mid-season), and 0.75 (late-season) [9].

Therefore, DT-Farms aims to optimize water consumption, considering initially low-frequency regimes while exploring potential strategies that lead to increased local productivity. The decision-making process determines the optimal irrigation strategy between DTs in the federation, considering shared resources and individual demands based on local conditions. The social interaction model [5] (Eq. 7) serves as a local utility, capturing spillovers and network effects, and guides weekly choices.

$$U_i(\omega_i, \omega_{-i}) = \left(\gamma x_i + z_i + \delta \sum_j c_{ij}(t)x_j\right)\omega_i - \frac{1}{2}\omega_i^2 + \phi \sum_j a_{ij}\omega_i\omega_j \tag{7}$$

Each DT_i selects a strategy ω_i from irrigation regimes that maximizes its local utility while accounting for peers' actions and influence. The variable x_i denotes the weekly predicted local yield ($TWSO$) and z_i the Water Use Efficiency (WUE). WUE represents the ratio of yield to the amount of water applied, expressed as kilograms per cubic meter $\left(\text{kg m}^{-3}\right)$ [7]. The term $\delta \sum_j c_{ij}(t)x_j$ captures *contextual effects* or direct influence from connected nodes while $\phi \sum_j a_{ij}\omega_i\omega_j$ reflects strategic complementarity [5]. A and C are weighted (normalized) sociomatrices, where a_{ij} and c_{ij} represent the peer and contextual effects. The middle term captures the convex costs of water consumption. Hence, farms with similar characteristics will most likely influence one another. For simplicity, parameters γ, δ and ϕ are set to 1.

A is derived using the *Haversine distance* based on geographical locations according to Eq. 8.

$$a_{i,j} = 2 \cdot r \cdot \arcsin\left(\sqrt{\sin^2\left(\frac{\phi_j - \phi_i}{2}\right) + \cos(\phi_i) \cdot \cos(\phi_j) \cdot \sin^2\left(\frac{\lambda_j - \lambda_i}{2}\right)}\right) \quad (8)$$

Where a_{ij} is the distance in meters, and r is the Earth's radius $= 6371\,\text{km}$. ϕ_j and ϕ_i are the latitudes of DT-Farms j, and i, respectively, while λ_j and λ_i are their longitudes.

C was obtained using the *Cosine distance* between connected farms as a metric of similarity according to Eq. 9.

$$c_{i,j} = \frac{\boldsymbol{u}_i \cdot \boldsymbol{u}_j}{\|\boldsymbol{u}_i\|\|\boldsymbol{u}_j\|} \quad (9)$$

Where $\boldsymbol{u}_i$, $\boldsymbol{u}_j$ are normalized vectors encoding weather conditions and soil attributes between DT-Farms i and j, respectively. Weather features include average temperature, evapotranspiration, solar radiation, wind speed, vapour pressure, and total precipitation. Soil features comprise field capacity, wilting point, and saturation point.

An iterative message-passing process is done weekly to coordinate irrigation strategies among DT-Farms, using the CSM to predict the expected productivity (TWSO) that maximizes joint utility. The utility function considered is *topology-dependent*, capturing network effects on farm utilities. Successful strategies are identified and exchanged between connected nodes. In this connectionist approach, communication is modulated by local spiking activity and event-based transmission across synaptic links. If the cumulative stimulation exceeds the threshold in the inter-twin LIF model, an interaction process is initiated, enabling a what-if analysis based on shared data among DTs.

If information from one farm leads to notable improvements in another, the connectivity weight between them is strengthened according to the STDP rule (LTP), indicating similarities and potential synergies. If no meaningful gains or no activity is detected, the connectivity weight decays (LTD). The dynamic evolution of these synaptic weights influences social utility across the network, ultimately shaping decision-making processes. The objective is to study the impact of static and emergent dynamic topologies derived from connectionism. The STDP parameters used were $A_+ = 0.01$, $A_- = 0.005$ and $\tau = 5ms$. Δt is the time difference between the pre- and postsynaptic spikes obtained as $\Delta t = 0.001 \times (t_{post} - t_{pre})$.

5   Experimental Results

Experiments were conducted under two scenarios: a fixed nearest neighbours topology and a dynamic connectionist network. Fixed topologies constrain information exchange to predefined links, limiting the emergence of interactions among DT-Farms. In contrast, the dynamic connectionist approach enables adaptive links and synergies modulated by spiking neural activity. Figures 3(a) and 3(b) illustrate these structural differences.

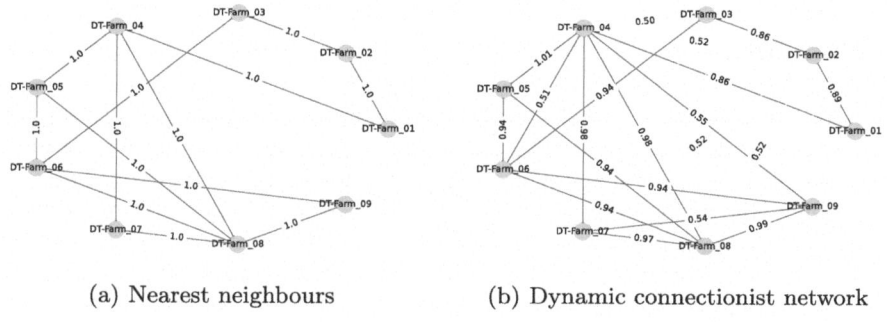

(a) Nearest neighbours   (b) Dynamic connectionist network

Fig. 3. FDT network topology for connected DT-Farms

Results revealed the emergence of *synaptic links* between synergistic DT-farms, such as DT-Farm 4 and DT-Farm 9 or DT-Farm 2 and DT-Farm 7 (Fig. 3(b)), formed probabilistically through DTs intermediaries based on environmental conditions and soil similarities. These connections, enable productive

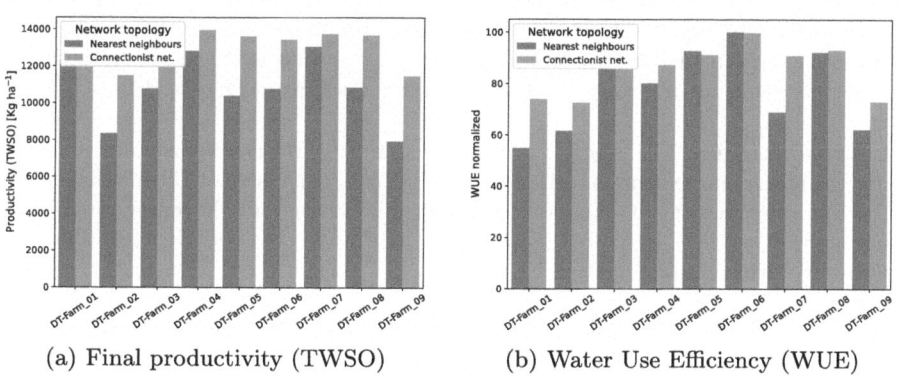

(a) Final productivity (TWSO)   (b) Water Use Efficiency (WUE)

Fig. 4. FDT performance comparison

DT-Farms to influence others through their intra- and inter-twin LIF models, fostering dynamic communication channels that promote interoperability and cooperative behavior within the federation.

Figure 4(a) illustrates increased productivity across all connected DT-Farms, benefiting farmers with improved profits. Simultaneously, the integrated social model for irrigation decision-making penalises excessive water use, mitigating environmental impact. Therefore, synaptic links in a connectionist federation facilitate information diffusion among DTs, while fostering sustainable agricultural decisions based on the social model in Eq. 7. Experimental results indicate a total productivity increase of up to 20.6% achieved by a connectionist network compared to the Nearest neighbours approach by the end of the 2023 growing season. Furthermore, Water Use Efficiency improved, ranging from 0.7% to 22.04% across all DT Farms, as illustrated in Fig. 4(b), due to information exchange and decentralized decision-making considering the emergent network topology.

6 Conclusions

This paper proposes a novel framework based on connectionism theory to support Federated Digital Twins, modelling the dynamic interrelationships among interconnected DTs in a federated environment. Drawing on principles from cognitive neuroscience, the proposed approach captures collaborative patterns for communication and decision-making, enabling cooperation and resource exchange across autonomous DTs. By incorporating temporal dynamics and adaptive connectivity, this methodology enhances the capability of federated ecosystems to handle complex, large-scale systems.

The proof of concept in the context of precision agriculture demonstrates the potential of the proposed framework as an innovative solution to enhance computation for sustainability, by monitoring and analysing pending global challenges such as food security.

Future research will focus on evaluating the scalability and broader implications of this approach within more complex network topologies and federated ecosystems of Digital Twins in different domains. It will also refine the underlying architecture by incorporating new patterns and cognitive principles, and integrating human-in-the-loop modalities and expert knowledge, contributing to an extended suite of reference architectures for Federated Digital Twins [32,33].

Acknowledgments. This work was supported by the Research Institute of Trustworthy Autonomous Systems (RITAS), the SUSTech-University of Birmingham Joint PhD Programme and the Guangdong Province Innovative and Entrepreneurial Team Programme (No. 2017ZT07X386). Christian Vergara-Marcillo and Georgios Theodoropoulos are corresponding authors.

Disclosure of Interests. The authors have no competing interests to declare that are relevant to the content of this article.

References

1. Allen, R.G., Pereira, L.S., Raes, D., Smith, M., et al.: Crop evapotranspiration-guidelines for computing crop water requirements-FAO irrigation and drainage paper 56. Fao Rome **300**(9), D05109 (1998)
2. Alves, R.G., et al.: A digital twin for smart farming. In: 2019 IEEE Global Humanitarian Technology Conference (GHTC), pp. 1–4. IEEE (2019)
3. Bennett, H., Birkin, M., Ding, J., Duncan, A., Engin, Z.: Towards ecosystems of connected digital twins to address global challenges (2023). https://doi.org/10.5281/zenodo.7840266
4. Berkeley, I.S.: The curious case of connectionism. Open Philos. **2**(1), 190–205 (2019)
5. Blume, L.E., Brock, W.A., Durlauf, S.N., Jayaraman, R.: Linear social interactions models. J. Polit. Econ. **123**(2), 444–496 (2015)
6. Bucaioni, A., et al.: Multi-partner project: a model-driven engineering framework for federated digital twins of industrial systems (MATISSE). In: Design, Automation and Test in Europe Conference (DATE 2025) (2025)
7. Cao, X., Xiao, J., Wu, M., Zeng, W., Huang, X.: Agricultural water use efficiency and driving force assessment to improve regional productivity and effectiveness. Water Resour. Manage **35**(8), 2519–2535 (2021)
8. De Wit, A., et al.: 25 years of the WOFOST cropping systems model. Agric. Syst. **168**, 154–167 (2019)
9. Doorenbos, J., Pruitt, W.O.: Guidelines for predicting crop water requirements (1977)
10. Eshraghian, J.K., et al.: Training spiking neural networks using lessons from deep learning. Proc. IEEE **111**(9), 1016–1054 (2023)
11. Fuller, A., Fan, Z., Day, C., Barlow, C.: Digital twin: enabling technologies, challenges and open research. IEEE Access **8**, 108952–108971 (2020)
12. Gerstner, W., Kistler, W.M., Naud, R., Paninski, L.: Neuronal Dynamics: From Single Neurons to Networks and Models of Cognition. Cambridge University Press (2014)
13. GOV.UK: National digital twin programme (NDTP). https://www.gov.uk/government/collections/the-national-digital-twin-programme-ndtp
14. Grieves, M., Vickers, J.: Digital twin: mitigating unpredictable, undesirable emergent behavior in complex systems. In: Kahlen, F.-J., Flumerfelt, S., Alves, A. (eds.) Transdisciplinary Perspectives on Complex Systems, pp. 85–113. Springer, Cham (2017). https://doi.org/10.1007/978-3-319-38756-7_4
15. Hebb, D.O.: The Organization of Behavior: A Neuropsychological Theory. Psychology Press (2005)
16. Hua, Z., Djemame, K., Tziritas, N., Theodoropoulos, G.: A framework for digital twin collaboration. In: 2024 Winter Simulation Conference (WSC), pp. 3046–3057. IEEE (2024)
17. Alan Turing Institute: Towards ecosystems of connected digital twins to address global challenges. https://www.turing.ac.uk/news/publications/towards-ecosystems-connected-digital-twins-address-global-challenges
18. Irmak, S., Djaman, K., Rudnick, D.R.: Effect of full and limited irrigation amount and frequency on subsurface drip-irrigated maize evapotranspiration, yield, water use efficiency and yield response factors. Irrig. Sci. **34**(4), 271–286 (2016). https://doi.org/10.1007/s00271-016-0502-z

19. Jeong, D.Y., et al.: Digital twin: technology evolution stages and implementation layers with technology elements. IEEE Access **10**, 52609–52620 (2022)
20. Maurer, H.: Cognitive Science: Integrative Synchronization Mechanisms in Cognitive Neuroarchitectures of Modern Connectionism. CRC Press (2021)
21. Medler, D.A.: A brief history of connectionism. Neural Comput. Surv. **1**, 18–72 (1998)
22. Monaco, J.D., Rajan, K., Hwang, G.M.: A brain basis of dynamical intelligence for AI and computational neuroscience. arXiv preprint arXiv:2105.07284 (2021)
23. Praharaj, L., Gupta, M., Gupta, D.: Hierarchical federated transfer learning and digital twin enhanced secure cooperative smart farming. In: 2023 IEEE International Conference on Big Data (BigData), pp. 3304–3313. IEEE (2023)
24. Rumelhart, D.E., McClelland, J.L., Group, P.R., et al.: Parallel distributed processing, volume 1: Explorations in the microstructure of cognition: Foundations. The MIT Press (1986)
25. Singh, A.K., et al.: Smart connected farms and networked farmers to improve crop production, sustainability and profitability. Front. Agron. **6**, 1410829 (2024)
26. Sjöström, J., Gerstner, W.: Spike-timing dependent plasticity. Scholarpedia **5**(2), 1362 (2010)
27. Thomas, M.S., McClelland, J.L.: Connectionist models of cognition. In: The Cambridge Handbook of Computational Psychology, pp. 23–58 (2008)
28. Thorpe, S.J.: Why connectionist models need spikes. In: Computational Modelling in Behavioural Neuroscience, pp. 35–58. Psychology Press (2009)
29. Tzachor, A., Richards, C.E., Jeen, S.: Transforming agrifood production systems and supply chains with digital twins. NPJ Sci. Food **6**(1), 47 (2022)
30. Van Diepen, C.V., Wolf, J.V., Van Keulen, H., Rappoldt, C.: Wofost: a simulation model of crop production. Soil Use Manag. **5**(1), 16–24 (1989)
31. Van Gelder, T.: The dynamical hypothesis in cognitive science. Behav. Brain Sci. **21**(5), 615–628 (1998)
32. Vergara, C., Bahsoon, R., Theodoropoulos, G., Yanez, W., Tziritas, N.: Federated digital twin. In: 2023 IEEE/ACM 27th International Symposium on Distributed Simulation and Real Time Applications (DS-RT), pp. 115–116. IEEE (2023)
33. Vergara, C.R., Theodoropoulos, G., Bahsoon, R., Yanez, W., Tziritas, N.: Federated digital twins as an enabling technology for collaborative decision-making. In: Proceedings of the 38th ACM SIGSIM Conference on Principles of Advanced Discrete Simulation, SIGSIM-PADS 2024, pp. 67–68. Association for Computing Machinery, New York (2024). https://doi.org/10.1145/3615979.3662152
34. de Wit, A.: PCSE documentation. Technical report (2019). https://app.readthedocs.org/projects/pcse/downloads/pdf/latest/
35. Wu, Y., Zhang, K., Zhang, Y.: Digital twin networks: a survey. IEEE Internet Things J. **8**(18), 13789–13804 (2021)
36. Yu, T., Li, Z., Hashash, O., Sakaguchi, K., Saad, W., Debbah, M.: Internet of federated digital twins: connecting twins beyond borders for society 5.0. IEEE Internet Things Mag. (2024)

Low Latency Recoding CORDIC Algorithm for FPGA Implementation

Pawel Poczekajlo[2] , Leonid Moroz[4] , Ewa Deelman[1] ,
Michela Taufer[3] , Pawel Gepner[4(✉)] , and Jerzy Krawiec[4]

[1] USC Information Sciences Institute, University of Southern California,
Marina del Rey, CA 90292, USA
deelman@isi.edu
[2] Faculty of Electronics and Computer Science, Koszalin University of Technology,
Koszalin, Poland
pawel.poczekajlo@tu.koszalin.pl
[3] Department of Electrical Engineering and Computer and Science,
University of Tennessee, Knoxville, TN, USA
mtaufer@utk.edu
[4] Faculty of Mechanical and Industrial Engineering,
Warsaw University of Technology, Warsaw, Poland
{leonid.moroz,pawel.gepner,jerzy.krawiec}@pw.edu.pl

Abstract. The Coordinate Rotation Digital Computer (CORDIC) algorithm is widely recognized for its fast real-time processing capabilities, making it highly suitable for hardware implementations in diverse applications such as signal processing, high-performance computing, and edge computing devices. Despite its advantages, the traditional CORDIC algorithm's iterative computational method introduces significant challenges, including a complex structure and high hardware resource consumption, which can limit its efficiency and scalability in certain applications.

In this article, we introduce an innovative and efficient variation of the CORDIC algorithm designed to address these challenges. Our proposed algorithm significantly reduces the number of required operations while maintaining computational accuracy, thereby optimizing performance. Furthermore, we demonstrate that this streamlined algorithm can be effectively implemented on Field-Programmable Gate Arrays (FPGAs), leveraging their reconfigurable hardware to achieve enhanced processing speeds and reduced resource utilization. This advancement not only improves the feasibility of using CORDIC in resource-constrained environments but also expands its applicability in modern computing contexts.

Keywords: CORDIC · algorithm · recoded · FPGA · latency

1 Introduction

The main drawback of the classical CORDIC method [5,17,18] is its low speed due to its linear convergence (only one correct bit of the result per iteration) and

M. H. Lees et al. (Eds.): ICCS 2025, LNCS 15905, pp. 75–89, 2025.
https://doi.org/10.1007/978-3-031-97632-2_6

the hardware complexity associated with the need to implement three simultaneous iteration equations for $x_{i+1}, y_{i+1}, z_{i+1}$ when applying a pipeline structure to the computation:

$$x_{i+1} = x_i - \sigma_i 2^{-i} y_i;$$
$$y_{i+1} = y_i + \sigma_i 2^{-i} x_i;$$
$$z_{i+1} = z_i - \sigma_i \arctan(2^{-i}); \tag{1}$$
$$\sigma_i = \text{sign}(z_i), \quad i = (0, ..., m)$$

that needs $m + 1$ elementary rotations.

To reduce the number of iterations, hybrid structures have been developed that use three steps sequentially: table-based + CORDIC + piecewise-linear multiplication (linear approximation) [1,16,19]. To simplify the hardware complexity, a CORDIC method with angle recoding has been proposed [9–11,13–15], which reduces the system CORDIC iterations (1) to only two equations for x_{i+1}, y_{i+1}. The simplest in terms of hardware implementation is the CORDIC angle recoding method [3,14]. Recently, there have been several new publications on the topic of angle transcoding in CORDIC, one of them being [15]. The theory described therein does not allow for flexible separation of iterations between memory table size and CORDIC. Therefore, it cannot lead to a significant reduction in latency and an increase in the operating frequency of the CORDIC. The algorithm in [3,14] exhibits a drawback as large size of memory (type LUT) is required for large values of m (a table of size not less than $2^{(m/3)} \times m$ bits is needed without the possibility of its reduction). Additionally, the output multipliers are implemented in the $\{-1,1\}$ basis, preventing the use of multipliers that are part of the DSP blocks in modern FPGA devices.

This work addresses the challenges described above by introducing a new approach to angle recoding that allows flexible adjustment of the memory table size and the number of CORDIC iterations. The performance issue is the main motivation of the work (primarily occupancy of the hardware structure and computation time).

2   Description of the Proposed Method

2.1   Known Approaches to Angle Recoding

Let's consider an arbitrary angle θ given in radians ($\theta \in \{0, 2 - 2^{-m}\}$), represented as

$$\theta = \sum_{i=0}^{m} a_i 2^{-i} \tag{2}$$

where a binary basis of coefficients $a_i \in \{0, 1\}$ is used with the weight of the corresponding digit 2^{-i}. In the CORDIC method, this angle is presented through the arctangent set of constant angles

$$\theta = \sum_{i=0}^{m} \sigma_i \arctan(2^{-i}) \tag{3}$$

where a signed basis of rotation direction $\sigma_i \in \{-1, 1\}$ is used with the weight of the corresponding angle $\arctan(2^{-i})$. This set acts as a new arctangent system when iteratively rotating the unit vector

$$\{x_0, y_0\} = \{1, 0\} \tag{4}$$

around the origin in the Cartesian coordinate system by an angle θ [19].

One of the drawbacks of the traditional CORDIC method is the sequential execution of all consecutive iterations, which results in its high latency. The latency has two main causes. First, it is necessary to maintain a constant value of the scaling factor P:

$$P = \prod_{i=0}^{m} \frac{1}{\sqrt{1 + 2^{-2i}}} \tag{5}$$

All iterations must be carried out; skipping them is not allowed.

Second, there is some ambiguity in the subsequent direction σ_i of vector rotation. To find the actual value of σ_i for any i, all previous iterations must be calculated before estimating σ_i.

Let's consider an approach that can eliminate the second reason, which is the bottleneck of the CORDIC method. In this case, the first reason remains valid: the constancy of the value P. To achieve this, instead of the binary basis of coefficients a_i, a signed basis $b_i \in \{-1, 1\}$ similar to σ_i can be applied, for example, $b_i = 2a_i - 1$ (other options for b_i are provided below). This substitution leads to a clear choice of the vector rotation direction because $b_i = -1$ when $a_i = 0$ and $b_i = 1$ when $a_i = 1$. Thus, the ambiguity in estimating σ_i in the next rotation step is eliminated, simplifying the method's structure by avoiding the calculation of this estimate. This technique is known as angle recoding θ [10,11, 13–15]. It is known that during the recoding of the angle θ into θ_r, any of the three formulas can be used [13–15]:

$$\theta_r = \sum_{i=1}^{m+1} b_i 2^{-i} \tag{6}$$

where $b_i = 2a_{i-1} - 1$ [14];

$$\theta_r = \sum_{i=0}^{m} b_i 2^{-i-1} \tag{7}$$

where $b_i = 2a_i - 1$ [13];

$$\theta_r = \sum_{i=0}^{m} b_i 2^{-i} \tag{8}$$

where $b_i = 2a_i - 1$ [15].

In this work, we use formula (5). If we perform the recoding θ_r in the case of the binary basis of coefficients a_i, we get

$$\theta_r = \sum_{i=0}^{m} b_i 2^{-i-1} = \sum_{i=0}^{m} (2a_i - 1) 2^{-i-1} \tag{9}$$

$$= -(1/2)^{m+1} + \sum_{i=0}^{m} a_i 2^{-i}. \tag{10}$$

The relationship between the angles θ and θ_r is as follows:

$$\theta = \theta_r - \theta_c \tag{11}$$

where

$$\theta_c = -(1/2)^{m+1} \tag{12}$$

2.2   Our Approach Recoding CORDIC

If we want to apply formula (5) for the CORDIC method, we need to encode (5) as an angle θ_{rr} in the arctangent system. Thus, we will have

$$\theta_{rr} = \sum_{i=0}^{m} b_i \arctan(2^{-i-1}) = \sum_{i=0}^{m} (2a_i - 1)\arctan(2^{-i-1})$$

$$= \sum_{i=0}^{m} 2a_i \arctan(2^{-i-1}) - \sum_{i=0}^{m} \arctan(2^{-i-1}) \tag{13}$$

or

$$\theta_{rr} = \theta_{rv} - \theta_{rc}; \tag{14}$$

where

$$\theta_{rv} = \sum_{i=0}^{m} 2a_i \arctan(2^{-i-1}); \tag{15}$$

$$\theta_{rc} = -\sum_{i=0}^{m} \arctan(2^{-i-1}). \tag{16}$$

Then the approximation of the angle θ (similar to (7)) for recoding CORDIC will be:

$$\theta_{rv} = \theta_{rr} - \theta_{rc} \tag{17}$$

To implement formula (12), the CORDIC method with the following iterative equations can be applied:

$$x_{i+1} = x_i - b_i y_i \cdot 2^{-i-1};$$

$$y_{i+1} = y_i + b_i x_i \cdot 2^{-i-1};$$

$$i = (0, 1, 2, ..., m);$$

$$x_0 = P\cos(-\theta_{rc}); \tag{18}$$

$$y_0 = P\sin(-\theta_{rc});$$

$$b_i = 2a_i - 1;$$

$$P = \frac{1}{\prod_{i=0}^{m} \sqrt{1 + 2^{-2i-2}}}.$$

As a result, we get $y_{m+1} = \sin(\theta_{rv})$ and $x_{m+1} = \cos(\theta_{rv})$. Obviously, the angle θ_{rv} will be smaller than the angle θ by the value del (lag angle)

$$del = \theta - \theta_{rv} = \sum_{i=0}^{m} a_i d_i =$$

$$= \sum_{i=0}^{m} a_i [2^{-i} - 2 \arctan(2^{-i-1})]. \tag{19}$$

To correctly determine the sine and cosine of the given θ angle, it is necessary to additionally rotate the vector $\{ x_{m+1}, y_{m+1} \}$ by the lag angle del.

Let's estimate the number of iterations within which we should consider the value del. For this, let's set the condition:

$$d_i \leq 2^{-m}, \tag{20}$$

then the inequality must hold:

$$d_i = 2^{-i} - 2 \arctan(2^{-i-1}) \leq 2^{-m}. \tag{21}$$

Using the Taylor series expansion for $\arctan(2^{-i-1})$, we obtain:

$$2^{-i} - 2 \left(2^{-i-1} - \frac{2^{-3i-3}}{3} \right) \leq 2^{-m}, \tag{22}$$

or

$$i = m_a = \lceil (m - 2 - \log_2 3) / 3 \rceil \tag{23}$$

- for this value of i, the condition $d_i \leq 2^{-m}$ will be satisfied.

Starting from this value $i = m_a$, the approximate equality holds (with accuracy to m bits):

$$2^{-m_a} - 2 \arctan(2^{m_a - 1}) \leq 2^{-m}. \tag{24}$$

For example if $m = 16$, then $m_a = 5$.

Hence, the value del for this m should be computed within only these limits of i:

$$del = \sum_{i=0}^{m_a} a_i d_i. \tag{25}$$

The use of the recoding approach allows excluding one of the three components of CORDIC - determining the next direction of micro-rotations σ_i (approximately one-third of hardware resources are saved in FPGA implementation), and also eliminates the need to continuously store the micro-rotation angles $\arctan(2^{-i-1})$.

Let's divide the input angle θ (1) into three angles:

$$\theta = \theta_1 + \theta_2 + \theta_3 \tag{26}$$

We process angle θ_1 to which we add a constant angle θ_{rc} using the look-up table method (where $m_1 + 1$ is the number of most significant bits of angle θ fed into the LUT, for example, $m_1 = 1...m_2$ for $m = 16$), angle θ_2 using the recoding CORDIC method (bits from $m_1 + 1$ to m_2, for example, $m_2 = 8...m$ for $m = 16$), and angle θ_3 using the method of output multiplication (bits from $m_2 + 1$ to m), where:

$$\theta_1 = \sum_{i=0}^{m_1} a_i 2^{-i};$$

$$\theta_2 = \sum_{i=m_1+1}^{m_2} a_i 2^{-i}; \tag{27}$$

$$\theta_3 = \sum_{i=m_2+1}^{m} a_i 2^{-i};$$

and $a_i \in \{0, 1\}$.

Applying the table method involves reading precomputed sine and cosine values of the angle $\theta_1 - \theta_{rc}$ from LUT. Here, the angle θ_{rc} is necessary to consider the recoding. These values are obtained using the formulas:

$$x_{m_1} = P\cos(\theta_1 - \theta_{rc});$$

$$y_{m_1} = P\sin(\theta_1 - \theta_{rc});$$

$$P = \prod_{i=m_1+1}^{m_2} \frac{1}{\sqrt{1 + 2^{-2i-2}}}; \tag{28}$$

$$\theta_{rc} = -\sum_{i=m_1+1}^{m_2} \arctan(2^{-i-1}).$$

For angle θ_2, we use the iterative equations of our recoding for CORDIC:

$$x_{i+1} = x_i - b_i y_i \cdot 2^{-i-1};$$

$$y_{i+1} = y_i + b_i x_i \cdot 2^{-i-1}; \tag{29}$$

$$i = (m_1 + 1, ..., m_2);$$

In the final stage, we use the method of output multiplication with the angle z,

$$x_{m+1} = x_{m_2+1} - z y_{m_2+1};$$

$$y_{m+1} = y_{m_2+1} + z x_{m_2+1}. \tag{30}$$

The formula defines the angle z:

$$z = \theta_3 + del, \tag{31}$$

where $del = \sum_{i=m_1+1}^{m_a} a_i[2^{-i} - 2\arctan(2^{-i-1})]$.

The uniqueness of our approach to angle conversion in the CORDIC method compared to existing ones lies in the following modifications:

- we propose to organise the computation of the sine and cosine of the input Theta angle with a scheme of LUT+ CORDIC+ linear interpolation, which gives the flexibility to choose the number of older bits of the angle (m- exponential of the Theta argument represented in binary code) that determine the size of the LUT type array and reduce the number of CORDIC iterations. This flexibility gives the possibility to dispense with a rigidly defined LUT size up to m/3 as in [14];
- this flexibility of our approach additionally gives the possibility to organise the implementation of the computation scheme: CORDIC+ linear interpolation [15], LUT + linear interpolation (the fastest computation scheme in the approach) according to the user's needs;
- we propose a new conversion approach taking into account the deformation of the input theta angle (an absolutely unique approach that did not exist before) when using a LUT, which gives the possibility to reduce the error significantly (which cannot be done in [14]).

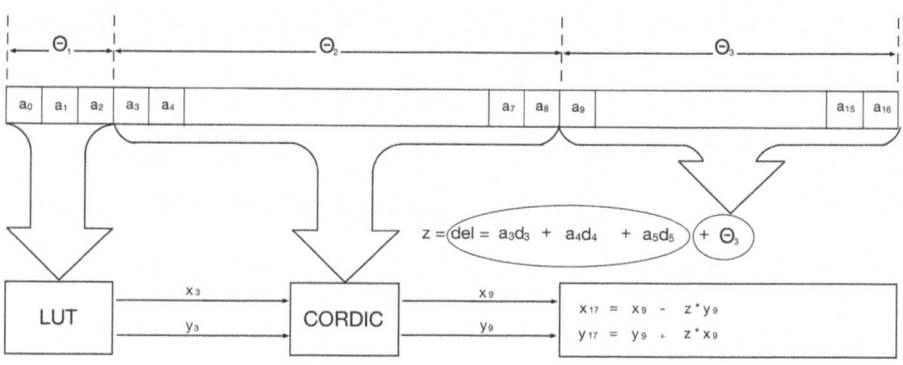

Fig. 1. Schematic view of our CORDIC algorithm

Our modified, recoded CORDIC algorithm is presented in the schematic view in Fig. 1 and its FPGA implementation block scheme in Fig. 2, where it is easy to observe the flow and relationships for signals and variables. This approach allows for the efficient implementation of an FPGA architecture aimed at achieving high computational throughput. The physical realization of the algorithm consists of three basic blocks: A lookup table (LUT), CORDIC, and Multiplier. The presented scheme is directly linked to the pseudo-code of the Algorithm 1. Also shown is the full cycle of calculating the output values for an example angle in Algorithm 2.

The LUT block has been implemented with a 3-bit input and two 20-bit outputs. The output of the LUT gives the values $x3$, $y3$ according to formula (11), which serve as input parameters for the CORDIC block. In our approach, we can arbitrarily choose the number of the most significant bits of the angle θ (the argument) in the binary code, which determines the size of the LUT and

Algorithm 1. Pseudocode of the proposed algorithm.

Input a in binary format

$a \leftarrow \{0, 1, 1, 1, 1, 0, 1, 1, 1, 1, 0, 0, 1, 0, 0, 1, 1\}$

$xx \leftarrow \{1037958, 973957, 849399, 672030,$
      $452877, 205567, -54525, -311226\}$

$yy \leftarrow \{128268, 381076, 610190, 801365, 942715,$
      $1025452, 1044432, 998473\}$

$x \leftarrow \{0, 0, 0, 0, 0, 0, 0, 0, 0, 0\}$

$y \leftarrow \{0, 0, 0, 0, 0, 0, 0, 0, 0, 0\}$

$j \leftarrow a[0] \times 4 + a[1] \times 2 + a[2]$

$x[3] \leftarrow xx[j]$

$y[3] \leftarrow yy[j]$

for $i \leftarrow 3$ **to** 8 **do**

  if $a[i] == 1$ **then**

    $x[i+1] \leftarrow x[i] - (y[i] >> (i+1))$

    $y[i+1] \leftarrow y[i] + (x[i] >> (i+1))$

  else

    $x[i+1] \leftarrow x[i] + (y[i] >> (i+1))$

    $y[i+1] \leftarrow y[i] - (x[i] >> (i+1))$

  end if

end for

$d3 \leftarrow 170$

$d4 \leftarrow 21$

$d5 \leftarrow 3$

$del \leftarrow a[3] \times d3 + a[4] \times d4 + a[5] \times d5$

$Theta3 \leftarrow a[9] \times 2048 + a[10] \times 1024 + a[11] \times 512 + a[12] \times 256 + a[13] \times 128 + a[14] \times 64 + a[15] \times 32 + a[16] \times 16$

$z \leftarrow del + Theta3$

$x17 \leftarrow x[9] - ((z \times y[9]) >> 20)$

$y17 \leftarrow y[9] + ((z \times x[9]) >> 20)$

return $x17, y17$

reduces the number of iterations. The table size can vary from 1 to m/2 at the designer's request (note that with m/2, there are no CORDIC iterations, and the algorithm's structure takes the form of LUT + linear interpolation).

In the CORDIC block, each iteration of the algorithm from $x3$ to $x9$ is implemented as a pipeline stage by logic elements, and it has two 20-bit outputs. This block has a pipelined structure consisting, in this case, of 5 stages. Each stage has two adders with an output register for each adder.

The Multipliers block performs hardware multiplication by the residual angle z, as described in theory and Algorithm 1.

The outputs of the Multipliers block are 20 bits, and they are implemented by dedicated blocks on the FPGA. The efficiency of the proposed approach is achieved by implementing a table-based computation method for the most significant bits of the input angle. It reduces the number of iterations of the CORDIC block. In addition, our approach allows for flexible resizing of the table (number of high-angle bits as an input LUT parameter). However, we have used only

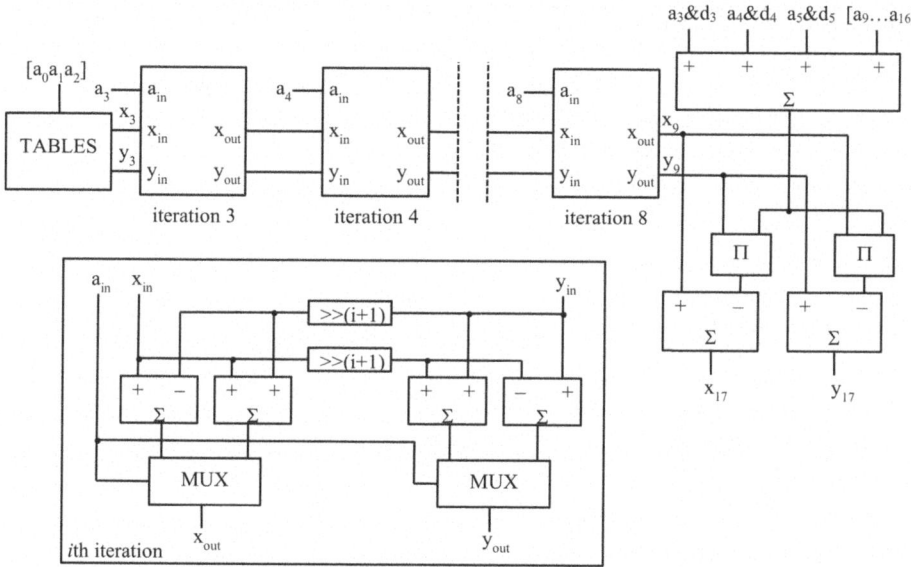

Fig. 2. Implementation block scheme of the algorithm in FPGA chip

the three oldest bits for this purpose. In the best-known FPGA implementation of the conventional CORDIC algorithm for high throughput, all iterations from x1 to x16 are unrolled, and each iteration is implemented as a pipelined stage. Such an implementation method, therefore, requires high latency and many logic elements. In contrast, our implementation of the CORDIC requires fewer logic elements. Moreover, it causes shorter latency due to fewer pipeline stages. It is worth mentioning that computing m-bit sin(x) and cos(x) in a pipeline, fashion requires $3(m+g)$ additions of size $m+g$ where g is a number of guard bits ensuring last-bit accuracy [4].

The remainder of this article examines the practical implementation of the proposed method on an FPGA circuit and compares it with the well-known built-in CORDIC implementation in the Altera library and other CORDIC implementations presented in the literature. A comprehensive analysis of the current advancement in CORDIC algorithms is presented in [15]. We compared our approach to the best implementation of the CORDIC algorithms presented in this article.

3   FPGA Implementation and Results

The hardware implementation is done in Verilog language in the Quartus Prime environment for the Cyclone EP4CE115 FPGA chip. The implementation of the algorithm in the FPGA chip is made of basic logic elements (gates, registers, flip-flops) - without DSP blocks (to have a fair comparison with other implemen-

Algorithm 2. Example of calculation of values according to the presented algorithm for a selected angle (step by step).

The vectors xx, yy, x, y a values d3, d4, d5 are completed as in Algorithm 1.

$Theta = 0.967086792(rad) \times 2^{16} = 63379 = (01111011110010011)BCD$

$a = \{0, 1, 1, 1, 1, 0, 1, 1, 1, 1, 0, 0, 1, 0, 0, 1, 1\}$

$j = 3$

$x[3] = 672030$

$y[3] = 801365$

for $i = 3$ **to** 8 (consecutive iterations) **do**

    for $i = 3$ and $a = 1$

        $x[4] = 621945$

        $y[4] = 843366$

    for $i = 4$ and $a = 1$

        $x[5] = 595590$

        $y[5] = 862801$

    for $i = 5$ and $a = 0$

        $x[6] = 609071$

        $y[6] = 853495$

    for $i = 6$ and $a = 1$

        $x[7] = 602404$

        $y[7] = 858253$

    for $i = 7$ and $a = 1$

        $x[8] = 599052$

        $y[8] = 860606$

    for $i = 8$ and $a = 1$

        $x[9] = 597372$

        $y[9] = 861776$

end for

$del = 191$

$Theta3 = 2352$

$z = 2543$

$x17 = 595283$

$y17 = 863224$

$x17/2^{20} = 0.567706108(\cos)$

$y17/2^{20} = 0.823234558(\sin)$

tations). Figure 2 shows a block diagram of the implementation of the algorithm in an FPGA.

The function calculating $sin()$ and $cos()$ values from Altera IP Core libraries (ALTERA_CORDIC) are also implemented for comparison. For the ALTERA_CORDIC libraries, implementations with outputs with 16 bits of fractional precision (fixed-point) are made. Furthermore, the implementations of the ALTERA_CORDIC library are made only on logic elements (without DSP blocks).

The Altera library supports fixed-point calculations with a computational core controlled by latency or frequency. Code for the VHDL or Verilog HDL description language can be generated. The returned results are rounded to one of the two closest numbers that can be represented in the selected format.

The ALTERA_CORDIC library supports several functions:

- determining trigonometric sine and cosine functions for a given angle (SinCos Function);
- determining the value for 2-argument arctangent (Atan2 Function);
- determining the angle and magnitude of the input vector for the given input coordinates (Vector Translate Function);
- determining the coordinates of a point after rotation by a given angle (Vector Rotate Function).

For the purposes of the conducted research, the focus was only on SinCos Function. This function can be configured in two ways, depending on the input angle range:

- for the angle range $[-\pi, +\pi]$ (signed), the values of the sine and cosine functions take the full range $[-1, +1]$ (this is important because some systems do not return the boundary values -1 and $+1$);
- for the angle range $[0, +\pi/2]$ (unsigned), the output sine and cosine values take the full range $[0, +1]$.

When configuring the library, it is possible to set additional parameters (format) of inputs/outputs. The following can be entered: number of fraction bits (1 to 64, but there must be no more for the output than for the input); width of fixed-point data and the sign of the fixed-point data (signed or unsigned).

Figure 3 shows the implementation model of both versions of the CORDIC algorithm.

3.1   Testing Scenario and Evaluation Methodology

The performance evaluation of CORDIC algorithms on FPGA devices differs from their evaluation on general-purpose computing platforms like CPUs and

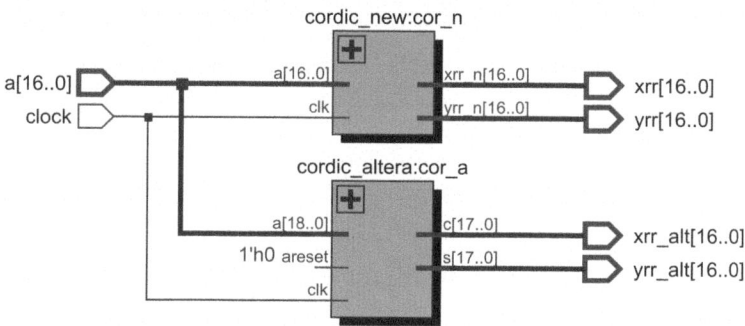

Fig. 3. Implementation model of both versions of the CORDIC algorithm (from RTL tool in Quartus Prime)

GPUs. While similar metrics such as speed, precision, and resource usage are considered, the architectural differences between FPGAs and these platforms lead to distinct evaluation criteria.

For CPUs and GPUs, performance is often measured in terms of execution time, computational accuracy, and resource efficiency (e.g., cycles, bandwidth, and cache use). Energy efficiency is particularly critical in high-performance and mobile computing. The performance evaluation of CPUs and GPUs and their key metrics are well described in various studies [2,6–8,12].

FPGA-based evaluations emphasize the unique flexibility and hardware configurability of these devices. FPGAs excel in implementing repetitive computations, such as those required by CORDIC algorithms, through parallelism and pipelining. Key metrics in this context include:

– **Throughput/Latency**: While both are critical, FPGA designs often optimize throughput—completing multiple operations concurrently—while maintaining acceptable latency levels.
– **Precision/Accuracy**: FPGAs enable customization of data bit-widths, optimizing computational precision for specific applications. This reduces resource consumption and power usage, offering a balance between accuracy and efficiency.
– **Logic Occupation**: Efficient use of FPGA resources, such as Look-Up Tables (LUTs) and Flip-Flops, is vital for achieving high performance without excessive costs.

For testing purposes, 104 input samples (θ angle) are prepared. They are determined with a fit to the bit representation (register lengths).

We measure the following metrics:

– Occupancy of the FPGA chip - number of LUTs (Lookup Tables) - values returned from the report after compilation in the Quartus Prime environment;
– Maximum and minimum absolute error - determined according to:

$$\max\{q(j) - q_{fp}(j)\}$$
$$\min\{q(j) - q_{fp}(j)\} \tag{32}$$

for $j = 1, 2, ..., 104$, where $q(j)$ is the value obtained from a given implementation, $q_{fp}(j)$ is the full precision value for a given angle (determined in the SciLab computer environment);
– Time required to return the result (latency) - determined as the minimum period (=1/clock frequency) for which all 104 input samples (given one after another) returned the expected result.

We compare [15] and aggregate the results in Tables 1, 2 and 3.

Table 1. Results of FPGA occupancy.

CORDIC Realisation	LUTs
Presented CORDIC algorithm	1597
ALTERA_CORDIC with 16-bits fractional outputs	2866
CORDIC algorithm [15]	1438
BBR-CORDIC [34] from [15]	1643
CORDIC II [38] from [15]	1433

Table 2. Results of response time (latency).

CORDIC Realisation	Time (ns)
Presented CORDIC algorithm	56
ALTERA_CORDIC with 16-bits fractional outputs	94
CORDIC algorithm [15]	60
BBR-CORDIC [34] from [15]	60
CORDIC II [38] from [15]	90

Table 3. Results of maximum and minimum absolute errors.

CORDIC Realisation	Max error	Min error
Presented CORDIC algorithm	4.78e-06 (sin) 4.84e-06 (cos)	−1.12e-05 (sin) −1.27e-05 (cos)
ALTERA CORDIC with 16-bit fractional outputs	1.33e-05 (sin) 1.37e-05 (cos)	−1.21e-05 (sin) −1.14e-05 (cos)
CORDIC algorithm [15]	3.04e-05 (sin) 3.04e-05 (cos)	−3.04e-05 (sin) −3.04e-05 (cos)

In addition, it is worth noting that the ALTERA_CORDIC library used does not get the configured parameters. Declaratively, the implemented version of ALTERA_CORDIC should return a result in less than 53ns. The results from ALTERA_CORDIC for the 104 samples used were unsatisfactory - most of the results were incorrect. As the clocking was lowered, the results improved. The ALTERA_CORDIC library returned correct results only when feeding samples with a period of 94ns - much more than the declared value. At the same time, ALTERA_CORDIC has the highest occupancy, which is due to the poor optimization of the ready-made library Table 1. The only advantage of this library is the accuracy (precision of calculations), which is comparable with other implementations Table 3. Presented CORDIC algorithm has occupancy (LUTs) at a very good level - only slightly higher than the lowest results Table 1. The advantage of the presented algorithm is the latency time (56ns) which allows higher sampling (and processing) frequencies Table 2. The aspect of smaller errors is also important (even an order lower than for other algorithms in the case of max errors) Table 3. Other errors are at similar levels (about 1.2e-05). Numerical accuracy is preserved while providing faster processing. CORDIC algorithm [15] has by far the largest errors - which may be related to almost the lowest hardware occupancy (LUTs). Typically, lower occupancy is obtained by reducing fractional bits, and this results in larger errors.

4   Conclusion

The use of our approach has allowed high performance and efficiency improvements of the CORDIC method, with up to 40% power consumption reduction and up to 21% delay reduction, offering overall occupancy reduction of up to 44% (compared to the presented implementation of ALTERA_CORDIC). The approximate CORDIC has been characterized by its absolute arithmetic error, which shows negligible average and maximal errors, i.e., RMS relative errors being only 0.0014%.

In this work, we give the basics of the recoding theory and formalize it in detail using algorithms suitable for hardware implementation on the FPGA platform. We propose a flexible algorithm for freely choosing the number of most significant bits of the angle θ that sets the size of the LUT and reduces the number of iterations. In addition, the difference from [3] is that the residual angle, by which the values from the CORDIC output are multiplied, is always positive. This simplifies the structure of the output multipliers (unsigned multiplication can be used). The proposed approach simplifies the angle recoding method's software and hardware implementation and gives it an advantage over the built-in library functions offered by FPGA developers.

Acknowledgments. This research was partly supported by PLGrid Infrastructure at ACK Cyfronet AGH, Krakow, Poland. This work was also partly supported by the National Science Foundation under grant #2331153.

References

1. Antelo, E., Bruguera, J., Zapata, E.: Unfolded redundant CORDIC VLSI architectures with reduced area and power consumption. IEEE Trans. Very Large Scale Integr. (VLSI) Syst. **31**(5), 872–880 (2023)
2. Ciznicki, M., Kopta, P., Kulczewski, M., Kurowski, K., Gepner, P.: Elliptic solver performance evaluation on modern hardware architectures. In: Wyrzykowski, R., Dongarra, J., Karczewski, K., Waśniewski, J. (eds.) PPAM 2013. LNCS, vol. 8384, pp. 155–165. Springer, Heidelberg (2014). https://doi.org/10.1007/978-3-642-55224-3_16
3. Curticapean, F., Niittylahti, J.: A hardware efficient direct digital frequency synthesizer. IEEE J. Solid-State Circuits **58**(4), 876–884 (2023)

4. Dinechin, F., Istoan, M., Sergent, G.: Fixed-point trigonometric functions on FPGAs. ACM SIGARCH Comput. Archit. News **41**, 83–88 (2014). https://doi.org/10.1145/2641361.2641375
5. Doe, J., Smith, J., Johnson, A.: Implementation of floating point CORDIC algorithm using 45 nm technology. IEEE Trans. Comput. **72**(3), 450–458 (2023)
6. Gepner, P.: Using AVX2 instruction set to increase performance of high performance computing code. Comput. Inform. **36**(5), 1001–1018 (2017)
7. Gepner, P., Fraser, D.L., Kowalik, M.F.: Second generation quad-core Intel Xeon processors bring 45 nm technology and a new level of performance to HPC applications. In: Bubak, M., van Albada, G.D., Dongarra, J., Sloot, P. (eds.) ICCS 2008. LNCS, vol. 5101, pp. 417–426. Springer, Heidelberg (2008). https://doi.org/10.1007/978-3-540-69384-0_47
8. Gepner, P., Gamayunov, V., Fraser, D.L.: Effective implementation of DGEMM on modern multicore CPU. Procedia Comput. Sci. **9**, 126–135 (2012). https://doi.org/10.1016/j.procs.2012.04.014
9. Hu, Y.H., Naganathan, S.: An angle recoding method for CORDIC algorithm implementation. IEEE Trans. Comput. **42**(1), 74–79 (1993)
10. Juang, T.: Low latency angle recoding methods for the higher bit-width parallel CORDIC rotator implementations. IEEE Trans. Circuits Syst. II Express Briefs **55**(11), 1139–1143 (2008)
11. Juang, T.B., Hsiao, S.F., Tsai, M.Y.: Para-CORDIC: parallel CORDIC rotation algorithm. IEEE Trans. Circuits Syst. I Regul. Pap. **51**(8), 1515–1524 (2004)
12. Kopta, P., et al.: Parallel application benchmarks and performance evaluation of the Intel Xeon 7500 family processors. Procedia Comput. Sci. **4**, 372–381 (2011). https://doi.org/10.1016/j.procs.2011.04.039
13. Kuhlmann, M., Parhi, K.K.: P-CORDIC: a precomputation based rotation CORDIC algorithm. EURASIP J. Adv. Signal Process. **2002**(9), 1–8 (2002). https://doi.org/10.1155/S1110865702205028
14. Madisetti, A., Kwentus, A.Y., Willson, A.N.: A 100 mhz, 16-b, direct digital frequency synthesizer with 100-DBC spurious-free dynamic range. IEEE J. Solid-State Circuits **34**(8), 1034–1043 (1999)
15. Qin, M., Liu, T., Hou, B., Gao, Y., Yao, Y., Sun, H.: A low-latency RDP-CORDIC algorithm for real-time signal processing of edge computing devices in smart grid cyber-physical systems. Sensors **22**(19) (2022). https://doi.org/10.3390/s22197489
16. Timmermann, D., Hahn, H., Hosticka, B.: Low latency time CORDIC algorithms. IEEE Trans. Comput. **41**, 1010–1015 (1992)
17. Volder, J.E.: The CORDIC trigonometric computing technique. IEEE Trans. Electron. Comput. **EC-8**(3), 330–334 (1959)
18. Walther, J.S.: A unified algorithm for elementary functions. In: Proceedings of AFIPS Conference, vol. 38, pp. 385–389 (1971)
19. Wang, S., Piuri, V., Swartzlander, E.E.: Hybrid CORDIC algorithms. IEEE Trans. Comput. **46**(11), 1260–1263 (1997)

Global Optimization of Microwave Circuits Using Dimensionality Reduction and Multi-fidelity EM Simulations

Slawomir Koziel[1,2]($\boxtimes$) (iD), Anna Pietrenko-Dabrowska[2] (iD), and Leifur Leifsson[3] (iD)

[1] Engineering Optimization and Modeling Center, Department of Engineering, Reykjavík University, Menntavegur 1, 102, Reykjavík, Iceland
koziel@ru.is
[2] Faculty of Electronics Telecommunications and Informatics, Gdansk University of Technology, Narutowicza 11/12, 80-233 Gdansk, Poland
anna.dabrowska@pg.edu.pl
[3] School of Aeronautics and Astronautics, Purdue University, West Lafayette, IN 47907, USA
leifur@purdue.edu

Abstract. The growing complexity of contemporary microwave circuits made numerical optimization imperative as a performance-boosting tool. Yet, it is intricate because of the high costs incurred by electromagnetic (EM) analysis required to evaluate the system's quality reliably. These expenses are particularly significant in global optimization, which is necessary in many situations. This study introduces an innovative strategy for high-efficacy globalized optimization of passive components. Our methodology leverages reduction of the problem dimensionality implemented using a rapid global sensitivity analysis and a custom-developed machine learning (ML) algorithm employing fast surrogate models established in the reduced domain. The designs rendered by the ML process are further refined in the local sense in the full-dimensionality parameter space using a gradient-based routine. Additional improvement in efficiency is obtained by employing multi-fidelity EM simulations with the low-fidelity models used for global search and high-fidelity ones only utilized in fine-tuning. The presented approach has been comprehensively validated utilizing two coupling circuits and juxtaposed against a pool of benchmark algorithms. The obtained results underscore the remarkable efficacy of our procedure. The typical running cost does not exceed a hundred high-fidelity EM analyses, corresponding to sizable savings over the benchmark. At the same time, the proposed method renders designs of competitive quality.

Keywords: Computer-aided design · microwave engineering · optimization · EM-driven design · dimensionality reduction · multi-fidelity simulations

1 Introduction

Microwave passive components have become increasingly complex in fulfilling performance demands associated with emerging application areas. On top of meeting strict requirements concerning electrical characteristics [1, 2], one of the crucial considerations nowadays is miniaturization [3]. Size reduction can be achieved through appropriate

M. H. Lees et al. (Eds.): ICCS 2025, LNCS 15905, pp. 90–103, 2025.
https://doi.org/10.1007/978-3-031-97632-2_7

geometry modifications (e.g., line folding and utilization of metamaterials [4, 5]), which lead to the further increase of topological sophistication. Accurate assessment of such circuits requires electromagnetic (EM) simulation because conventional methods, such as equivalent network modeling, cannot quantify phenomena such as cross-coupling, dielectric losses, or the effects of the system's environment (e.g., connectors, housing) [6].

Although EM-driven design is imperative, at least at the later stages of the circuit development process [7], it is also intricate. The major issues include handling multiple decision variables, design objectives, and constraints. At the same time, repetitive EM simulations incur considerable computational expenses, which is especially problematic when using numerical optimization methods. Employing the latter also requires an appropriate background and familiarity with optimization algorithms, which microwave engineers often lack. Consequently, traditional approaches relying on combining engineering insight and parametric studies are still widely used even though they cannot yield optimum results. Nonetheless, such interactive techniques are immensely laborious. Another challenge is that in a growing number of cases, global optimization is necessary to address the design problem's multimodality (e.g., design of metasurfaces, array pattern synthesis, etc. [8, 9]), unavailability of a good starting point [10], or optimization-driven miniaturization [11].

Today, global optimization is mainly conducted with nature-inspired algorithms [12, 13]. These methods leverage the exchange of information between sets of candidate solutions processed during the optimization run and the employment of stochastic components [14] to allow escaping from local optima. The literature is replete with specific procedures (e.g., [15–17]). Their popularity stems from straightforward handling and accessibility. Unfortunately, direct optimization of EM simulation models through population-based methods incurs tremendous computational expenses. In practice it is typically attempted if cheaper (e.g., analytical) models are available [18]. Cost-related difficulties may be alleviated with the help of surrogate modeling [19], often incorporated into machine learning (ML) frameworks [20, 21]. The ML process renders candidate solutions (infill points) using a metamodel as a fast predictor, subsequently refined using accumulated EM data [22]. Despite its potential benefits, the construction of a reliable surrogate is the most severe bottleneck of ML, mainly due to the curse of dimensionality. Mitigation methods include domain confinement [23], variable-fidelity approaches [24], feature-based methods [25, 26], and physics-based frameworks (e.g., space mapping [27]). Unfortunately, many of these solutions lack the generality and are challenging to integrate with global search engines.

This research introduces a new approach to reduced-cost globalized optimization of microwave circuits. We focus on enhancing the efficiency and reliability of the search procedure. To pursue this goal, a fast machine learning algorithm is introduced that incorporates kriging metamodels built in a dimensionality-reduced subspace. The latter is determined with the help of a rapid global sensitivity analysis (RGSA), which identifies directions corresponding to the most prominent changes in the system's frequency characteristics. The candidate solutions are rendered by a particle swarm optimizer (PSO) used as a core search algorithm. Cost efficiency is further enhanced by involving low-resolution EM analysis for global search. Meanwhile, the dependability is ensured by employing high-fidelity EM models at the final (gradient-based) tuning stage. Extensive numerical verification demonstrates the exquisite efficacy of our method. It is superior over several benchmark techniques, among others nature-inspired and machine-learning routines. The average cost does not exceed a hundred high-fidelity EM analyses. Meanwhile, multiple runs of the algorithm corroborate the ability of the framework to yield satisfactory results in each instance.

2 Fast Microwave Optimization by Dimensionality Reduction

Here, we elucidate the operation of the suggested global optimization procedure. The design task statement is recalled in Sect. 2.1. Variable-resolution EM models and rapid sensitivity analysis are discussed in Sects. 2.2 and 2.3. Sections 2.4 and 2.5 elucidate the global and local search stages. Finally, Sect. 2.6 puts together the operation of the entire algorithm.

2.1 Problem Statement

We aim to reduce a merit function $U(x)$ encoding the design quality. Here, $x = [x_1 \ldots x_n]^T$ represents the decision variables (here, geometry parameters of the circuit). The task is stated as

$$x^* = \arg \min_{x \in X} U(x) \tag{1}$$

in which X is the search domain. The circuit outputs (scattering parameters versus frequency) are evaluated using EM analysis and denoted as $S_{ij}(x, f)$, with i and j being the indices of the port indices; f stands for the frequency. A representative example is a microwave coupler designed to enhance return loss and port isolation at a target frequency f_0. At the same time, we aim to maintain a target power division ratio K_P at f_0. The merit function may take the form of

$$U(x) = \max\{|S_{11}(x, f_0)|, |S_{41}(x, f_0)|\} + \beta\big[|S_{21}(x, f_0)| - |S_{31}(x, f_0)| - K_P\big]^2 \tag{2}$$

In (2), the second part is a regularization factor that enforces the power division condition.

2.2 Multi-fidelity EM Models

Multi-fidelity models have been utilized in microwave engineering for over two decades to accelerate design procedures [28]. The idea is to trade-off accuracy for speed under controlled conditions, i.e., appropriate enhancement of the low-resolution model. Although the less reliable representation is typically an equivalent network, it is of limited generality. In this work, we implement the low-resolution model $R_c(x)$ by reducing the resolution of the EM analysis (i.e., using coarse discretization of the simulated circuit). This versatile approach ensures a sufficient correlation with the high-resolution model $R_f(x)$ [29]. A representative example and typical relationship between R_c and R_f have been showcased in Fig. 1.

This study employs R_c to execute sensitivity analysis, construct the initial surrogate model (Sect. 2.3), and carry out the global search stage (Sect. 2.4). In contrast, R_f is for final tuning (Sect. 2.5). Note that due to a good correlation between the models (cf. Fig. 1), R_c can be used uncorrected, also because possible inaccuracies will be rectified at the last stage.

2.3 Global Sensitivity Analysis. Dimensionality-Reduced Search Domain

This study uses a surrogate-assisted ML scheme to conduct the global search stage. The major challenge is building a reliable data-driven metamodel. Here, it is facilitated through dimensionality reduction implemented using a rapid global sensitivity analysis (RGSA) adopted from [30] and outlined in Fig. 2. RGSA produces a set of orthonormal directions e_j ordered regarding their effect on the circuit response variability. These effects are quantified by the eigenvalues $\lambda_1 \geq \lambda_2 \geq \ldots \geq \lambda_n$. The restricted domain is then established using a small number N_d of the most relevant vectors collectively accounting for most of the system's response variability. Given the threshold C_{min}, N_d is set as the smallest integer such that [30]

$$\sqrt{\sum_{j=1}^{N_d} \lambda_j^2} \Big/ \sqrt{\sum_{j=1}^{n} \lambda_j^2} \geq C_{min} \tag{3}$$

Here, $C_{min} = 0.9$ so that the vectors e_j determining the domain are responsible for 90% of the overall variability. The reduced domain X_d is determined as

$$X_d = \left\{ x \in X : x = x_c + \sum_{j=1}^{N_d} a_j e_j \right\} \cap X \tag{4}$$

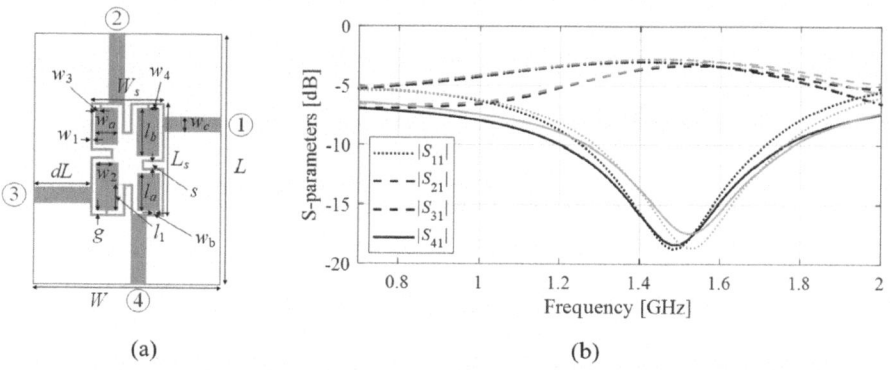

(a)   (b)

Fig. 1. Multi-fidelity models: (a) exemplary miniaturized microstrip coupler, (b) frequency characteristics obtained with the low- and high-fidelity EM models R_c (gray) and R_f (black). The evaluation times for R_f and R_c are 210 and 90 s, respectively.

The vector $x_c = [l + |u|]/2$ (decision variable space center); $a_j, j = 1, \ldots, N_d$ are real numbers.

2.4 Global Search Stage

The first search step is global optimization using R_c and conducted in the reduced space X_d. It is a machine learning (ML) process with the infill criterion involving the improvement of the merit function. The underlying surrogate is kriging interpolation, which is optimized using the particle swarm optimizer (PSO).

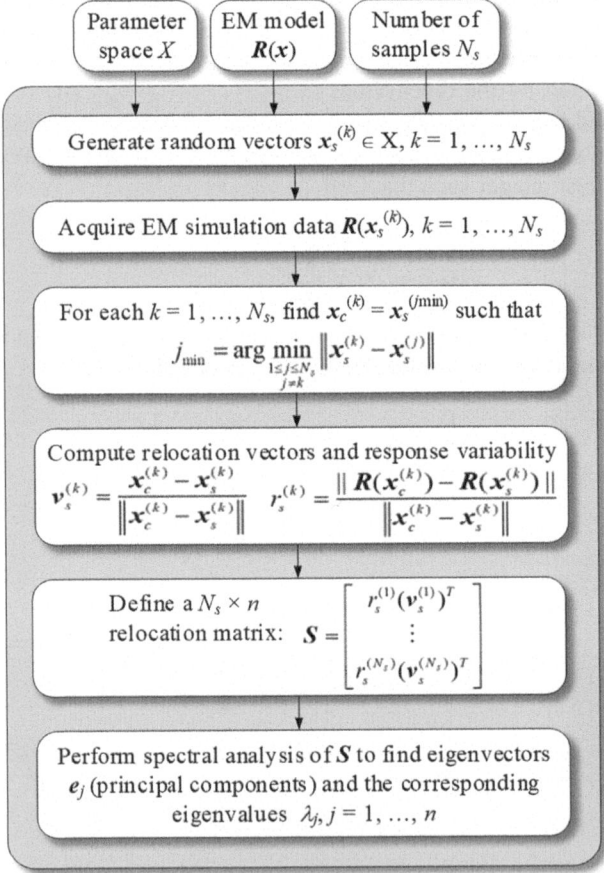

Fig. 2. The outline of RGSA [30]. The principal vectors e_j are associated with the directions significantly affecting the circuit outputs as quantified by the corresponding eigenvalues λ_j.

The initial surrogate model is constructed using N_iN_d samples, where $N_i = 20$ in the verification experiments discussed in Sect. 3. The training points $x_B^{(k)} = k = 1, \ldots, N_iN_d$, are allocated uniformly in the reduced domain X_d. A temporary surrogate $s_{tmp}(x)$ is first built using $\left\{ x_B^{(k)}, R_c(x_B^{(k)}) \right\}_{k=1,\ldots,N_iN_d}$. Next, s_{tmp} is iteratively enhanced using the points produced by maximizing the mean square error (MSE) of the metamodel for $j = 1, 2, \ldots$

$$x_B^{(N_iN_d+j)} = \arg \max_{x \in X_d} MSE(s_{tmp}(x)) \tag{5}$$

The infill vector is incorporated into the training dataset so that we have $\{x_B^{(k)}, R_c(x_B^{(k)})\}_{k=1,\ldots,NiNd+j}$. Maximization of MSE (here, using PSO) promotes the enhancement of the surrogate's accuracy. The model construction is terminated if the cross-validation-estimated relative RMS error becomes smaller than E_{max} (here, set to 20%) or the overall number of generated samples exceeds $2N_iN_d$. The final version of $s_{tmp}(x)$ is renamed as the initial surrogate $s^{(0)}(x)$.

Having $s^{(0)}(x)$, the global search stage is launched within X_d, which is a machine learning (ML) algorithm. It starts by optimizing $s^{(0)}$. Subsequent models, $s^{(j)}, j = 1, 2, \ldots$, are obtained using the EM simulation results generated during the search. ML generates candidate solutions $x^{(i+1)}$, $i = 0, 1, 2, \ldots$, by optimizing the cost function $U_S(x, s^{(i)}(x))$, i.e., using predicted improvement of the merit function. U_S coincides with the original merit function U (cf. Sect. 2.1) but it is computed based on $s^{(i)}(x)$ rather than EM-simulated outputs. We have

$$x^{(i+1)} = \arg \min_{x \in X_d} U_S(x, s^{(i)}(x)) \tag{6}$$

Again, PSO acts as the core search engine, although any bio-inspired routine may be used. Note that because of the low cost of evaluating $U_S(x, s^{(i)}(x))$, solving of (6) may be executed using a high computational budget (here, 10,000 cost function calls). The candidate designs and the accumulated EM data are used to refine the surrogate: $s^{(i)}(x)$ is built based on $\left\{ x_B^{(k)}, R_c(x_B^{(k)}) \right\}_{k=1,\ldots,2N_iNd+i}$, with $x_B^{2N_iNd+i} = x^{(i)}$ for $i = 1, 2, \ldots$. The termination conditions are: $\left\| x^{(i+1)} - x^{(i)} \right\| < \varepsilon$ (convergence in argument), or no improvement in cost function for the last $N_{no\_improve}$ iterations. The parameter values utilized in the verification experiments are $\varepsilon = 10^{-2}$ and $N_{no\_improve} = 20$.

2.5 Fine Tuning

The global search stage is conducted in X_d using R_c. Both factors enable significant computational savings. The same factors contribute to a degradation of reliability. This is rectified by adding a fine tuning stage, executed in the original decision variable space, and using high-fidelity EM analyzes. The underlying search routine is the trust-region (TR) algorithm [31]. The circuit response gradients are computed using finite differentiation (FD) [32]. Because the process starts from an already good design rendered by ML, the cost of final tuning is low. The algorithm is further expedited by replacing FD with a Broyden updating scheme [33], when the process approaches convergence, i.e., when $\left\| x^{(i+1)} - x^{(i)} \right\| < 10\varepsilon_{TR}$, where $\varepsilon_{TR} = 10^{-3}$ is the termination threshold.

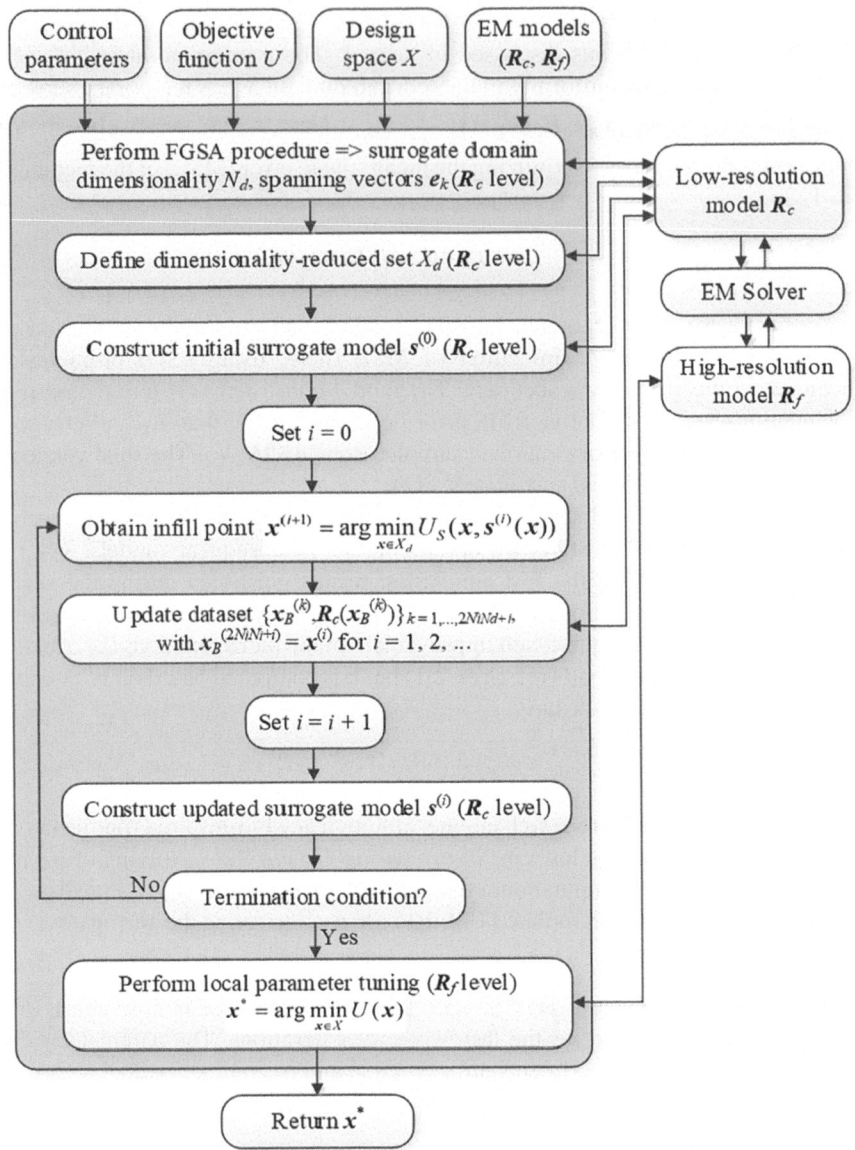

Fig. 3. Proposed machine learning procedure for global microwave optimization: flow diagram.

2.6 Complete Algorithm

The complete optimization framework employs the algorithmic components discussed in Sects. 2.2 through 2.5. Our approach utilizes fast surrogates built in a dimensionality-reduced space defined using RGSA. It should be reiterated that sensitivity analysis (RGSA) and the global search stage are conducted using the faster but less accurate low-fidelity EM model R_c. Fine tuning is conducted using the high-fidelity model R_f. There are few control parameters ($N_r, N_i, E_{\max}, \varepsilon, N_{no\_improve}, \varepsilon_{TR}$), all of which were discussed in detail in the preceding sections. None of these parameters is critical, and most control the resolution of the search process. The algorithm's flow diagram is illustrated in Fig. 3.

3 Results

This section showcases the operation of the suggested technique, demonstrated with two microstrip circuits and extensive comparisons to four benchmark methods, nature-inspired, gradient-based, and machine learning.

3.1 Test Circuits

Consider the test circuits illustrated in Fig. 4, and referred to as Circuits I and II. Their important parameters are listed in Table 1. CST Microwave Studio [36] is used to implement the computational models. The low-fidelity R_c model is a coarse-discretization version of the high-fidelity representation R_f (cf. Sect. 2.2). The resolution of R_f is determined through a grid convergence study. Note that both verification problems are challenging and require handling four distinct responses (matching, transmission, and isolation characteristics), several objectives, and carrying out the search in vast parameter spaces.

3.2 Setup

The setup of our framework and the benchmark procedures have been included in Table 2. There are four methods, including particle swarm optimizer (PSO) (Algorithm I) executed in two versions (500 and 1000 objective function evaluations), random-initialization gradient-based search (Algorithm II), and two ML routines. Among these, Algorithm III operates in the original parameter space. In contrast, Algorithm IV uses dimensionality reduction similar to that employed in this work, but the entire search process employs high-fidelity EM simulations. Observe that the budget assigned to Algorithm I is lower for bio-inspired methods yet considerable in absolute terms (a few days of CPU time). On the other hand, Algorithm II is incorporated to demonstrate the multimodality of our verification tasks.

3.3 Results

The results are displayed in Tables 3 and 4. The data encapsulates the mean value of the merit function and the optimization cost evaluated for ten independent executions of each method. Additionally, a success rate is displayed, which is the fraction of runs leading to satisfactory outcomes. The expenses are expressed in the equivalent number of R_f evaluations. Figures 5 and 6 show the circuit frequency characteristics for the selected runs of the suggested technique.

The main performance factors are reliability and computational efficiency. The reliability is assessed using the success rate, which is perfect (10/10) for our methodology and both test cases. In contrast, the benchmark techniques perform much worse. In particular, the fraction of successful runs for the random-start gradient search is only 5/10 and 8/10 for Circuit I and II, respectively, which underscores multimodality of the considered verification problems.

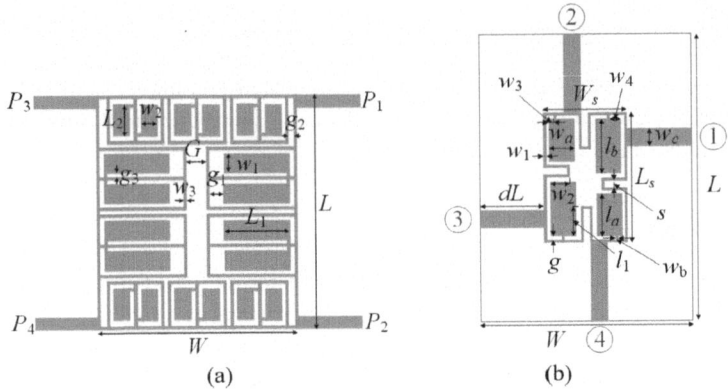

(a)   (b)

Fig. 4. Test cases: (a) Circuit I [34], (b) Circuit II [35].

Table 1. Essential data and design requirements for Circuits I and II

Parameter	Test case	
	Circuit I	Circuit II
Substrate	FR4 ($\varepsilon_r = 4.4$, $h = 1.0$ mm)	AD300 ($\varepsilon_r = 2.97$, $h = 0.76$ mm)
Design parameters$	$x = [G\ g_1\ g_2\ g_3\ w_1\ w_3\ L_1\ L_2]^T$	$x = [g\ l_{1r}\ l_a\ l_b\ w_1\ w_{2r}\ w_{3r}\ w_{4r}\ w_a\ w_b]^T$
Other parameters$	$L = 4w_1 + 10w_3 + 15g_3 + 2L_2$, $W = 4w_3 + 2L_1 + G + 2g_1 + 2g_3$	$L = 2dL + L_s$, $L_s = 4w_1 + 4g + s + l_a + l_b$, $W = 2dL + W_s$, $W_s = 4w_1 + 4g + s + 2w_a$, $l_1 = l_b l_{1r}$, $w_2 = w_a w_{2r}$, $w_3 = w_{3r} w_a$, $w_4 = w_{4r} w_a$, $w_c = 1.9$ mm
EM model	CST Microwave Studio	CST Microwave Studio
R_c (low-fidelity)	~ 200,000 mesh cells Simulation time 70 s	~ 30,000 mesh cells Simulation time 90 s
R_f (high-fidelity)	~ 600,000 mesh cells Simulation time 320 s	~ 170,000 mesh cells Simulation time 210 s
Target frequencies [GHz]	$f_0 = 1.0$ GHz	$f_0 = 1.5$ GHz
Design goals	• Minimize $|S_{11}|$ and isolation $|S_{41}|$ at the operating frequency f_0 • Ensure power split ratio $K_P = 0$ dB	• Minimize matching $|S_{11}|$ and isolation $|S_{41}|$ at the operating frequency f_0 • Ensure power split ratio $K_P = 0$ dB
Parameter space X	$l = [0.2\ 0.2\ 0.2\ 0.2\ 1.0\ 0.2\ 4.0\ 2.0]T$ $u = [1.5\ 1.5\ 1.5\ 1.0\ 3.5\ 0.5\ 12.0\ 8.0]^T$	$l = [0.4\ 0.6\ 3\ 9\ 0.6\ 0.4\ 0.1\ 0.6\ 4.0\ 0.6]^T$ $u = [0.5\ 0.85\ 6.5\ 11\ 0.95\ 0.7\ 0.4\ 0.9\ 5\ 0.7]^T$

$ Dimensions in mm except for the relative ones (with subscript r), which are unitless.

The performance of PSO is much better (9/10), yet not perfect, which is indicative of insufficient computational budget. On the other hand, the ML-based techniques match ours; however, they exhibit significantly higher computational expenses. The design quality measured by the objective function value is also highly competitive for the proposed approach. It is comparable to Algorithms III and IV but significantly better than Algorithms I and II. Also, one can note that increasing the budget for Algorithm I translates into a noticeable improvement in the results (by a few dB), which is another argument against direct EM-driven parameter tuning by means of bio-inspired techniques.

Computational efficiency is another advantage of the suggested technique. The average expenses incurred by the optimization process are below a hundred R_f evaluations, corresponding to 91% relative speedup over Algorithm I, 67% acceleration over Algorithm III, and 45% acceleration over Algorithm IV. At the same time, our method incurs costs comparable to gradient-based search (94 versus 54 EM simulations), which is remarkable given the local nature of Algorithm II. These benefits are the results of the mechanisms incorporated into the proposed framework, especially dimensionality reduction, the two-stage optimization process, and multi-fidelity EM analyses.

Table 2. Benchmark techniques

Algorithm	Algorithm type	Setup
This work	ML framework with dimensionality reduction	Control parameters: $N_r = 50, N_i = 20, E_{\max} = 20\%, \varepsilon = 10^{-2}, N_{no\_improve} = 20, \varepsilon_{TR} = 10^{-3}$ (the parameter's meaning has been explained in Sect. 2)
I	Particle swarm optimizer (PSO)	Swarm size $N = 10$, standard control parameters ($\chi = 0.73, c_1 = c_2 = 2.05$); the number of iterations set to 50 (ver. I) and 100 (ver. II)
II	Trust-region gradient-based optimizer	Random initial design, response gradients estimated using finite differentiation, termination criteria based on convergence in argument, and reduction of the trust region size [121]
III	Machine-learning procedure	Algorithm setup: • Initial surrogate set up to ensure relative RMS error not higher than 20% with the maximum number of training samples equal to 400; • Algorithm operates in the original parameter space (no dimensionality reduction); • Infill criterion: minimization of the predicted objective function
IV	Machine-learning procedure	Algorithm setup: • The method is the same as the proposed one; however, the algorithm operates at the level of high-resolution EM models; • Control parameters: default values as in Table 1

Table 3. Optimization results for Circuit I

Optimization algorithm	Performance figure		
	Average objective function value [dB]	Computational cost$	Success rate#
Algorithm I: PSO (50 iterations)	−20.8	500	9/10
Algorithm I: PSO (100 iterations)	−22.2	1,000	9/10
Algorithm II: Trust-region gradient-based algorithm	−7.5	49.0	5/10
Algorithm III: Machine learning operating in the original parameter space X	−26.2	449.8	10/10
Algorithm IV: Machine learning operating in the reduced space X_d; high-resolution model only	−24.5	183.6	10/10
Proposed algorithm	−25.0	66.0	10/10

$ The cost expressed in terms of the number of EM simulations of the circuit under design.
# Number of runs at which the operating frequencies were allocated near the target frequency.

Table 4. Optimization results for Circuit II

Optimization algorithm	Performance figure		
	Average objective function value [dB]	Computational cost$	Success rate#
Algorithm I: PSO (50 iterations)	−25.2	500	9/10
Algorithm I: PSO (100 iterations)	−29.1	1,000	9/10
Algorithm II: Trust-region gradient-based algorithm	−10.7	57.4	8/10
Algorithm III: Machine learning operating in the original parameter space X	−30.2	238.4	10/10
Algorithm IV: Machine learning operating in the reduced space X_d; high-resolution model only	−25.7	165.4	10/10
Proposed algorithm	−35.7	122.2	10/10

$ The cost expressed in terms of the number of EM simulations of the circuit under design.
# Number of runs at which the operating frequencies were allocated near the target frequency

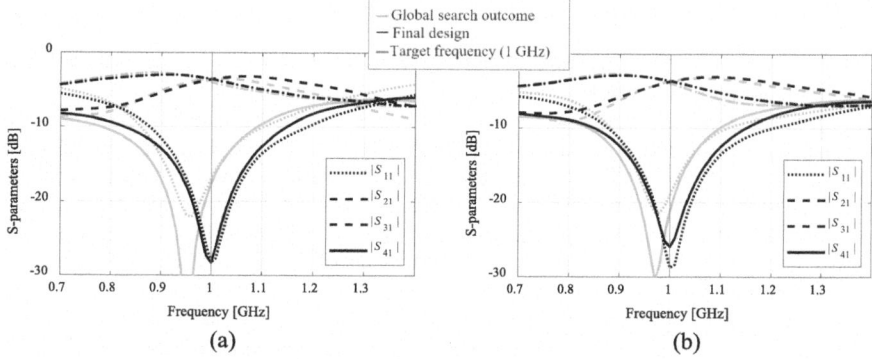

Fig. 5. Circuit I: designs found by the proposed multi-fidelity ML algorithm: (a) selected run 1, (b) selected run 2.

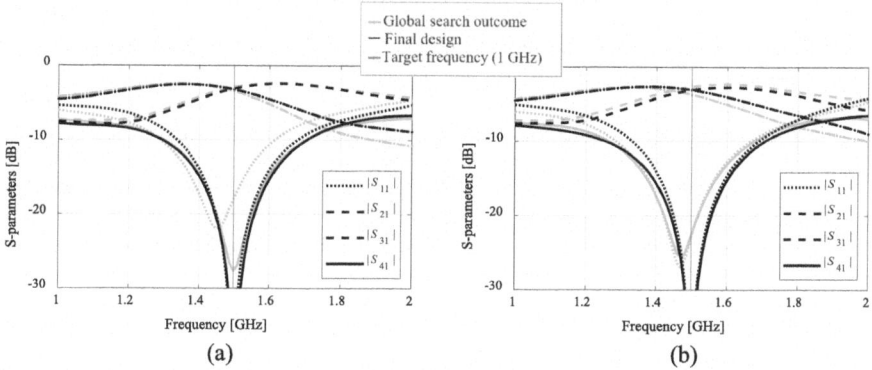

Fig. 6. Circuit II: designs found by the proposed multi-fidelity ML algorithm: (a) selected run 1, (b) selected run 2.

4   Conclusion

In this study, we developed an innovative methodology for the high-efficacy optimization of microwave passives. Our technique employs rapid global sensitivity analysis (RGSA) to determine a confined search domain (spanned by vectors responsible for the most significant changes in the system outputs) and a machine learning (ML) algorithm to execute global stage, complemented by gradient-based fine tuning. The efficiency is boosted by incorporating multi-fidelity EM simulations, low fidelity (faster but less accurate) for RGSA and ML, and high fidelity for final tuning. Combining these mechanisms results in competitive computational efficiency and reliability, as demonstrated by using two planer circuits and benchmarking against several state-of-the-art techniques.

Acknowledgment. The authors thank Dassault Systemes, France, for making CST Microwave Studio available. This work is partially supported by the Icelandic Centre for Research (RANNIS) Grant 239858 and by the National Science Centre of Poland Grant 2020/37/B/ST7/01448.

References

1. Liu, M., Lin, F.: Two-section broadband couplers with wide-range phase differences and power-dividing ratio. IEEE Microw. Wirel. Comput. Lett. **31**, 117–120 (2021)
2. Li, Q., Chen, X., Chi, P., Yang, T.: Tunable bandstop filter using distributed coupling microstrip resonators with capacitive terminal. IEEE Microw. Wirel. Comput. Lett. **30**, 35–38 (2020)
3. He, Z., Liu, C.: A compact high-efficiency broadband rectifier with a wide dynamic range of input power for energy harvesting. **30**, 433–436 (2020)
4. Zhou, J., Rao, Y., Yang, D., Qian, H.J., Luo, X.: Compact wideband BPF with wide stopband using substrate integrated defected ground structur. IEEE Microw. Wirel. Comput. Lett. **31**, 353–356 (2021)
5. Singh, H., Gupta, A., Kaler, R.S., Singh, S., Gill, J.: Designing and analysis of ultrathin metamaterial absorber for W band biomedical sensing application. IEEE Sens. J. **22**, 10524–10531 (2022)
6. Koziel, S., Pietrenko-Dabrowska, A., Plotka, P.: Reduced-cost microwave design closure by multi-resolution EM simulations and knowledge-based model management. IEEE Access **9**, 116326–116337 (2021)
7. Zhang, Z., Liu, B., Yu, Y., Cheng, Q.S.: A microwave filter yield optimization method based on off-line surrogate model-assisted evolutionary algorithm. IEEE Trans. Microw. Theory Techn. **70**, 2925–2934 (2022)
8. Esmail, B.A.F., Koziel, S., Szczepanski, S., Majid, H.A.: Overview of approaches for compensating inherent metamaterials losses. IEEE Access **10**, 67058–67080 (2022)
9. Liang, S., Fang, Z., Sun, G., Liu, Y., Qu, G., Zhang, Y.: Sidelobe reductions of antenna arrays via an improved chicken swarm optimization approach. IEEE Access **8**, 37664–37683 (2020)
10. Abdullah, M., Koziel, S.: A novel versatile decoupling structure and expedited inverse-model-based re-design procedure for compact single-and dual-band MIMO antennas. IEEE Access **9**, 37656–37667 (2021)
11. Jin, H., Zhou, Y., Huang, Y.M., Ding, S., Wu, K.: Miniaturized broadband coupler made of slow-wave half-mode substrate integrated waveguide. IEEE Microw. Wirel. Comp. Lett. **27**, 132–134 (2017)
12. Li, X., Luk, K.M.: The grey wolf optimizer and its applications in electromagnetics. IEEE Trans. Ant. Prop. **68**, 2186–2197 (2020)
13. Oyelade, O.N., Ezugwu, A.E.-S., Mohamed, T.I.A., Abualigah, L.: Ebola optimization search algorithm: a new nature-inspired metaheuristic optimization algorithm. IEEE Access **10**, 16150–16177 (2022)
14. Liu, F., Liu, Y., Han, F., Ban, Y., Jay, G.Y.: Synthesis of large unequally spaced planar arrays utilizing differential evolution with new encoding mechanism and Cauchy mutation. IEEE Trans. Antennas Propag. **68**, 4406–4416 (2020)
15. Li, X., Guo, Y.-X.: Multiobjective optimization design of aperture illuminations for microwave power transmission via multiobjective grey wolf optimizer. IEEE Trans. Ant. Prop. **68**, 6265–6276 (2020)
16. Li, W., Zhang, Y., Shi, X.: Advanced fruit fly optimization algorithm. IEEE Access **7**, 165583–165596 (2019)
17. Jiang, Z.J., Zhao, S., Chen, Y., Cui, T.J.: Beamforming optimization for time-modulated circular-aperture grid array with DE algorithm. IEEE Ant. Wirel. Propag. Lett. **17**, 2434–2438 (2018)
18. Cui, L., Zhang, Y., Jiao, Y.: Robust array beamforming via an improved chicken swarm optimization approach. IEEE Access **9**, 73182–73193 (2021)
19. Zhang, Z., Cheng, Q.S., Chen, H., Jiang, F.: An efficient hybrid sampling method for neural network-based microwave component modeling and optimization. IEEE Microw. Wirel. Comp. Lett. **30**, 625–628 (2020)

20. Wu, Q., Wang, H., Hong, W.: Multistage collaborative machine learning and its application to antenna modeling and optimization. IEEE Trans. Ant. Propag. **68**, 3397–3409 (2020)
21. Yu, X., Hu, X., Liu, Z., Wang, C., Wang, W., Ghannouchi, F.M.: A method to select optimal deep neural network model for power amplifiers. IEEE Microw. Wirel. Comp. Lett. **31**, 145–148 (2021)
22. Forrester, A.I.J., Keane, A.J.: Recent advances in surrogate-based optimization. Prog. Aerospace Sci. **45**, 50–79 (2009)
23. Koziel, S., Pietrenko-Dabrowska, A.: Performance-driven surrogate modeling of high-frequency structures. Springer, New York (2020)
24. Pietrenko-Dabrowska, A., Koziel, S.: Expedited gradient-based design closure of antennas using variable-resolution simulations and sparse sensitivity updates. IEEE Trans. Ant. Propag. **70**, 4925–4930 (2022)
25. Pietrenko-Dabrowska, A., Koziel, S.: Design centering of compact microwave components using response features and trust regions. Energies **14**, 1–15 (2021)
26. Zhang, C., Feng, F., Gongal-Reddy, V., Zhang, Q.J., Bandler, J.W.: Cognition-driven formulation of space mapping for equal-ripple optimization of microwave filters. IEEE Trans. Microw. Theory Techn. **63**, 2154–2165 (2015)
27. Bandler, J.W., Rayas-Sánchez, J.E.: An early history of optimization technology for automated design of microwave circuits. IEEE J. Microw. **3**, 319–337 (2023)
28. Cervantes-González, J.C., Rayas-Sánchez, J.E., López, C.A., Camacho-Pérez, J.R., Brito-Brito, Z., Chávez-Hurtado, J.L.: Space mapping optimization of handset antennas considering EM effects of mobile phone components and human body. Int. J. RF Microw. CAE **26**, 121–128 (2016)
29. Ogurtsov, S., Koziel, S.: Model management for cost-efficient surrogate-based optimization of antennas using variable-fidelity electromagnetic simulations. IET Microw. Ant. Prop. **6**, 1643–1650 (2012)
30. Koziel, S., Pietrenko-Dabrowska, A., Leifsson, L.: Improved efficacy behavioral modeling of microwave circuits through dimensionality reduction and fast global sensitivity analysis. Sc. Rep. **14**, paper no. 19465 (2024)
31. Conn, A.R., Gould, N.I.M., Toint, P.L., Trust Region Methods, MPS-SIAM Series on Optimization (2000)
32. Levy, H., Lessman, F.: Finite Difference Equations. Dover Publications Inc., New York (1992)
33. Broyden, C.G.: A class of methods for solving nonlinear simultaneous equations. Math. Comp. **19**, 577–593 (1965)
34. Letavin, D.A., Shabunin, S.N.: Miniaturization of a branch-line coupler using microstrip cells. In: International Scientific-Technical Conference Actual Problems of Electronics Instrument Engineering (APEIE), pp. 62–65 (2018)
35. Tseng, C., Chang, C.: A rigorous design methodology for compact planar branch-line and rat-race couplers with asymmetrical T-structures. IEEE Trans. Microw. Theory Techn. **60**, 2085–2092 (2012)
36. CST Microwave Studio, ver. 2023, Dassault Systemes, France, 2023

Understanding the Limitations of Deep Transformer Models for Sea Ice Forecasting

Julia Borisova$^{(\boxtimes)}$ ⓘ, Andrey Kuznetsov ⓘ, Gleb Solovev ⓘ,
and Nikolay O. Nikitin ⓘ

ITMO University, St. Petersburg 197101, Russia
yulashka.htm@yandex.ru

Abstract. It would not be an exaggeration to say that we live in the era of transformers. Due to the great results of generative models for video prediction, spatio-temporal data of various kinds are usually treated as video-like sequences - and this is a good assumption for many problems. However, we want to argue that transformer-based prediction is not the best option for some spatio-temporal cases with regular grid and strong periodicity (since most discussions about the limitations of transformer applicability focus only on time series).

In the paper, we considered the task of sea ice forecasting and analyzed two transformer-based architectures (TimeSformer and SwinLSTM) against the proposed baseline - a lightweight convolutional network with different setups of convolutional layers (2D and 3D). Experiments for long-term forecasting of Arctic seas show that transformers do not reproduce the annual dynamics of sea ice. At the same time, the CNN-based solutions allow to outperform the existing state-of-the-art numerical (SEAS5) and data-driven (IceNet) forecasts, with a quality improvement of up to 30% in the mean absolute error and up to 10% in the structural similarity index. A similar experiment is provided for the synthetic example of video data. Due to the analysis of the obtained results, this problem is caused by the nature of the model and the data and can be faced in many scientific and industrial tasks outside sea ice.

Code and supplementary materials for this research are available on GitHub: https://github.com/ITMO-NSS-team/sea_ice_transformers.

Keywords: sea ice concentration · transformers · CNN · spatio-temporal data · long-term forecasting

1 Introduction

The discussion on the limitations of the applicability of transformers is widely presented in the literature [25], especially for the forecasting of time series. However, for the broad class of spatio-temporal tasks (e.g. video prediction), transformers are considered as the basis of almost any state-of-the-art model. One of the most challenging spatio-temporal problems for AI is sea ice forecasting [12].

M. H. Lees et al. (Eds.): ICCS 2025, LNCS 15905, pp. 104–118, 2025.
https://doi.org/10.1007/978-3-031-97632-2_8

Since spatio-temporal data have similarities with a video sequence and its forecasting can be considered as a video prediction task, it looks promising to apply more complex architectures with the attention mechanism - transformers - that have proven successful for processing video [14]. While the application of transformers for long-term, high-resolution forecasting of the entire area of the Arctic Ocean is challenging (due to the amount of computational resources required to train a model), it appears suitable for the task of regional modelling of ice dynamics in specific water areas, which is also quite important [5].

We found that *self-designed baseline models based on simple convolutional architecture significantly outperform deep transformers for long-term regional sea ice forecasting* for all regions considered. Similar problems with the applicability of transformers are widely discussed for univariate and multivariate time series forecasting [25]. However, for tasks similar to video prediction, the limits of the applicability of transformers are still not well discussed. Therefore, we conducted a detailed investigation using different models and setups.

In the experimental part of the paper, we provide a comparison of four approaches to predictive modelling of ice conditions. Two transformer-based decisions: TimeSformer [3], which adapts the architecture for sequence prediction, and SwinLSTM [20]. Shallow two- and three-dimensional convolutional networks have been proposed as strong baselines. The experimental setup includes regional prediction of ice concentration for five Arctic seas.

Experiments show the inability of transformers to reproduce the annual dynamics of ice concentration due to incorrect periodicity components. At the same time, CNN-based baselines provide adequate results for long-term forecasting. We compare them with two state-of-the-art (SOTA) solutions for sea ice forecasting (physics-based system SEAS5 [10] and neural ensemble-based model IceNet [1]). *The improvement of the CNN-based baseline over domain-specific SOTA is up to 30% in mean absolute error, up to 10% in structural similarity index and up to 6% in ice edge reconstruction accuracy.*

We can conclude that it is important to extend the existing benchmarks for video prediction by novel tasks with strong periodic component to represent the limits of transformers for a wide class of real-world "periodic" tasks (from remote sensing to the analysis of production systems [6]). We have also provided a synthetic example of periodic data for video forecasting task as an empirical proof that the considered problem is not specific only to sea ice.

2   Related Works

Sea ice forecasts can be obtained using global physical models based on systems of differential equations have been developed to simulate ice conditions. However, despite the scale of the system, the simulation is global and roughly reproduces local processes, which is a serious limitation of its applicability For this reason, regional physical models such as SI^3 are widely used for real-world tasks.

The accumulation of remote sensing data has made it possible to use fully data-driven models to forecast ice conditions. The task is known as spatio-temporal forecasting, so image-based deep learning models are actively used. Disadvantages of most deep learning models for forecasting ice conditions include a short forecasting horizon (several days or weeks [8]), while long-term forecasting (>3 months) is especially important for planning industrial work for the next seasons.

Convolutional Neural Networks (CNN) [13] are widely used in sea ice modelling. The most popular architecture is the U-net [1,8]. There are examples of the use of U-net models for long-term forecasts up to one year [11]. However, the limiting factor for the use of such solutions is the need for a large amount of additional input data on the atmosphere (temperature, pressure, solar radiation, etc.) for training and inference. To improve the quality of forecasts, simple models are often combined into ensembles, which allows a probabilistic modelling component to be introduced, taking into account the confidence of each of the ensemble models [1].

Since the spatio-temporal data is similar to a video sequence, video prediction methods can be applied to sea ice concentration forecasting. The first group of methods are recurrent networks. There are many architectures from ConvLSTM [18] to the more recent (e.g. CrevNet [24]), which proposes a CNN-based recurrent network for learning spatio-temporal dependencies. The PhyDNet [9] model introduces physical knowledge into a CNN-based model to improve the quality of prediction. For video prediction, these models perform reasonably well due to their ability to account for spatial and temporal dependencies.

The Transformer [22] architecture has also been widely applied to video processing. The ViT [7] model was the first to use Transformers directly for image classification and achieved impressive results. However, the performance of the ViT model is highly dependent on the size of the training sample. There is also a promising SwinLSTM [20] model for video prediction based on Swin Transformer [14] blocks with a simplified LSTM. This model performs well in analysing temporal and spatial dependencies in video files.

Limitations of transformers is a topic that is widely discussed in the literature [25]. There is even a repository *Transformers And LLM Are What You Dont Need*[1], which contains the examples where simple models overcome deep transformers for different tasks. However, the papers in this repository focus on time series data (both univariate and multivariate). Even if the data has a spatiotemporal nature (e.g. the task of predicting traffic at different spatial points, it is not represented as a regular grid and solved by other methods (e.g. graph neural networks). Thus, the limitation of the applicability of transformers to video-like sequences is still an under-discussed topic.

3   Problem Statement

We consider the problem of regional sea ice concentration prediction not only from a domain-specific point of view. In this paper, we discuss the applicability of state-of-the-art computer vision models to the data with specific properties that are characteristic of the environmental case considered.

---

[1] https://github.com/valeman/Transformers_And_LLM_Are_What_You_Dont_Need.

3.1   Nature of the Data and Models

The main difference between metocean forecasting and conventional spatio-temporal forecasting is the different nature of the data. First, the dynamic processes in environmental systems are multiscale and non-stationary. In addition, they contain an irremovable stochastic component.

However, state-of-the-art computational methods still perform well on a large part of natural systems forecasting tasks - for example, the transformer-based basic model can outperform both state-of-the-art classical simulation tools and specialised deep learning models for weather forecasting [4]. So what is the problem? Why can we not apply state-of-the-art CV tools directly to sea ice data? What is so special about this?

One issue is the non-differentiable nature of sea ice data - it is not a smooth field, but data with a clear distinction between concentrated ice and clean water (the so-called "ice edge"). In addition, sea ice has very complex periodic patterns (e.g. annual periodicity for sea ice), the reproduction of which is crucial for current forecasting.

In this paper we aimed to prove or reject the **hypothesis**: *the practical applicability of regular-grid transformer-based models for spatio-temporal data with specific periodic properties is very limited.* We use regional sea ice concentration prediction as a real-world case study to empirically confirm it.

The theoretical basis of this hypothesis is as follows: artificial neural networks are combinations of several simple mathematical functions that implement more complex functions from one real data value to another. The spaces of multivariate functions that can be implemented by a network are determined by the structure of the network and its parameters. Ice concentration data is non-linear, it is a time series with pronounced periodicity. Since neural network architectures based on transformers have a linear nature [17], the main hope for improving the quality of prediction is achieved through a large number of parameters.

It is known that adding the ReLU activation function allows to increase the efficiency of networks on linear layers by transforming the model architecture [23]. Therefore, in the process of adapting the transformers to the task, ReLU activation functions were added to the architecture to improve the quality of data approximation based on a large number of parameters.

3.2   Benchmarks for Spatio-Temporal Tasks

As the task of spatio-temporal prediction is not new, there are many well-known open benchmarks against which the model can be compared. For example, the OpenSTL[2] [19] benchmark for spatio-temporal predictive learning covers several tasks (including weather prediction from WeatherBench [16]). However, tasks similar to sea ice forecasting are not included in these benchmarks. For this reason, we cannot base our experimental setup on existing benchmarks and prepare our own dataset.

---

[2] https://github.com/chengtan9907/OpenSTL.

4   Proposed Approaches

We propose a strong baseline for the task of sea ice prediction based on a convolutional architecture. As typical examples of transformer models, we choose TimeSformer and SwinLSTM. The mean absolute error was used as the loss function for all models. The technical details of the model implementations and their adaptation to the sea ice forecasting task are given below.

4.1   Baseline

CNN-2D was implemented as a CNN with an encoder-decoder architecture. It consists of 5 convolutional 2D layers with ReLU activation function and its transposed mirror. The values in the input images range from 0 to 1 due to the nature of the ice concentration data. As input data, the model receives a multichannel image with the history of the parameter; the output of the model is a multichannel image with a prediction n steps ahead.

The training sample was formed by a sliding window along the space-time series. The scheme illustrating the dataset formation is shown in Fig. 1. This approach allows the starting point of the model to be varied and a forecast to start on any day of the year. This is important for applying the model to real industrial problems as the forecast can be based on the most recent data.

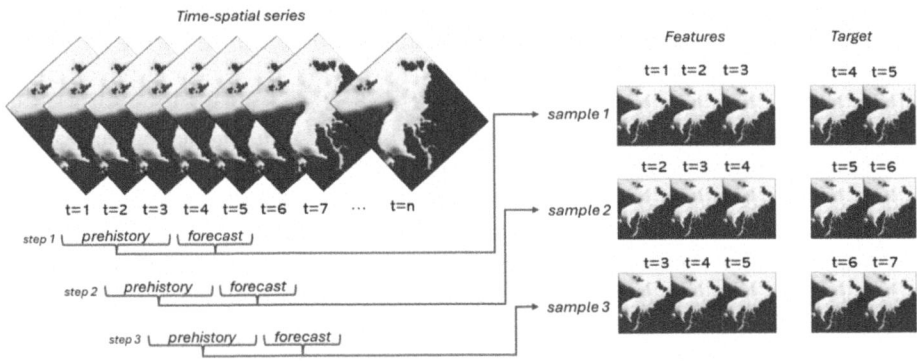

Fig. 1. The preprocessing of training set of sea ice forecasting

Models trained with the L1 loss function tend to produce grain artefacts during inference. To solve this problem and make the model lighter, we reduced the spatial resolution of the input images (by a factor of 2).

Baseline CNN-3D uses a time component sensitive CNN encoder-decoder architecture with 3D convolution. Each of the encoder and decoder parts consists of 2 layers, forming a symmetric structure. Otherwise, the architecture and training process are identical to the previous model.

The model has fewer layers and parameters than the baseline because 3d convolution is asymptotically more complex. This increased complexity means that convolution operations with a 3d kernel can be more time-consuming than those with a 2d kernel, even when the number of parameters is reduced. For example, with almost the same number of parameters for 2D and 3D convolutions (2234 and 2529 estimated with software), their total number of multi-adds is 42.83 million and 216.32 million respectively. The number of parameters for TimeSformer is 33 million and 20 million for SwinLSTM. The impact of such an increase on the runtime is shown in the Table 2.

The third dimension of the kernel acts as a temporal dimension to the input, which consists of 2D images over time. By tuning this third dimension, the model can effectively extract temporal components such as seasonality or trends, thereby improving its ability to detect and predict time-dependent patterns in the data. For these reasons, 52 was chosen as the third dimension of the kernel, corresponding to one year of prediction, with each time step representing one week. Choosing a higher frequency or increasing the number of layers becomes challenging because with the input size halved and a history of 104 timesteps, the feature maps can degenerate to zero after convolution.

4.2   Transformers

The dynamics of ice concentration changes can be represented as spatio-temporal data closest to the video of ice melt and ice intrusion. In this video series, not only neighbouring images are linked, but there can also be a link between images related by seasonality. For example, the data for January of each year are linked, and taking this into account it is possible to predict February more effectively. An appropriate attention mechanism can be used to identify and take account of this relationship. Although the original Transformer was developed for NLP (natural language processing) tasks, there are now solutions for processing images, such as ViT [2] and Swin Transformer [14]. For comparison with the proposed baseline solutions, two transformer-based models were applied to the ice concentration forecasting problem: TimeSformer [3] and SwinLSTM [20].

TimeSformer. Processing frames alone is not enough to create an effective approach to sea ice forecasting. ViViT [2] and TimeSformer can provide a more in-depth method for processing data such as video. These models are designed for the task of video series classification and are encoder-only models.

TimeSformer implements the Divided Space-Time attention mechanism, which we believe is the promising basis for the sea ice prediction task. Thus, our experiments with transformers are based on the developments of TimeSformer.

The model architecture had to be refined because the original TimeSformer was designed for video classification, and the task at hand requires the prediction of ice concentration changes over multiple frames. To this end, the transformer head responsible for classification was replaced by a convolutional decoder. This decoder translates the hidden state of the output data from the TimeSformer

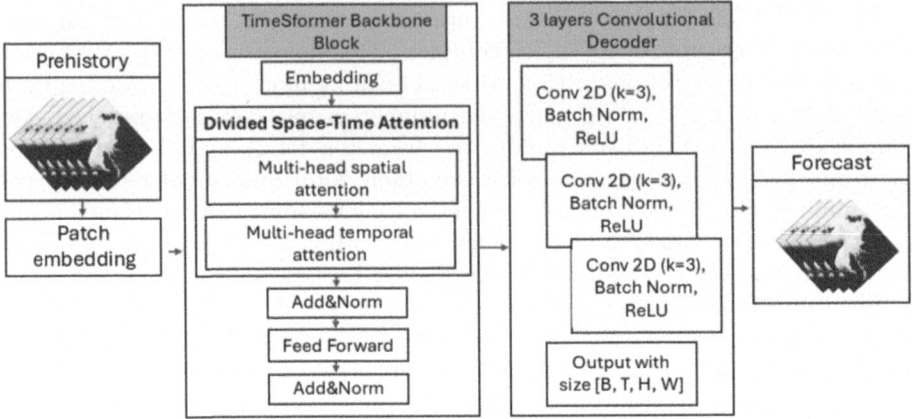

Fig. 2. Application of the TimeSformer to the long-term sea ice forecasting

backbone into the $[B,T,H,W]^3$ dimension of the sea ice data set. Three convolution layers with ReLU activation function and BatchNorm normalization were used to unlock the decoder. A schematic illustration of the transformer for ice concentration forecasting is shown in Fig. 2.

Two NVIDIA Tesla P100 GPUs were used to train the model, and the time spent is shown in the Table 2. The total number of epochs was set at 120, in accordance with the estimates used by the authors of the architecture (the original work trains the model for 15 epochs). We also performed the additional experiments and made sure that increasing the number of epochs did not improve the results. To ensure the adequacy of the chosen number of epochs, convergence curves were constructed for the training and test samples (presented in the supplementary material).

SwinLSTM. This approach has performed well for video sequence prediction on well-known datasets such as Moving MNIST, TaxiBJ, Human3.6 m, and KTH. The SwinLSTM architecture is based on Swin Transformer blocks and the simplified LSTM. This approach is successful in extracting spatio-temporal representations.

This model was used in both SwinLSTM-D and SwinLSTM-B without significant architectural changes. The only change in our approach was to change the resolution of the input data. Since the original SwinLSTM used data with resolutions of 32×32, 64×64 and 128×128, it is expected that this model can be successfully applied to our data resolutions. The learning process, optimizer, loss function and learning rate values have not been changed. The only limitation of this model is the frame prediction range. The authors of the paper conducted experiments to predict from 4 to 14 consecutive frames. Our experiment requires the prediction of 52 frames (52 weeks in a year). The time taken to train the

---

[3] (B - batch, T - time, H - height, W - width).

model in this statement for the different test areas shown in the Table 2. For the 6-frame (monthly temporal resolution) prediction training took $13 \pm 3\,h$ for 90 epochs. However, this setup does not allow the intra-month dynamics of sea ice to be represented.

5   Experiments Studies

The experimental setup in the paper is focused on comparing the performance of the proposed baseline model and transformer architectures in the sea ice forecasting task.

We use the OSI SAF Global Sea Ice Concentration [21] product as training data. The spatial resolution of the images is reduced to 14 km. To test the generalisability of the developed models for different water areas, five Arctic seas were selected as test areas. The spatial position of each sea is shown in the supplementary materials.

The forecast horizon for the predictive models was set at one year ahead in order to produce long-term forecasts. For inference, the pre-history length was set to two years. The time resolution of the series was set to 7 days. Models were trained over the period 1979 to 2020 years. Dates from 01/01/2020 to 31/12/2023 are used as a test set.

In order to compare the predictive capabilities of the models, forecasts were made on the test sample starting on 1 January of each year. The quality metrics chosen were the mean absolute error (MAE) for each prediction step and the structural similarity index (SSIM). Ice edge product can be computed with ice concentration through binarization. The choice of threshold is due to studies [1] as a marker for the presence of ice in remote sensing data. Binary accuracy on predicted ice edge was calculated to indicate the quality of thick ice position prediction. Averaged metrics for the test sample are presented in Table 1. For convenience, expanded tables with metrics averaged by quarters of each test year are presented in the supplementary materials.

Architectures based on 2D and 3D convolutional layers differ significantly in the complexity of the operations performed. Measurements of the time taken to train 1000 epochs on NVIDIA GeForce RTX 4080 for each of the architectures, depending on the size of the input area, are presented in the Table 2.

According to the metrics, the TimeSformer has a lower quality compared to simpler models. To understand which period of the year makes the largest contribution to the average error, the course of the metric over the year is shown in the Fig. 3 for the Kara Sea test area. Vertical lines mark the beginning of each year. There is a pattern in the plot of the TimeSformer error that differs from other models - the error increases significantly in the summer months.

To understand the reason for the increase in TimeSformer error in the summer period, we look at the ice concentration maps for each prediction time step. Example of each model prediction in July compared to the ground truth map shown in Fig. 4. We also plot the time series of each prediction at one point to make sure that the pattern of error does not change over the years. As we can

Table 1. Quality metrics for implemented models (averaged over 2020–2023), forecast horizon - 1 year (bold are the best)

Metric	Mean Absolute Error (MAE)			Structural Similarity Index (SSIM)			Accuracy (0.2 threshold)		
Model	2D CNN	3D CNN	Time Sformer	2D CNN	3D CNN	Time Sformer	2D CNN	3D CNN	Time Sformer
Kara Sea	**0.080**	0.082	0.111	**0.673**	0.655	0.530	**0.929**	0.928	0.892
Barents Sea	0.065	**0.061**	0.134	**0.679**	0.670	0.482	0.935	**0.937**	0.839
Laptev Sea	**0.075**	0.079	0.161	**0.720**	0.700	0.591	0.933	**0.934**	0.875
East-Siberian Sea	0.087	**0.081**	0.176	**0.710**	0.705	0.688	0.923	**0.931**	0.870
Chukchi Sea	0.084	**0.083**	0.153	0.696	**0.700**	0.567	**0.936**	0.935	0.899

see from the maps and plot, TimeSformer does not predict ice melt during the summer period correctly.

TimeSformer. Results are related to the way data are transformed when fed into the spatial attention and temporal attention blocks. In the original work this solution performs well for the video series classification task, however, for predicting ice concentration this solution is not optimal. Perhaps, this solution requires modernization of the input data patching, for example, using 3D convolution as implemented in the ViViT model, as well as applying a different approach in the attention blocks. However, these changes may lead to higher computational complexity, which will eventually require much more computational resources to achieve the quality of models based on 2D convolutions.

SwinLSTM. This model demonstrated low efficiency in forecasting of long sequences. In the considered formulation of the problem of predicting one year from two years of prehistory, SwinLSTM could not achieve the results of TimeSformer. This model predicts all 52 weeks with one coarse value of ice concentration. This prediction is even worse than predicting each week with the average ice concentration for the whole year. In the original experiments of SwinLSTM developers, this model did not predict more than 14 frames. Therefore, the failure in predicting 52 frames of ice concentration is not so surprising.

Comparison with SOTA for Sea Ice. To evaluate the absolute values of the errors of the implemented models, we compare them with the SOTA solution SEAS5 forecast system. SEAS5, ECMWF's fifth generation seasonal forecasting

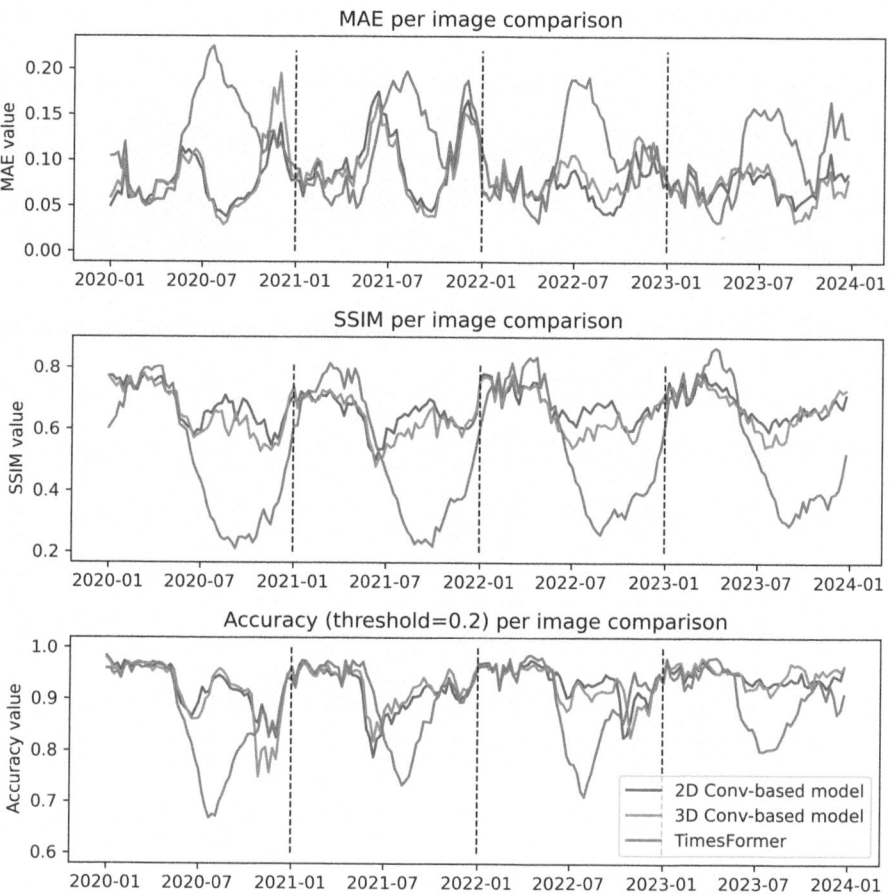

Fig. 3. Metrics for each time step of prediction for Kara sea

system, is physics-based and uses systems of differential equations. It provides a global Arctic forecast 7 months ahead and includes 51 ensemble elements. Due to differences in forecast horizons, in the generalised Table 3 the forecast of all models is limited to the SEAS5 horizon, detailed tables can be found in the supplementary material.

To assess the quality of ice edge prediction, IceNet, a forecasting system based on an ensemble of neural networks, was chosen as a data-driven SOTA. IceNet consists of 25 ensemble members, each of which is a U-net architecture model. Eleven climate and ice cover variables are used as input parameters. As the solution provides a monthly probabilistic forecast for 6 months, we used a confidence threshold of 0.8 for the probabilistic model. In the generalized Table 3 for each of the seas the forecast is limited to a 6 month horizon, a detailed table can be found in the supplementary.

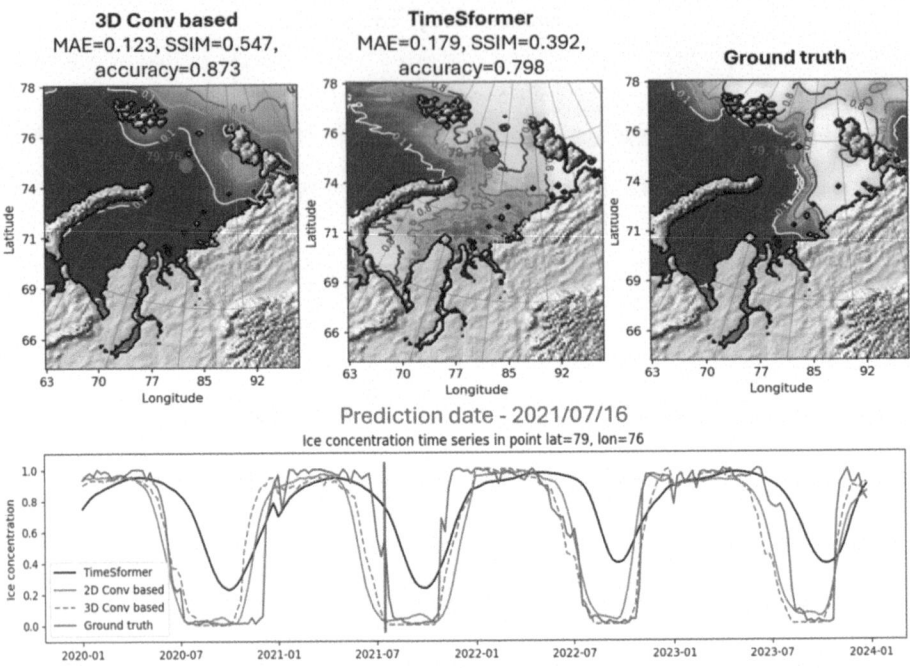

Fig. 4. Comparison of spatial distribution of values on prediction for 2021/07/16 in Kara Sea for different models

Table 2. Time spent on training models on test areas

Model architecture		2D CNN model	3D CNN model	SwinLSTM	TimeSformer
Sea	Image size	Train runtime in hours (1000 epochs)*		Train runtime in hours (120 epochs)*	
Kara	70 × 60	1.2	2.0	26.2	105.4
Barents	80 × 75	2.4	3.0	28.7	105.6
East-Siberian	50 × 62	0.8	1.5	24.6	105.3
Laptev	55 × 65	1.1	1.7	25.4	105.3
Chukchi	42 × 73	0.9	1.4	22.1	105.1

* 2D, 3D Conv-based trained on NVIDIA GeForce RTX 4080, SwinLSTM and TimeSformer on NVIDIA Tesla P100 GPU

Table 3. Comparison of averaged metrics with SOTA-solutions (SEAS5 and IceNet), 7 month ahead forecast (bold are the best)

Metric	Mean Absolute Error (MAE)				Structural Similarity Index (SSIM)				Accuracy (comparison with ice mask from IceNet)			
Model	SEAS5	2D CNN	3D CNN	TimeS former	SEAS5	2D CNN	3D CNN	TimeS former	Ice Net	2D CNN	3D CNN	TimeS former
Kara Sea	0.093	**0.076**	0.076	0.109	0.653	**0.683**	0.663	0.581	0.918	**0.945**	0.943	0.929
Barents Sea	0.073	0.063	**0.060**	0.129	0.634	**0.684**	0.672	0.489	0.906	0.922	**0.944**	0.916
Laptev Sea	0.101	**0.068**	0.072	0.146	0.703	**0.722**	0.706	0.608	0.967	**0.982**	0.980	0.966
East-Siberian Sea	0.098	0.074	**0.069**	0.177	**0.723**	0.718	0.714	0.685	0.980	**0.990**	0.990	0.988
Chukchi Sea	**0.067**	0.075	0.073	0.147	**0.780**	0.713	0.719	0.588	0.974	0.979	**0.981**	0.962

As can be seen from the tables, both the absolute values of ice concentration and the ice edge position predicted by convolution-based models are of better quality than SOTA. Statistical significance was confirmed using the non-parametric one-sided Mann-Whitney test. IceNet, SEAS5, 2D CNN and TimeSformer all have difference with p-value <0.05. 3D CNN and 2D CNN are not different with p-value 0.91. These models can therefore form a foundation for solutions that go beyond the current state-of-the art.

Toy Example on Periodic Video Data. To ensure that the problem of transformers in modelling periodic spatio-temporal data is not specific for analyzed case only, we performed an additional experiment on a 10-frame video (gif animation) based on the manga character "Menhera Shoujo Kurumi-chan" [15]. The animation was divided into frames, scaled to 45 × 45 resolution, transformed from RGB to 1-channel gray scale with values from 0 to 1. To imitate spatio-temporal data, 10 frames were repeated 5 times, a train set was formed with a slide window on this time series. As a pre-history 20 images were used, the prediction horizon was 10 images of the series ahead. Experiment run with TimeSformer architecture and 2D CNN architecture. The prediction results are shown in Fig. 5. Statistical significance of models errors difference confirmed with Mann-Whitney test (p-value for MAE - 0.002, for SSIM - 0.001).

Due to the small training set, we were able to run TimeSformer for 4000 epochs, the CNN was trained for 100000 epochs, the stopping criterion is the number of epochs without L1loss improvement. Detailed convergence plots are described in the supplementary materials (*Media data convergence*).

As the data has an explicit periodicity and no stochastic component, it is expected that the models will be able to approximate the training sample with near-zero error. However, TimeSformer reaches a plateau at 0.02 and produces artifacts in the center of the image. The CNN model converges asymptotically to zero error. Both models capture the temporal dynamics of contour changes well,

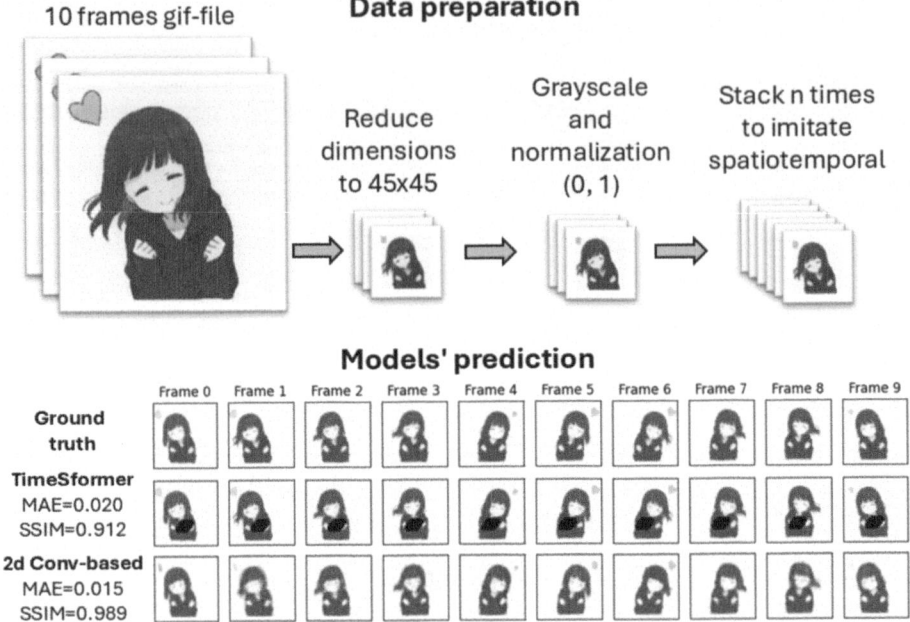

Fig. 5. Media images pre-processing and prediction result (for TimeSformer, CNN-2D)

but the Transformer reproduces the distribution of values within the contour poorly. This behavior is similar to the results obtained with ice data - the model tends to reproduce particularities while losing sight of the more general trend, or vise versa.

6   Conclusion

The results of the experiments confirmed the hypothesis about the limited applicability of transformers for spatio-temporal data with strong periodicity.

For the sea ice concentration forecasting task, the adaptation of the TimeSformer architecture showed a weak reproduction of the time component, due to which the ice concentration in the water area in the summer period did not fall below 0.3, making such a forecast inapplicable. SwinLSTM, aimed at the video prediction task, proved to be helpless in reproducing the annual dynamics. It showed a tendency to self-repeat - for a 7-day forecast a year ahead, summer ice conditions were indistinguishable from the model's initial conditions. It is also worth noting the grainy artifacts in the predictions of transformer-based models.

At the same time, shallow baseline models based on CNN showed reasonable quality compared to SOTA solutions in the field of sea ice prediction. For the forecasts of ice concentration we achieve a quality improvement of up to 30% (Laptev Sea) against SEAS5 system. For ice edge position prediction we achieved comparable results against data-driven system IceNet in terms of accuracy (at the 0.2 threshold) and quality improvements of up to 4–5% in certain water areas (Kara Sea, Barents Sea).

The video prediction experiments for synthetic periodic data also confirm the existence of highlighted problem - convolutional baseline outperforms TimeSformer by 8% for SSIM and 25% for MAE. While further evaluation of the limitations of transformers is still required, we can claim to have provided the solid empirical conformations on the previously poorly discussed problem.

Acknowledgments. The research was carried out within the state assignment of Ministry of Science and Higher Education of the Russian Federation (project No. № FSER-2024-0004).

Disclosure of Interests. The authors have no competing interests to declare that are relevant to the content of this article.

References

1. Andersson, T.R., et al.: Seasonal arctic sea ice forecasting with probabilistic deep learning. Nat. Commun. **12**(1), 5124 (2021)
2. Arnab, A., Dehghani, M., Heigold, G., Sun, C., Lučić, M., Schmid, C.: Vivit: a video vision transformer. In: Proceedings of the IEEE/CVF International Conference on Computer Vision, pp. 6836–6846 (2021)
3. Bertasius, G., Wang, H., Torresani, L.: Is space-time attention all you need for video understanding? In: ICML, vol. 2, p. 4 (2021)
4. Bodnar, C., et al.: Aurora: a foundation model of the atmosphere. arXiv preprint arXiv:2405.13063 (2024)
5. Bushuk, M., et al.: Regional arctic sea-ice prediction: Potential versus operational seasonal forecast skill. Clim. Dyn. **52**, 2721–2743 (2019)
6. Destro, M., Gygli, M.: Cyclecl: self-supervised learning for periodic videos. In: Proceedings of the IEEE/CVF Winter Conference on Applications of Computer Vision, pp. 2861–2870 (2024)
7. Dosovitskiy, A., et al.: An image is worth 16x16 words: transformers for image recognition at scale. arXiv preprint arXiv:2010.11929 (2020)
8. Grigoryev, T., et al.: Data-driven short-term daily operational sea ice regional forecasting. Remote Sens. **14**(22), 5837 (2022)
9. Guen, V.L., Thome, N.: Disentangling physical dynamics from unknown factors for unsupervised video prediction. In: Proceedings of the IEEE/CVF Conference on Computer Vision and Pattern Recognition, pp. 11474–11484 (2020)
10. Johnson, S.J., et al.: SEAS5: the new ECMWF seasonal forecast system. Geosci. Model Dev. **12**(3), 1087–1117 (2019)
11. Kim, Y.J., Kim, H.C., Han, D., Stroeve, J., Im, J.: Long-term prediction of arctic sea ice concentrations using deep learning: effects of surface temperature, radiation, and wind conditions. Remote Sens. Environ. **318**, 114568 (2025)
12. Li, W., Hsu, C.Y., Tedesco, M.: Advancing arctic sea ice remote sensing with AI and deep learning: now and future. EGUsphere **2024**, 1–36 (2024)
13. Liu, Y., Bogaardt, L., Attema, J., Hazeleger, W.: Extended-range arctic sea ice forecast with convolutional long short-term memory networks. Mon. Weather Rev. **149**(6), 1673–1693 (2021)

14. Liu, Z., et al.: Video swin transformer. In: Proceedings of the IEEE/CVF Conference on Computer Vision and Pattern Recognition, pp. 3202–3211 (2022)
15. Pom: Menhera shoujo kurumi-chan (2018). https://pom-official.jp/menhera_kurumichan/
16. Rasp, S., Dueben, P.D., Scher, S., Weyn, J.A., Mouatadid, S., Thuerey, N.: Weatherbench: a benchmark data set for data-driven weather forecasting. J. Adv. Model. Earth Syst. **12**(11), e2020MS002203 (2020)
17. Razzhigaev, A., et al.: Your transformer is secretly linear. arXiv preprint arXiv:2405.12250 (2024)
18. Shi, X., Chen, Z., Wang, H., Yeung, D.Y., Wong, W.K., Woo, W.C.: Convolutional LSTM network: a machine learning approach for precipitation nowcasting. In: Advances in Neural Information Processing Systems, vol. 28 (2015)
19. Tan, C., et al.: Openstl: a comprehensive benchmark of spatio-temporal predictive learning. Adv. Neural. Inf. Process. Syst. **36**, 69819–69831 (2023)
20. Tang, S., Li, C., Zhang, P., Tang, R.: Swinlstm: improving spatiotemporal prediction accuracy using swin transformer and LSTM. In: Proceedings of the IEEE/CVF International Conference on Computer Vision, pp. 13470–13479 (2023)
21. Tonboe, R., Lavelle, J., Pfeiffer, R.H., Howe, E.: Product user manual for OSI SAF global sea ice concentration. Danish Meteorological Institute, Copenhagen, Denmark (2016)
22. Vaswani, A., et al.: Attention is all you need. In: Advances in Neural Information Processing Systems, vol. 30 (2017)
23. Yarotsky, D.: Error bounds for approximations with deep ReLU networks. Neural Netw. **94**, 103–114 (2017)
24. Yu, W., Lu, Y., Easterbrook, S., Fidler, S.: Efficient and information-preserving future frame prediction and beyond. In: International Conference on Learning Representations (2020)
25. Zeng, A., Chen, M., Zhang, L., Xu, Q.: Are transformers effective for time series forecasting? In: Proceedings of the AAAI Conference on Artificial Intelligence, vol. 37, pp. 11121–11128 (2023)

Microscopic Binary Engagement Model

Marco Lemos⬤, Pedro J. S. Cardoso(✉)⬤, and João M. F. Rodrigues⬤

NOVA LINCS & ISE, Universidade do Algarve, Faro, Portugal
{a72178,pcardoso,jrodrig}@ualg.pt

Abstract. Tracking audience engagement in real-time offers numerous benefits. For instance, event planners can make dynamic adjustments to presentations or activities to maintain high levels of interest and participation. This enhances the overall experience for attendees by ensuring the content remains engaging and relevant. This paper proposes a model for computing the binary engagement within groups. The model does this by identifying individuals' engagement during the events' time frames, which are then combined, i.e., the engagement of the group is computed by aggregating the engagement of each individual. For each individual of the group, the engagement model incorporates the computation over time of the gaze direction, valence, and arousal, classifying the engagement into two primary levels: not-engaged and engaged. The engaged category is further divided into two sublevels: positive and negative engagement. Experimental results confirm the model's effectiveness, showcasing reliable identity tracking and accurate assessment of engagement states in dynamic scenarios.

Keywords: Engagement · Affective Computing · HCI · Group Engagement · Real-time Engagement

1 Introduction

Detecting audience engagement in real-time during events is a cutting-edge approach that leverages advanced technologies to measure and analyze how attendees interact and respond throughout an event. This process may involve various data collection methods, such as video analysis, audio cues, physiological sensors, and social media monitoring, to capture real-time feedback.

Machine learning and computer vision advancements have paved the way for engagement understanding by combining human emotions through automated analysis of visual and behavioral cues. Emotional states, many times represented in a valence-arousal space (e.g., [3]), provide valuable insights into individual and group behaviors. Likewise, engagement levels (e.g. [15, 24]) offer an understanding of a person or group focus and participation in activities.

Although several studies emphasize personal or group engagement online, such as social media engagement [5, 8] or learner engagement with virtual educational events [4], in the context of real-world and real-time (live) engagement detection during indoor and outdoor events, very few models have been proposed [15].

M. H. Lees et al. (Eds.): ICCS 2025, LNCS 15905, pp. 119–134, 2025.
https://doi.org/10.1007/978-3-031-97632-2_9

At this point, it is important to define the terms crowd and group [13]. A *group* is a collection of individuals, ranging in size from two to hundreds, who are present together at any given time and engaging in social contact. Its members move in a similar direction and at a similar speed, making them near to one another. Multiple groups can cohabit during an event. Conversely, a *crowd* (or mass) is a special huge gathering of people who are physically present in the same place. It typically arises when individuals who have a common objective unite as a single entity, losing their individuality and assuming the characteristics of the crowd entity.

The engagement analysis in groups and crowds can be divided into two main methodologies [15, 20]: *Microscopical* (or bottom-up) methods, typically applied to groups, where individuals in the "video streaming" are analyzed and the resulting data is then used to extrapolate information at the collective level, i.e., a group analysis is considered as a collection of individuals analysis; *Macroscopical* (or top-down) methods, typically applied to crowds, are made up of comprehensive processes that view the crowd as a single cohesive unit, rather than requiring the tracking and segmenting of every individual. Macroscopic approaches are (more) suited when population density increases and tracking quality drastically decreases.

It is also important to define *instantaneous engagement*, which corresponds to the engagement detected at each instant t (or frame f) of the stream or video. Similarly, the *period engagement* corresponds to the engagement for a specific time period, while *event engagement* accounts for the engagement throughout the entire event. For more details see [15] and Sect. 3.

This paper focuses on engagement in groups, i.e., when there are few or no occlusions, low density, and a clear view of people. In such cases, microscopic approaches frequently perform best. By using a microscopical approach, this paper focuses on presenting a modular and scalable model for instantaneous, period, and event engagement detection in groups during real-world events – Microscopic Binary Engagement Model (MiBE). The model integrates person tracking, with the combination of two dimensions: (i) emotion (valence and arousal level estimation) and (ii) attention (focus-head pose estimation). More dimensions can be integrated in the future [15].

The main contribution of the paper is a scalable binary engagement detection model for groups, where engagement can be classified as positive if the person is "appreciating/liking" the event, or negative if the person despite being engaged, is not "appreciating/liking" the event. A secondary contribution is the introduction of an initial model for valence-arousal computation.

In the present section, the subject and goals of the paper are presented. Section 2 briefly summarizes the state of the art. Section 3 introduces the proposed model – MiBE, and Sect. 4 presents the initial tests and results achieved. The final section outlines some conclusions and future work.

2   Related Work

As already mentioned, MiBE is based in two dimensions: emotion and attention. Here we will not go into detail on the different emotion and attention models, we will rather briefly enumerate some recent models. For emotion computation the valence-arousal (VA) predictions can be used. Valence is a measure of the emotional intensity, ranging from negative to positive, while arousal indicates the emotional intensity, ranging from low to high. Nguyen et al. [14] present an approach to affective behavior analysis, focusing on VA prediction within the Affective Behavior Analysis in the Wild (ABAW3) challenge. Leveraging deep learning (DL) techniques, the authors propose a two-stage model for continuous emotion estimation. Experimental results on the Aff-Wild2 dataset demonstrate significant improvements over baseline methods, achieving a Concordance Correlation Coefficient core of 0.507 for VA estimation and an F1-score of 0.533 for action unit detection. Stephen et al. [11] presented a DL method for predicting continuous affect from facial expressions (FE) in the VA space. The method maps discrete emotion labels and FE to this space, outperforming existing methods on the AffectNet dataset [13] and showing strong generalization. Andrew [16] introduced a real-time video-based algorithm for predicting FE, VA, and action units on mobile devices. Lorenzo et al. [1] explore VA estimation from neuromorphic vision data using event cameras, which excel at capturing subtle and rapid facial micro-movements. Other models also exist, such as the one proposed in [3].

For attention detection, head pose estimation (HPE) can be used. The Wide Headpose Estimation Network (WHENet) [25] is a model designed for HPE using single RGB images. It excels in predicting Euler angles—yaw, pitch, and roll—over a full 360-degree yaw range, which is critical for applications like autonomous driving and augmented reality. Built on the EfficientNet-B0 backbone, WHENet combines regression and classification objectives for robust and fine-grained pose prediction. Evaluation on BIWI and AFLW2000 datasets shows WHENet achieving a mean absolute error as low as $3.81°C$. Hempel et al. [9] introduce a method from single images using a continuous 6D rotation matrix representation. Later, the same authors used a geodesic loss function within the Special Orthogonal Group to stabilize learning and ensure precise predictions [10]. The model, named 6DRepNet360, is open-sourced to facilitate further research and application development.

Finally, there are models designed to detect engagement. Gupta et al. [7] introduce a real-time DL-based learner engagement detection system that leverages facial emotion recognition (FER). Addressing the challenges of online education, it measures student engagement by analyzing facial expressions captured via webcams during online sessions. Lasri et al. [12] detect the engagement levels of deaf and hard-of-hearing students through FER. More recently, Zhao et al. [24] present a model designed to detect student engagement through FE in real-time classroom settings.

According to the literature that has been presented and examined, no model has been found that can handle actual events that take place both indoors and outdoors and that can aggregate the engagement of various cameras, groups, and time periods; in other words, it cannot drill down the information from the individual's engagement with each object to the group, to the period, to the entire event.

3   Binary Engagement Model

Before going into the detail of the model, let us define C as the combined information from different dimensions, $D_1, D_2, \ldots, D_n$, where n is the number of dimensions. A dimension refers to "emotion", "sentiment", "scene dynamics", "attention" etc. (for further details see [15]). With this in mind, let us also define *instantaneous engagement* as $IE(t, G) = C\{D_1, \ldots, D_n\}$, which corresponds to the engagement detected at time t (or frame f) of the streaming/movie for a non-empty set G of individuals. If G is a set with a single individual, $G = \{h\}$, then it will be the engagement of the person h at time t. If G is a set with more than one individual, $G = \{h_1, h_2, \ldots, h_n\}$, then it will be the engagement of the group at time t. The *P-period engagement* is given by $PE(P, G) = \uplus_{t \in P} IE(t, G)$, where $P = \{t_i, t_{i+1}, \ldots, t_f\}$ is a period of time, t_i and t_f (with $t_i < t_f$) are two different times in the event timeline, and $\uplus$ is the combination of the information retrieved from the different instants. Finally, the *event engagement* is given by $E(G) = PE(I, G) = \uplus_{t \in I} IE(t, G)$, i.e., it accounts for the entire event, with duration interval I.

In the present model, the $P-$period engagement for a group (G) is computed as the mean engagement of all persons in the group, i.e., $PE(P, G) = \frac{1}{|G||P|} \sum_{h \in G} \sum_{t \in P} IE(t, \{h\})$, where $|.|$ is the number of elements of the set. Similarly, the $E(G)$ is computed as the mean engagement of all persons in the group, i.e., $E(G) = \frac{1}{|G||I|} \sum_{h \in G} \sum_{t \in I} IE(t, \{h\})$. Both $PE(P, G)$ and $E(G)$ can be computed for a single person h by setting $G = \{h\}$.

Furthermore, we defined binary levels of engagement that are determined based on valence, arousal, and gaze direction, whether the person is looking or not to the point of interest/scene (PoI). These levels are represented as pairs, $IE = [x, y]$, of binary (0-1) values, as follows. (i) **Not Engaged** ($IE = [0, \times]$) is distinguishable in (i.1) $IE = [0, 0]$ if the person is not looking at the PoI, regardless of their valence and arousal; or (i.2) $IE = [0, 1]$ if the person display a negative arousal (low emotional intensity) but is looking at the PoI. The latter suggests that the person is disinterested in the activity and not engaged, although their gaze being directed at the PoI. (ii) **Engaged** ($IE = [1, \times]$) which encompasses all other situations, divided in: (ii.1) **Negative Engaged**, $IE = [1, 0]$, if a person has a negative valence (expressing a negative sentiment) but a positive arousal (indicating a strong emotional intensity), and are looking (gaze) at the PoI. In this case, even though the person feels negatively about the activity, the high arousal indicates its engagement or that the reaction to the activity is strong. (ii.2) **Positive/True Engaged**, $IE = [1, 1]$, if a person displays a positive valence (expressing a positive sentiment) and a positive arousal (indicating a strong emotional intensity) while looking (gaze) directly at the PoI. This combination suggests that the person is actively and positively engaged with the activity. Therefore, the instantaneous engagement $IE(t, G)$ will be equal to 1 if the person is engaged negatively or positively, and 0 if the person is not engaged at all.

The model's global block diagram is shown in Fig. 1, and operates as follows. The first block is (a) *Head Box Detection*, which identifies bounding boxes

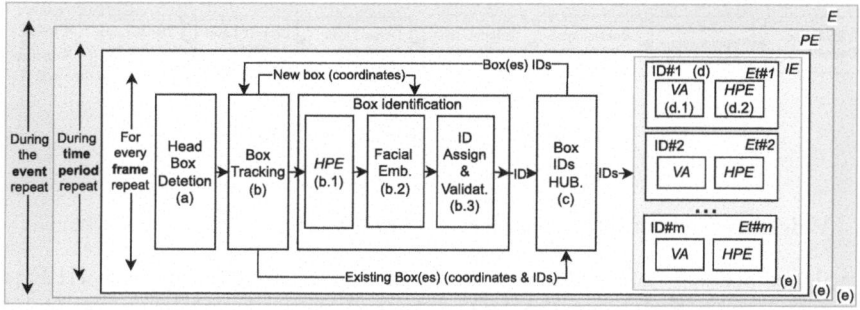

Fig. 1. Block Diagram of the MiBE model.

corresponding to heads in each frame. Next, the model performs (b) *Box Tracking*, where it tracks the previous identified boxes. For all boxes that matches an existing ID, the ID and box coordinates are passed to step (c). For any new bounding boxes (i.e., new heads), the (b) *Box Identification* is initiated and computes the (b.1) *HPE*, to determine head orientation. Then (b.2) *Facial Embeddings* are computed, to check if the face within the box resembles one from a previous frame. Finally, in step (b.3) *Box ID Assign & Validate*, the face is either assigned a new unique identity or an existing ID is validated and maintained.

Next, the (c) *Box ID HUB* is performed for both existing and newly processed bounding boxes, functioning as a central hub for managing and updating the box IDs. (d) For each of these bounding boxes (and for each dimension, $ID\#1$ to $ID\#n$) the model computes: (d.1) valence-arousal estimation to measure the individual's emotional state and (d.2) HPE for head orientation, computed once *per* frame (used also in step (b.1)) for each specific ID. (e) These computed dimensions are then used to calculate the values of engagement of the individuals and groups ($IE(t, G), PE(P, G)$, and $E(G)$), incorporating information from all groups involved, if more than one group exists.

Before going in details with each mentioned block, let us explain in more detail the HPE and VA estimation blocks.

3.1 Head Pose Estimation Block (HPE)

For each frame, the HPE was computed using WHENet[1] model [25], which allows to predict *yaw, pitch*, and *roll* values. In this context, *yaw* (ψ) indicates the degree of head turn to the left or right, with positive values indicating a turn to the right and negative values indicating a turn to the left. The range of yaw is $\psi \in [-180°, 180°]$, with $\psi = 0°$ indicating that the person is looking directly at the camera. *Pitch* (θ) represents the degree of head tilt up or down, with positive values indicating a tilt up and negative values indicating a tilt down. The range of pitch is $\theta \in [-90°, 90°]$. *Roll* (φ) represents the degree of head tilt to the left or right, with positive values indicating a tilt to the right and negative values indicating a tilt to the left. Roll values range in $\varphi \in [-180°, 180°]$.

---

[1] Model available at: https://tinyurl.com/yc47z9w3, accessed on 2025/01/16.

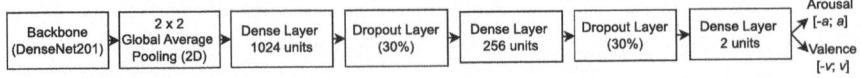

Fig. 2. Valence-Arousal model block diagram.

3.2   Valence-Arousal Estimation Block (VA)

The valence-arousal computation is performed concurrently with the HPE process through a newly developed VA Deep Neural Network (DNN) model - EVAm (see Fig. 2). The primary goal of the new model is to be seamlessly incorporated into the engagement framework, i.e., a model that can operate under real-world circumstances, using various cameras, positioned at various facial angles, in various lighting situations etc.

The initial model, still in its early steps, uses DenseNet201 pre-trained weights from ImageNet as the backbone of the DNN architecture. In the head of the DNN, the (i) first layer is a 2 × 2 *Global Average Pooling (2D)*, this choice preserved spatial information across channels while significantly reducing the number of trainable parameters compared to a Flatten operation [22]. This not only improved computational efficiency but also mitigated the risk of overfitting. The next layer (ii) is a *Dense Layer* with 1024 units (neurons), incorporated to learn complex feature interactions from the pooled features, followed by a (iii) *Dropout Layer*, to enhance regularization and prevent overfitting. The Dropout Layer, with a 30% dropout rate, is repeated after each dense layer in the architecture. The next layer (iv) is a *Dense Layer* with 256 neurons for further feature abstraction, and finally, the output (v) is a *Dense Layer* with 2 neurons used to predict respectively the Valence and Arousal values simultaneously.

Training, Results and Discussion (Partial Results). We used the Affect-Net dataset [13] for training, validation, and testing. The dataset contains over 400,000 facial images that have been manually labeled for the presence of eight different facial expressions. Additionally, the dataset includes annotations for VA intensity. For training, we utilized 288,000 images annotated with valence and arousal. The validation and testing sets each comprised 2,000 images.

In the training phase, Early Stopping was introduced with a 10 epochs patience, to prevent overfitting and reduce training time by halting the process once performance stagnated. Learning Rate Scheduling was employed with a 5 epochs patience and a factor of 0.5, allowing the optimizer to reduce the learning rate when progress slowed. For the dense layers, the Rectified Linear Unit (ReLU) activation was employed. To optimize the model, we employed the Adam optimizer and a mean squared error (MSE) loss function, as it is well-suited for regression tasks, penalizing larger deviations more heavily. Training was conducted with a batch size of 64, balancing memory efficiency and gradient stability for effective optimization. Finally, the DenseNet-201 model architecture requires images to be of size 224 × 224 pixels.

Table 1. Valence-Arousal model results, with MAE - Mean Absolute Error, RMSE - Root Mean Square Error, PCC - Pearson Correlation Coefficient, CCC - Concordance Correlation Coefficient, and SAGR - Sign Agreement Ratio.

Model/Metrics	Valence					Arousal				
	MAE	RMSE	PCC	CCC	SAGR	MAE	RMSE	PCC	CCC	SAGR
EAVm (ours)	0.290	0.390	0.630	0.610	0.760	0.280	0.360	0.560	0.470	0.760
Mollahosse et al. [13]	-	0.370	0.660	0.600	0.740	-	0.410	0.540	0.340	0.650
Stephen et al. [11]	0.146	0.179	0.952	0.948	-	0.121	0.164	0.952	0.950	-
Andrey [16]	-	-	-	0.429	-	-	-	-	0.496	-

To facilitate efficient batch processing and ensure consistent model performance, the ImageDataGenerator for both data preprocessing and augmentation was employed. A key preprocessing step involved normalizing pixel values to the range [0, 1] by dividing by 255. This normalization accelerates convergence and stabilizes the training process by standardizing the input, as highlighted in [6].

As for results, while our model presents good performance, see Table 1, it still falls short when compared to other models. When compared with the results in [13], the baseline, our model shows superior performance across all metrics for arousal. For valence, our model outperforms the model in two metrics, while the other two remain close to the baseline values. When comparing with the results in [11], the authors presented better results than us, but they only used images corresponding to seven of the eight existing emotions in the AffectNet dataset, they exclude the neutral emotion. In the case of Andrey [16], which uses the Aff-Wild2 dataset, our results demonstrate better performance in valence and similar results in arousal. However, since Andrey's work utilized a different dataset, a direct and fair comparison is not feasible.

Finally, it is important to stress that this is a first version of the model. Future work will involve testing different backbones and applying various approaches to the DNN head. Additionally, the model will be trained on a combination of datasets to have a better generalization.

We will now explain in more details the remaining blocks of the model.

3.3   Head Detection, Tracking and Identification Blocks

Head Box Detection. This is the first module where the streaming input is processed. The focus it to detect heads, rather than retain any facial image information. In compliance with the General Data Protection Regulation (GDPR), only the bounding boxes containing heads are of interest. This initial detection is performed using the YOLOv4 [2] model[2], trained on the Hollywood Heads [21] and CrowdHuman [19] datasets. More recent models exists but, for this initial prototype, YOLOv4 presented a good solution, with a good balance between accuracy and speed.

---

[2] Model available at: https://tinyurl.com/2pbydwn7, accessed on 2025/01/16.

Box Tracking. For each frame, in the *Box Tracking* module, the centroid of each bounding box is computed and compared with the centroid(s) of the previous frame. The Euclidean distance between centroids is tested against a threshold defined as half the box's width (w_{box}), i.e., $th_{bt} = w_{box}/2$. If the distance is less than th_{bt} then (i) the box is considered the same as the one in the previous frame, and the box coordinates along with its corresponding ID are sent to the *Box ID HUB* module. If the distance is greater than th_{bt} then (ii) the box is considered as a new box, and the box coordinates are sent to the *Box Identification* module.

Box Identification. In this module, the initial/new boxes are identified. To account for variations in head orientation, the *HPE* (see Sect. 3.1) is computed for each box. Faces within the boxes are "acknowledged" using the MTCNN model [23] from the DeepFace library [18] and each box is then processed by the FaceNet model [17], which computes facial embeddings – vector representations that capture the unique features of the face – *Facial Embedding* module.

Any new boxes detected in subsequent frames are processed by computing embeddings with FaceNet. These embeddings are compared to those in the stored database using cosine similarity. If the similarity score between a new embedding and the stored embeddings exceeds a predefined threshold (0.4, given by the FaceNet model), the system at *ID Assign & Validation* module assigns the box to an existing ID. Otherwise, a new ID is generated. This approach ensures consistent recognition of individuals, even when they temporarily leave and re-enter the scene, ensuring that no image is stored.

By using the HPE information, as the head orientation changes, the system computes multiple embeddings for each individual, representing different angles such as *yaw* and *pitch*, ensuring the embeddings remain robust against changes in orientation, lighting, and facial expressions. By storing multiple embeddings per box (person), the system reduces mismatches caused by these variables.

Box ID HUB. This module works like a Hub, receiving IDs and boxes coordinates from *Box Tracking* and from *Box Identification*, routing that information back to the *Box Tracking* module for the new position of the box, and at the same time to the *Instantaneous Engagement – IE* module, to the respective $IE\#1, \ldots, IE\#m$ (sub-)processes that are directly related with each box ID ($ID\#1, \ldots, ID\#m$).

3.4   Instantaneous Engagement (IE)

The next step is to compute the instantaneous engagement for each ID and frame ($IE(t, \{h_{ID\#i}\}), i = \{1, \ldots, m\}$), which are then combined to compute the G group instantaneous engagement $IE(t, G)$ (see Sect. 3).

First, let us define the Gaze (Ga) in relation to a PoI. The model operates, at the moment, under the assumption of static PoIs, a constraint dictated by the present design of our mapping approach. Initially, we have to generate a map

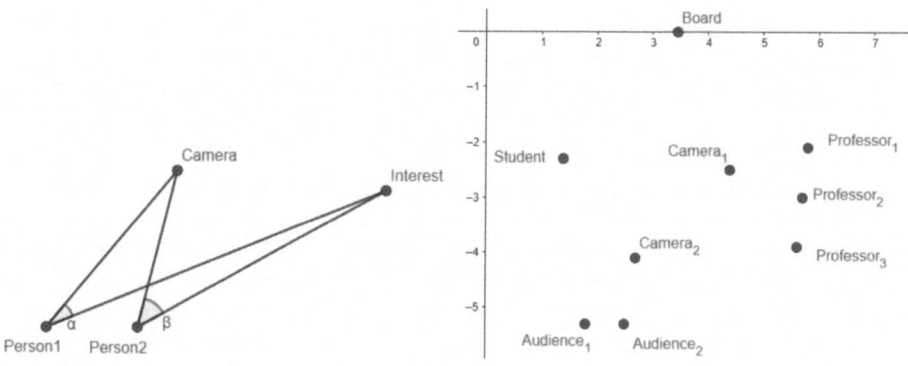

Fig. 3. On the left, a sketch of the angles between persons, PoI, and a camera. On the right, a room setup (see details in Sect. 4).

of the room, setting the (x, y) coordinates of the individuals, cameras, and PoI. Using this map, we compute the angles between each person, the PoI, and the camera, to determine whether a person is looking at a specific PoI or not. Then, we compare those angles with the *yaw* values predicted by the HPE model (see Sect. 3.1). To account for minor positional variations (e.g., leaning or the size of the PoI), we accept a ±25 cm (empirically chosen) shift of the coordinates of the persons, the cameras and the PoI, calculating maximum and minimum angles accordingly. This ensures a robust determination of gaze despite small positional shifts.

Figure 3 (left) illustrates an example of a mapping for two persons, a PoI, and a camera positioned within a room. The angles α and β, for each individual, are computed using standard trigonometric formulas. For instance, for $Person_1$, if d_p be the distance between $Person_1$ and the Camera, d_i the distance between $Person_1$ and the PoI, and d_c the distance between the Camera and the PoI then $\alpha = \arccos\left((d_i^2 + d_p^2 - d_c^2)/(2 \cdot d_i \cdot d_p)\right)$. Figure 3 (right) illustrates a real setup for a student presentation (more details are presented in Sect. 4).

If the predicted *yaw* and *pitch* angles fall within the calculated range, the person is classified as looking at the point of interest, $Ga = 1$. For example, considering Person1, the *yaw* angle, ψ_{p1}, returned by HPE, must lie within the interval $\psi_{p1} \in [\alpha - 10°, \alpha + 10°]$. Additionally, the *pitch* value must be greater than $-10°$ to ensure the person is looking at the PoI. If the *yaw* angle lies outside the range or the *pitch* value is less than $-10°$, the person is deemed not to be looking at the PoI, $Ga = -1$.

It is important to note that the *yaw* value of $0°$ represents alignment with the camera's optical axis, while the *pitch* value of $0°$ reflects no upward or downward tilt of the head, independent of the camera's vertical alignment. This distinction becomes crucial in future scenarios where the PoI change its elevation, as our model does not currently account for such vertical movements. In addition, this model only considers the gaze direction based on the orientation of the head, without accounting for eye movements. In other words, it assumes that the eyes are looking straight ahead and aligned with the head's direction. This simplification has not yet been addressed in the current implementation.

Now, let us define the engagement for each frame (f) in a bi-dimensional space, namely as: $E_{level,\pm} = (Ga.(A + 1)/2, V)$, where Ga is the gaze (computed as presented above), A is the arousal, and V the valence (A and V are estimated with the model presented in Sect. 3.1). The first coordinate shows the level of engagement and the second establishes if the engagement was generated by a positive/"good" or a negative/"bad" emotion. In the formula, the arousal is normalized to the interval $[0, 1]$, before being multiplied by the gaze. This normalization step means that if the arousal is negative, the resulting value will be lower than if the arousal were positive. Consequently, for a negative arousal value, the engagement level - calculated by multiplying the normalized arousal by gaze - will be smaller than for a positive arousal value. The multiplication reflects how gaze intensity and arousal level together determine the overall engagement level. The second coordinate, representing valence, indicates the person's emotional state, i.e., a positive valence indicates that the person experiences positive engagement whereas a negative valence indicates negative engagement.

Thus far, the instantaneous engagement of each individual has been computed as, $IE(t, \{h_i\})$, where $i = \{1, \ldots, m\}$. As defined in Sect. 3, the instantaneous engagement of the group is computed as the mean instantaneous engagement of all individuals in the group, i.e., $IE(t, G) = \frac{1}{m} \sum_{i=1}^{m} IE(t, \{h_i\})$.

3.5   Period Engagement (PE) and Event Engagement (E)

Following the above, repeating the instantaneous engagement module for each frame, and using the formulae presented in the beginning of this section, it is now possible to compute the group's P-period engagement $PE(P, \{h_i\})$, $i = \{1, \ldots, m\}$ (for each person) and event engagement $E(\{h_i\})$, $i = \{1, \ldots, m\}$).

Using the same reasoning, it can be computed the engagement in the 4 binary segments presented initially: $[0, 0]$ - no engagement, $[0, 1]$ - disinterested, $[1, 0]$ - negative engagement, and $[1, 1]$ - true engagement.

It is important to stress, the MiBE is completely scalable in terms of individuals and groups, and can cope with the information of more than 1 camera.

4   MiBE Operational Tests and Assessment

To illustrate the functionality of the MiBE, we present two tests. In **Test#1**, the simplest setup is considered, involving only one person and a PoI which is the same as the camera positioned directly in front of the person. Figure 4 (left) shows a frame extracted from the video, with the person looking directly at the camera, resulting in $Ga = 1$. The person exhibits negative valence and arousal, indicating a negative emotional response toward the scene. The middle plot shows the valence and arousal values per frame (each frame is represented by a point in the VA plane), while the right plot shows the instantaneous engagement, $IE(t, \{h\})$, during the full stream.

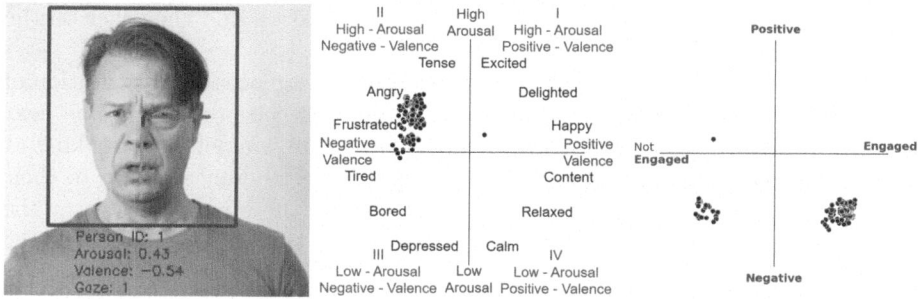

Fig. 4. Illustration of Test#1, see text.

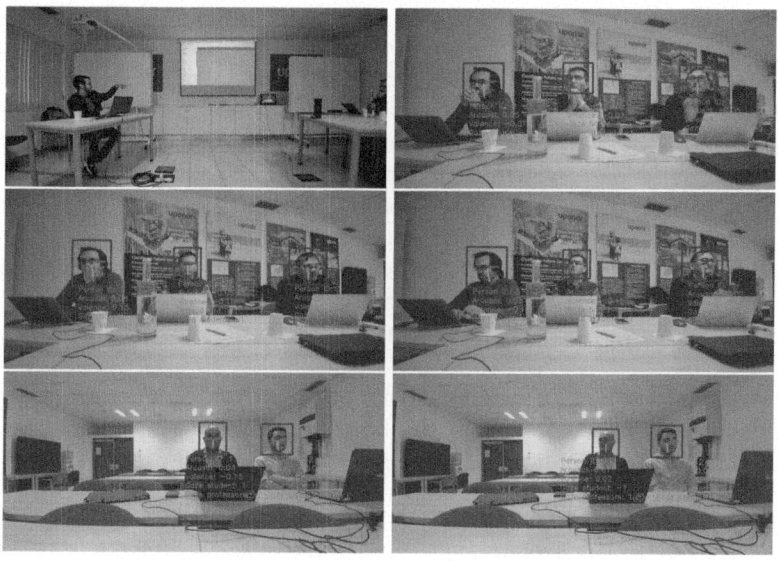

Fig. 5. Illustration of Test#2, see text.

The right plot reveals that the individual is not engaged during certain frames, which corresponds to moments when he is not looking at the PoI (camera). Additionally, when engaged, the person's engagement is negative, as indicated by the negative valence and low arousal values. Finally, $E(\{h\}) = 80\%$ for the length of the video, which has 6 s.

The second test – **Test#2** – was done in real indoor environment, during a student's master thesis presentation. Figure 3 (right) illustrates the layout setup (coordinates in meters). In the room, the student is positioned on the left and the professors, three (Group 1), on the right, being the presentation projected in the "board". Two pairs of cameras are positioned in the room. Camera$_1$ consists of a pair of identical cameras positioned back-to-back, with one focusing on the student and the other on the professors. Similarly, Camera$_2$ features the same setup, with one camera focusing on the presentation and the other on the audience, which comprises two individuals (Group 2).

Figure 5 showcases top to bottom, left to right: the general representation of the room, showing the student, the board and one of the professors; The next 3 images illustrate the professors in different situations, namely, one looking at the student and two at the board, one looking to the student and two to their computers, and one looking at the computer, one at the board and one at the student; The last two images illustrates the audience, two persons looking at the student, and one looking at the student and one at the professors.

Figure 6 (top-left) shows the valence and arousal values per frame for the individual with ID#3 (Professor$_1$). This person exhibits consistently low arousal, indicating weak emotional intensity, while the valence fluctuates from negative to positive. These dynamics suggest an overall neutral emotional state, as the valence shows low absolute values despite its polarity shifts. The top-middle plot displays the valence and arousal data for the individual with ID#2 (Audience$_1$). Similar to ID#3, this person demonstrates low arousal, signifying weak emotional intensity. However, the valence remains predominantly negative, indicating an overall mild negative emotional state.

In the top-right figure is depicted the engagement between the person with ID#3 and the student (PoI), while bottom-left shows the engagement of the person with ID#3 with the board. These plots reveal that this person tends to focus more on the student than the board. The valence transition observed in the top-left plot, from negative to positive, is consistent with the engagement patterns, as the engagement also transitions from low to high values.

The bottom-middle plot highlights the engagement from person with ID#2 with the student (as PoI), and in the bottom-right the engagement with the professors. These plots indicate that this person, as ID#3, directs its attention more frequently to the student. Furthermore, the negative valence observed in the top plots aligns with the engagement trends in the bottom, as the engagement values remain consistently low. The above-mentioned plots reinforce the previously discussed engagement dynamics, as defined by the engagement formula outlined earlier. Engagement occurs only when the individual is looking at the point of interest, with the engagement level modulating in accordance with arousal intensity. This highlights the interplay between gaze, attention, and emotional engagement, underscoring the importance of arousal in driving changes in interaction focus.

Figure 7 illustrates the engagement trends over time for the three professors during a 2-minute presentation followed by a 2-minute arguing. Green dots represent a positive engagement, yellow a negative engagement, and red no engagement. The top plot shows the professors engagement with the board, while the bottom depicts their engagement with the student. The first 3,600 frames correspond to the student's presentation, while the remaining frames correspond to the arguing.

In terms of results, for the same 4-minutes period mentioned before, for the group 1 (professors) the engagement to the board (b) was $PE_b(P, \text{Group 1}) =$

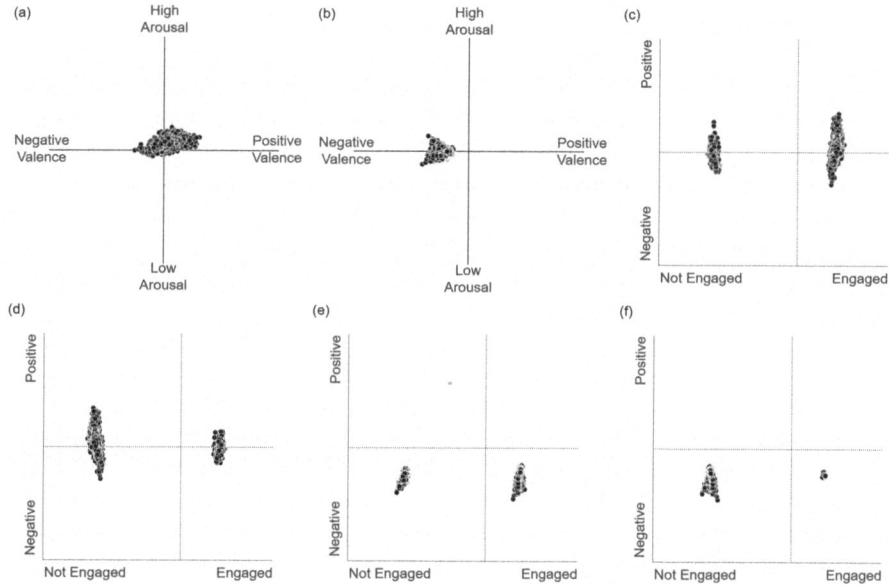

Fig. 6. (a) Frame-by-frame graph of valence and arousal for person ID#3 and (b) ID#2, engagement for person ID#3 towards the (c) student and (d) board, followed by the engagement towards the (e) student and the (f) professors for person ID#2.

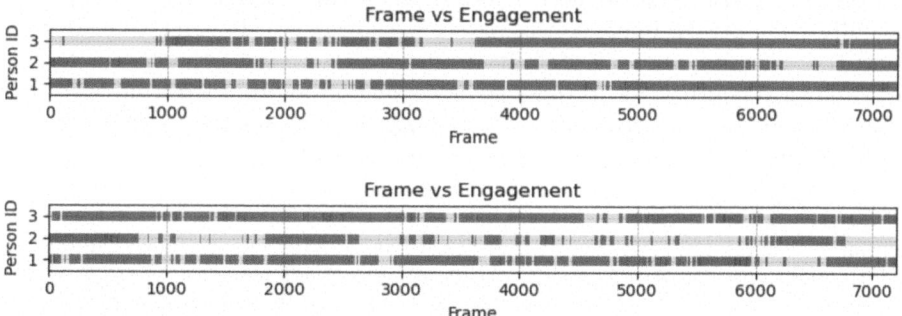

Fig. 7. Engagement of each professor, on the top, with the board, and on the bottom, with the student. See details in the text.

23% of the time, i.e., 1,656 frames of engagement for a maximum of 7,200 (4 min × 60 s × 30 frames). To the student (s), the period engagement was $PE_s(P, \text{Group 1}) = 27\%$. Finally, the $E(\text{Group 1} \cup \text{Group 2}) = 41\%$, i.e., the groups were engaged 41% of the event, counting both the engagement from the student and board.

5   Conclusion

This paper presents a modular and scalable model for real-time emotion analysis and engagement detection, combining advanced deep learning models with multimodal data processing. By integrating valence-arousal prediction, head pose estimation, and individual identity tracking, the system achieves robust performance in diverse scenarios, such as educational and behavioural studies. The framework demonstrates effective classification of engagement levels, leveraging gaze, emotional states, and head orientation to provide detailed insights into individual and group behaviours.

When the camera focuses the sole PoI (e.g., Test#1), the system operates effectively even with positional changes in the room, as the camera's optical axis provides a fixed reference for yaw detection. However, challenges arise with vertical movement since pitch values change with head tilts but are not currently linked to the camera's vertical alignment. When the PoI is not the main focus of the camera (e.g., Test#2), our model relies on fixed positions for individuals, PoI, and cameras. This limitation arises because accurate angle estimation depends on known spatial relationships. Enhancing the system to handle dynamic scenarios (e.g., moving individuals or PoI) is a future goal. This may involve integrating real-time positional tracking and incorporating changes in pitch due to vertical shifts.

Future work, in addition to the aspects already mentioned, includes integrating additional dimensions or sources such as speech and physiological data, which hopefully will further improve the model performance. However, the primary future goal is to improve adaptability to dynamic scenarios, thereby increasing the system's versatility. With these advancements, the proposed framework can become a valuable tool in fields such as human-computer interaction, offering deeper insights into human emotions and engagement.

Acknowledgments. This work is supported by UID/04516/NOVA Laboratory for Computer Science and Informatics (NOVA LINCS) with the financial support of FCT.IP, and by the project AI.EVENT: Monitor Live Audience with AI (ALGARVE-FEDER-01180500, Ref. 17325) co-financed by ALGARVE 2030, Portugal 2030 and by the European Union.

References

1. Berlincioni, L., Cultrera, L., Becattini, F., Bimbo, A.D.: Neuromorphic valence and arousal estimation. J. Ambient Intell. Humaniz. Comput. 1–11 (2024)
2. Bochkovskiy, A., Wang, C.Y., Liao, H.Y.M.: Yolov4: optimal speed and accuracy of object detection. arXiv preprint arXiv:2004.10934 (2020)
3. Bruin, J., et al.: Detection of arousal and valence from facial expressions and physiological responses evoked by different types of stressors. Front. Neuroergonomics **5**, 1338243 (2024). https://doi.org/10.3389/fnrgo.2024.1338243

4. Dickinson, K., et al.: Assessing learner engagement with virtual educational events: development of the virtual in-class engagement measure (VIEM). Am. J. Surg. **222**(6), 1044–1049 (2021)

5. Einsle, C.S., Escalera-Izquierdo, G., García-Fernández, J.: Social media hook sports events: a systematic review of engagement. Commun. Soc. **36**(3), 133–151 (2023)

6. Fanelli, G., Dantone, M., Gall, J., Fossati, A., Van Gool, L.: Random forests for real time 3D face analysis. Int. J. Comput. Vision **101**, 437–458 (2013)

7. Gupta, S., Kumar, P., Tekchandani, R.K.: Facial emotion recognition based real-time learner engagement detection system in online learning context using deep learning models. Multimedia Tools Appl. **82**(8), 11365–11394 (2023)

8. Harrison, E., Kwon, W.S.: Brands talking on events? Brand personification in real-time marketing tweets to drive consumer engagement. J. Prod. Brand Manag. **32**(8), 1319–1337 (2023)

9. Hempel, T., Abdelrahman, A.A., Al-Hamadi, A.: 6D rotation representation for unconstrained head pose estimation. In: 2022 IEEE International Conference on Image Processing (ICIP), pp. 2496–2500. IEEE (2022)

10. Hempel, T., Abdelrahman, A.A., Al-Hamadi, A.: Toward robust and unconstrained full range of rotation head pose estimation. IEEE Trans. Image Process. **33**, 2377–2387 (2024)

11. Hwooi, S., Othmani, A., Sabri, A.: Deep learning-based approach for continuous affect prediction from facial expression images in valence-arousal space. IEEE Access **10**, 96053–96065 (2022)

12. Lasri, I., Riadsolh, A., Elbelkacemi, M.: Facial emotion recognition of deaf and hard-of-hearing students for engagement detection using deep learning. Educ. Inf. Technol. **28**(4), 4069–4092 (2023)

13. Mollahosseini, A., Hasani, B., Mahoor, M.H.: AffectNet: a database for facial expression, valence, and arousal computing in the wild. IEEE Trans. Affect. Comput. **10**(1), 18–31 (2017)

14. Nguyen, H.H., Huynh, V.T., Kim, S.H.: An ensemble approach for facial expression analysis in video. arXiv preprint arXiv:2203.12891 (2022)

15. Rodrigues, J., Cardoso, P., Lemos, M., Cherniavska, O., Bica, P.: Engagement monotorization in crowded environments: a conceptual framework. In: 11th International Conference on Software Development and Technologies for Enhancing Accessibility and Fighting Info-exclusion (DSAI 2024), Abu Dhabi, UAE (2024). https://doi.org/10.1145/3696593.3696632

16. Savchenko, A.V.: Frame-level prediction of facial expressions, valence, arousal and action units for mobile devices. arXiv preprint arXiv:2203.13436 (2022)

17. Schroff, F., Kalenichenko, D., Philbin, J.: FaceNet: a unified embedding for face recognition and clustering. In: Proceedings of the IEEE Conference on Computer Vision and Pattern Recognition, pp. 815–823 (2015)

18. Serengil, S., Özpınar, A.: A benchmark of facial recognition pipelines and co-usability performances of modules. Bilişim Teknolojileri Dergisi **17**(2), 95–107 (2024)

19. Shao, S., et al.: CrowdHuman: a benchmark for detecting human in a crowd. arXiv e-prints pp. arXiv-1805 (2018)

20. Veltmeijer, E.A., Gerritsen, C., Hindriks, K.V.: Automatic emotion recognition for groups: a review. IEEE Trans. Affect. Comput. **14**(1), 89–107 (2021)

21. Vu, T.H., Osokin, A., Laptev, I.: Context-aware CNNs for person head detection. In: Proceedings of the IEEE International Conference on Computer Vision, pp. 2893–2901 (2015)

22. Zhang, N., Luo, J., Gao, W.: Research on face detection technology based on MTCNN. In: 2020 International Conference on Computer Network, Electronic and Automation (ICCNEA), pp. 154–158. IEEE (2020)
23. Zhang, Z., Luo, P., Loy, C.C., Tang, X.: From facial expression recognition to interpersonal relation prediction. Int. J. Comput. Vision **126**, 550–569 (2018)
24. Zhao, Z., Li, Y., Yang, J., Ma, Y.: A lightweight facial expression recognition model for automated engagement detection. SIViP **18**(4), 3553–3563 (2024)
25. Zhou, Y., Gregson, J.: WHENet: real-time fine-grained estimation for wide range head pose. In: Proceedings of the 31st British Machine Vision Virtual Conference, pp. 1–13 (2020)

Dead Gate Elimination

Yanbin Chen$^{(\boxtimes)}$, Christian B. Mendl , and Helmut Seidl

School of CIT, Technical University of Munich, Garching 85748, Germany
{yanbin.chen,christian.mendl,helmut.seidl}@tum.de

Abstract. Hybrid quantum algorithms combine the strengths of quantum and classical computing. Many quantum algorithms, such as the variational quantum eigensolver (VQE), leverage this synergy. However, quantum circuits are executed in full, even when only subsets of measurement outcomes contribute to subsequent classical computations. In this manuscript, we propose a novel circuit optimization technique that identifies and removes dead gates. We prove that the removal of dead gates has no influence on the probability distribution of the measurement outcomes that contribute to the subsequent calculation result. We implemented and evaluated our optimization on a VQE instance, a quantum phase estimation (QPE) instance, and hybrid programs embedded with random circuits of varying circuit width, confirming its capability to remove a non-trivial number of dead gates in real-world algorithms. The effect of our optimization scales up as more measurement outcomes are identified as non-contributory, resulting in a proportionally greater reduction of dead gates.

Keywords: Quantum compilation · Dynamic circuit optimization

1 Introduction

In recent efforts to address complex real-world problems, researchers are increasingly integrating quantum and classical computing to use the unique strengths of both paradigms [18]. In such interdisciplinary development, domain specialists are exploring ways to implement or even accelerate specific subroutines through quantum circuits tailored to quantum processing units ($QPUs$) [2,3,13,17,26]. Concurrently, quantum experts may incorporate classical computing procedures, given the wealth of sophisticated classical computing procedures that have been developing over decades [8,31,32]. A popular algorithm framework that allows to take advantage of the strength of both quantum and classical computers is *hybrid programs* [19]. In hybrid programs, quantum circuits are embedded as subroutines into programs from a classical host language. Usually, the classical host program handles optimization, control, and data processing, while the quantum circuits are used for specific calculations that may benefit from quantum speedups.

However, this integration can present challenges [10,11,28]. When researchers work beyond their core expertise, the interplay between classical and quantum

M. H. Lees et al. (Eds.): ICCS 2025, LNCS 15905, pp. 135–150, 2025.
https://doi.org/10.1007/978-3-031-97632-2_10

components may be suboptimal. This imperfect coupling risks inefficient resource utilization.

An implicit assumption is often made that the circuits are executed as external entities and all qubits are measured in the end and their outcomes are collected for purposes that are not of interest to circuits. So, circuits are fully executed even if not all measurement outcomes contribute to later calculations. However, in the following example, we see the potential for circuit simplification when knowing that some measurement outcomes are not needed.

Example 1. In the hybrid program in Fig. 1, the measurement outcome o_0 does not contribute to final results: in **Proc**$_a$, the initial value of variable a, i.e. o_0, gets canceled out in the expression $z - 2t$; in **Proc**$_b$ the initial value of variable a has no impact on the return value, because $0 \leq \eta a \leq 0.5$ and thus the ηa part is always rounded down to 0 by **int**$(\cdot)$ operator. So, if we execute the circuit Fig. 2, where the measurement outcome from q_0 is always discarded, and we assign an arbitrary value from $\{0, 1\}$ to o_0, the results of the both **Proc**$_a$ and **Proc**$_b$ will not be influenced. Then, we could optimize the program by running the simplified circuit Fig. 3 instead of circuit Fig. 2. We call gates removed by this analysis dead gates. We will formally justify that this simplification will never influence calculation results of hybrid programs in Sect. 3.

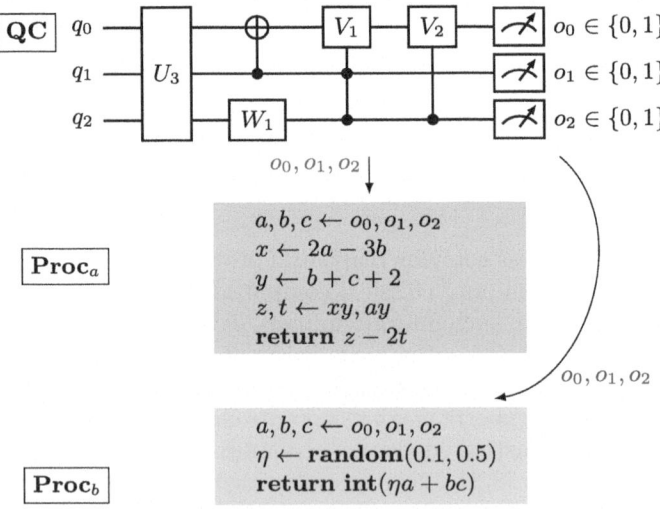

Fig. 1. An example of a hybrid program, where a quantum circuit **QC** of 3 qubits are first executed and then the measurement outcomes o_0, o_1, o_2 from qubits q_0, q_1, q_2, respectively, are dispatched to one of the two classical computing procedures, **Proc**$_a$, or **Proc**$_b$.

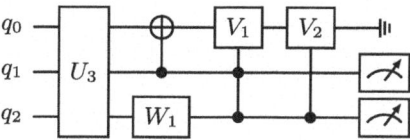

Fig. 2. A 3-qubit circuit. The measurement outcome of the top qubit is discarded.

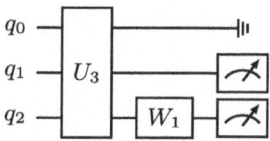

Fig. 3. Simplification of the circuit in Fig. 2. The probability distribution on measurement outcomes that are not discarded remains unchanged.

Non-contributory measurement outcomes can also occur in scenarios where the classical computing procedure only queries a subset of the available measurement results. This could make the same simplification possible, as we will demonstrate in the VQE and QPE examples in Sect. 4.

In addtion, qubits that are not explicitly measured, such as ancilla qubits used in intermediate computations can also be interpreted in the same manner. While not explicitly measured at the end of circuits, these qubits could be considered as implicitly measured, with their outcomes being discarded immediately. In this perspective, such qubits fit naturally into the consideration of our paper, as their measurement outcomes do not influence subsequent classical computations.

Several existing works address certain aspects of the matter discussed in this paper. The partial equivalence checking proposed in [4], verifies whether two circuits yield the same probability distribution for a given set of measurement outcomes, but it does not provide a way to simplify circuits while preserving these distributions. Moreover, it requires explicit global unitary operators, the computation of which is infeasible for large-scale circuits due to the inherent complexity of circuit simulation. QuTracer proposed in [16] optimizes circuits by eliminating gates that do not affect a subset of measured qubits, but it lacks a formal framework for this process and may fail to recognize redundant operations—such as SWAP gates that merely permute qubits without altering measurement distributions. In [1], it is mentioned that a measurement outcome depends only on its causal light cone, yet it does not provide a systematic method to exploit this insight for circuit simplification. Crucially, existing approaches overlook a key optimization opportunity: in hybrid quantum-classical workflows, some measurement outcomes become non-contributory to subsequent classical computations.

In this manuscript, we introduce a novel approach to simplify circuits that uses context information from the classical computing components of the hybrid program. By propagating the contextual information that some measurement outcomes are not contributory, our method identifies and removes dead gates in circuits without changing the semantics of the entire hybrid program, leading to more resource-efficient circuits and quantum-classical integration. We evaluate our method by running it on instances of VQE and QPE algorithm, and on random circuits in Sect. 4.

2 Preliminaries

This manuscript assumes that readers are familiar with the basics of quantum computing. For a detailed introduction to quantum computing, we recommend the following literature [14,22,27]. In this section, we explain some notations that we will use later.

For an n-qubit circuit C, we use $C.\textbf{gates}()$ to denote the set of all gates in C. Each gate in $C.\textbf{gates}()$ is an object storing information, including gate type, the set of qubits it acts on, and the set of gates it depends on. We use C to also denote the unitary matrix of circuit C if no ambiguity is produced. For example, when applying C to a n-qubit state S, the resulting state is CS. We use the following $-$ as an operator to remove one gate from a circuit.

Definition 1 ($-$ operator). *For an n-qubit circuit C and a gate $g \in C.\textbf{gates}()$, $C-g$ represents a n-qubit circuit obtained by removing the gate g from C, namely $(C-g).\textbf{gates}() := C.\textbf{gates}()\backslash\{g\}$.*

We use the following notation to describe the probability of the measurement outcomes of a subsystem of a quantum state being a given binary string.

Definition 2 (Probability distribution of subsystem measurement outcomes). *For an n-qubit state S_n on a set of qubits $Q_n = \{q_0, \ldots, q_{n-1}\}$, and a binary string k of length $|k| = |Q|$, where Q is a subset of qubits $\{q_{i_0}, \ldots, q_{i_{|k|-1}}\} = Q \subseteq Q_n$ where $i_0 < \cdots < i_{|k|-1}$, $\mathcal{P}^k_{i_0 \ldots i_{|k|-1}}[S_n]$ denotes the probability of q_{i_j} measuring $k[j]$ for all $j \in \{0, \ldots, |k|-1\}$, where $k[j]$ is the j-th element of k.*

Example 2. Consider a 2-qubit state $|\Phi\rangle$ on qubits q_0 and q_1, where $|\Phi\rangle = \alpha_0|00\rangle + \alpha_1|01\rangle + \alpha_2|10\rangle + \alpha_3|11\rangle$. $\mathcal{P}^{01}_{01}[|\Phi\rangle]$ represents the probability of measuring 0 on q_0 and 1 on q_1, namely the probability of the state collapsing to $|01\rangle$, which is $|\alpha_1|^2$. Similarly, $\mathcal{P}^{10}_{01}[|\Phi\rangle] = |\alpha_2|^2$, $\mathcal{P}^{00}_{01}[|\Phi\rangle] = |\alpha_0|^2$. $\mathcal{P}^1_0[|\Phi\rangle]$ represents the probability of measuring 1 on q_0, which is $|\alpha_2|^2 + |\alpha_3|^2$, because when q_0 is measured 1, q_1 could be measured either 0 or 1.

Definition 3 (Frontier). *Given a circuit C, its frontier is a set $\mathcal{F}_C$ satisfying: (a) $\mathcal{F}_C \subseteq C.\textbf{gates}()$; (b) for any gate $g \in \mathcal{F}_C$, any output wire of g is no input of any other gates.*

Example 3. The frontier of the following circuit only consists of V_5 and U_2.

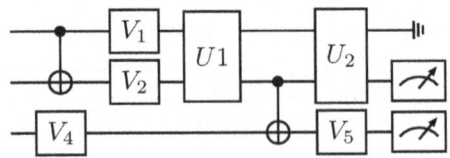

3 Method

In this work, We restrict our discussion to circuits that contain no mid-circuit measurements or resets. We start by introducing some concepts that we will use in later discussions.

For a quantum circuit C, we assume that for the outcomes we collect by measuring all qubits, a subset of them has no contribution to the classical computing procedures that come later. We explicitly mark such measurement outcomes as discarded, and we call them *discarded measurement outcomes*. The following notation is put at the end of a qubit wire to denote that the measurement outcome on that qubit is discarded: ──╫ . On the contrary, a *valid measurement outcome* is the one that is not discarded.

From now on, if a measurement outcome is valid, we omit the symbol of measurement at the end of the qubit wire in the circuit diagram for conciseness.

Definition 4 (Dead/Valid qubit). *A qubit is a* dead qubit *if its measurement outcome is discarded. A qubit is a* valid qubit *if it is not dead.*

Next, we establish an equivalent relation among circuits that is based on measurements performed only on valid qubits. That is, in this equivalence, we consider two circuits to be equal if their probability distributions of valid measurement outcomes are identical.

Definition 5 (Equivalence relative to valid outcomes). *Given two circuits C_1 and C_2 applied on the same set of qubits $Q_n = \{q_0, \ldots, q_{n-1}\}$, for a subset $D \subseteq Q_n$ where all qubits in D are dead, C_1 and C_2 are equivalent relative to D, denoted by $C_1 \equiv_D C_2$, if and only if for any n-qubit state S and any binary string k of length $|k| = |Q_n \backslash D|$, $\mathcal{P}_{i_0 \ldots i_{|k|-1}}^k[C_1 S] = \mathcal{P}_{i_0 \ldots i_{|k|-1}}^k[C_2 S]$, where $Q_n \backslash D = \{q_{i_0}, \ldots, q_{i_{|k|-1}}\}$ and $i_0 < \cdots < i_{|k|-1}$.*

Example 4. The following two circuits are equivalent relative to their valid outcomes, because the probability of measuring 0 on the valid qubit, q_1, is the same in both circuits.

Then, we move on to concepts of dead gates, which are essential to our method. Given the knowledge that some measurement outcomes do not influence subsequent calculations and we discard them explicitly, we define a gate as dead if removing it only affects the probability distribution of these discarded measurement outcomes.

Definition 6 (Dead gate). *Given a circuit C and a gate g in C, and a set of dead qubits D, g is a* dead gate *if and only if $C \equiv_D C' := C - g$, where $-$ is defined by Definition 1.*

Since removing dead gates does not change the probability distribution on valid measurement outcomes, we could simplify circuits by removing such dead gates. By Theorem 1, Theorem 2, and Theorem 3, we present our approach to identify dead gates and prove that removing these dead gates does not influence the results of calculation, therefore justifying the correctness of our method.

Theorem 1. *Given any operator U acting on $n+1$ qubits and any operator V acting on a single qubit q_i, and q_i is dead, it holds that*

$$\tag{1}$$

Proof. This is a special case of Theorem 2. □

Remark 1. By Definition 6, gate V in Eq. (1) is a dead gate, so we could optimize the circuit by removing it.

Theorem 2. *Given any operator U acting on $n+1$ qubits and any operator V acting on a single qubit q_i, and V is controlled by n_c qubits, where $n_c + n_r = n$, and q_i is dead, it holds that*

$$\tag{2}$$

Proof. Let the circuit on the left be C_1, and the circuit on the right be C_2. W.l.o.g, we assume that $i = 0$, and the gate V is controlled by qubits $q_1, \ldots, q_{n_c}$. Suppose the gate V is defined by $V|0\rangle = \alpha_{v_0}|0\rangle + \beta_{v_0}|1\rangle$ and $V|1\rangle = \alpha_{v_1}|0\rangle + \beta_{v_1}|1\rangle$. For any input state S, we assume that $|\Phi\rangle = C_2 S = \sum_{j=0}^{N-1} c_j|j\rangle$, where $N = 2^{n+1}$, $c_j \in \mathbb{C}$. Then, for any n-bit binary string k, we have $\mathcal{P}_{1\ldots n}^k[|\Phi\rangle] = |c_{0\oplus k}|^2 + |c_{1\oplus k}|^2$, where $\oplus$ is string concatenation (E.g., $00 \oplus 11 = 0011$ and $110 \oplus 1 = 1101$). In fact, $|\Phi\rangle$ could be rewritten as

$$
|\Phi\rangle = \sum_{j=0}^{N-1} c_j|j\rangle = \sum_{\substack{|s|=n_c,\, |t|=n_r \\ 0 \in s}} \sum c_{0\oplus s\oplus t}|0 \oplus s \oplus t\rangle +
$$

$$
\sum_{\substack{|s|=n_c,\, |t|=n_r \\ 0 \notin s}} \sum c_{0\oplus s\oplus t}|0 \oplus s \oplus t\rangle + \sum_{\substack{|s|=n_c,\, |t|=n_r \\ 0 \in s}} \sum c_{1\oplus s\oplus t}|1 \oplus s \oplus t\rangle + \tag{3}
$$

$$
\sum_{\substack{|s|=n_c,\, |t|=n_r \\ 0 \notin s}} \sum c_{1\oplus s\oplus t}|1 \oplus s \oplus t\rangle
$$

So, by applying C_1 to S, the output state $|\Psi\rangle = C_1 S = (C^{n_c} V)C_2 S$ is

$$|\Psi\rangle = C^{n_c} V |\Phi\rangle = \sum_{\substack{|s|=n_c, |t|=n_r \\ 0 \notin s}} \sum \sum_{b=0}^{1} c_{b\oplus s\oplus t}\alpha_{v_b} |0 \oplus s \oplus t\rangle +$$

$$\sum_{\substack{|s|=n_c, |t|=n_r \\ 0 \notin s}} \sum \sum_{b=0}^{1} c_{b\oplus s\oplus t}\beta_{v_b} |1 \oplus s \oplus t\rangle + \sum_{\substack{|s|=n_c, |t|=n_r \\ 0 \in s}} \sum \sum_{b=0}^{1} c_{b\oplus s\oplus t} |b \oplus s \oplus t\rangle \tag{4}$$

where $C^{n_c} V$ denotes the multi-controlled gate V.

If $\exists l \in \{1, \ldots, n_c\}$: $k[l] = 0$, then $\mathcal{P}_{1\ldots n}^k[C_1 S]$ is calculated by

$$\sum_{b=0}^{1} \sum_{s\oplus t=k} |c_{b\oplus s\oplus t}|^2 = |c_{0\oplus k}|^2 + |c_{1\oplus k}|^2 = \mathcal{P}_{1\ldots n}^k[C_2 S] \tag{5}$$

If $\forall l \in \{1, \ldots, n_c\}$: $k[l] = 1$, then $\mathcal{P}_{1\ldots n}^k[C_1 S]$ is calculated by

$$\sum_{b=0}^{1} \sum_{s\oplus t=k} |c_{b\oplus s\oplus t}\alpha_{v_b}|^2 + |c_{b\oplus s\oplus t}\beta_{v_b}|^2 = \sum_{b=0}^{1} \sum_{s\oplus t=k} |c_{b\oplus s\oplus t}|^2(|\alpha_{v_b}|^2 + |\beta_{v_b}|^2)$$

$$= |c_{0\oplus k}|^2(|\alpha_{v_0}|^2 + |\beta_{v_0}|^2) + |c_{1\oplus k}|^2(|\alpha_{v_1}|^2 + |\beta_{v_1}|^2) = \mathcal{P}_{1\ldots n}^k[C_2 S] \tag{6}$$

Since our choice of S is arbitrary, by Definition 5, Eq. (2) holds. $\qquad\square$

It could happen that removing some gate makes some dead qubits valid and some valid qubits dead, while the probability distribution of valid measurement outcomes is unchanged. For our analysis to encompass this case, we need to extend the equivalence in Definition 5 and the dead gate in Definition 6.

Definition 7 (Extended equivalence relative to valid outcomes). *Given circuits C_1 and C_2 applying on the set of qubits $Q_n = \{q_0, \ldots, q_{n-1}\}$, for $D1, D2 \subseteq Q_n$ where $|D_1| = |D_2|$ and all qubits in $D1$ and D_2 are dead, $C_1 \equiv_{D_2}^{D_1} C_2$ iff for any n-qubit state S and any binary string k of length $|k| = |Q_n \backslash D_1| = |Q_n \backslash D_2|$, $\mathcal{P}_{i_0 \ldots i_{|k|-1}}^k[C_1 S] = \mathcal{P}_{[e_1/f_1, \ldots, e_m/f_m](i_0 \ldots i_{|k|-1})}^k[C_2 S]$, where $Q_n \backslash D_1 = \{q_{i_0}, \ldots, q_{i_{|k|-1}}\}$, $i_0 < \cdots < i_{|k|-1}$, $D_1 \backslash (D_1 \cap D_2) = \{e_1, \ldots, e_m\}$, $D_2 \backslash (D_1 \cap D_2) = \{f_1, \ldots, f_m\}$, and $[b_1/a_1, \ldots, b_p/a_p]s$ denotes a string obtained by for each $l \in \{1, \ldots, p\}$ replacing a_l in string s with b_l (E.g., $[1/4, 2/5, 3/6]456 = 123$).*

Definition 8 (Extended dead gate). *Given a circuit C and a gate g in C acting on a set of qubits Q, g is a dead gate if and only if there exist subsets $D_1, D_2 \subseteq Q$ such that $C \equiv_{D_2}^{D_1} C' = C - g$, where $-$ is defined by Definition 1.*

Theorem 3. *Given any operator U acting on $n + 2$ qubits and a SWAP gate, it holds that*

$$\tag{7}$$

Proof. It follows directly the definition of SWAP gates and Definition 7.   □

Remark 2. The SWAP gate in Eq. (7) is a dead gate by Definition 8 and can be removed. After removing a SWAP gate, we also need to adapt the qubit mapping, if the qubit mapping/routing is performed at an earlier stage.

Our optimization algorithm is shown in Algorithm 1, of which the asympotic bound is given in Theorem 4.

Algorithm 1: Dead gates removal

Data: $C \in circuits$
Result: C_{opt}
$C_{opt} \leftarrow C$, $terminate \leftarrow$ **False**;
while $terminate \neq True \wedge \emptyset \neq \mathcal{F}_C \leftarrow C_{opt}.\textbf{\textit{frontier}}()$ **do**
   $terminate \leftarrow$ **True**;
   for $g \in \mathcal{F}_C$ **do**
      if g *is a dead gate by Theorem 1, Theorem 2, or Theorem 3* **then**
         | $C_{opt} \leftarrow C_{opt} - g$, $terminate \leftarrow$ **False**;
      end
   end
end

Theorem 4 (Algorithm 1 is polynomial). *The time complexity of 1 is* $\mathcal{O}(|C.\textbf{\textit{gates}}()|^2)$.

Proof. In each **while** iteration, $\mathcal{O}(|C.\textbf{gates}()|)$-many gates are checked to see whether they are dead, and since at least one gate is removed in every iteration (except the last iteration), there are at most $\mathcal{O}(|C.\textbf{gates}()|)$-many iterations.   □

One should be cautious when considering circuit simplification based on the knowledge about non-contributory measurement outcomes. There are cases where it seems that some gate only "writes" on dead qubits and looks like a dead gate, but it is not a dead gate and thus cannot be removed, as demonstrated in the following example.

Example 5. The following simplification would, in general, lead to a changed probability distribution on valid measurement outcomes.

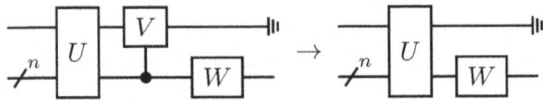

To see this, we consider a special case of it shown as follows:

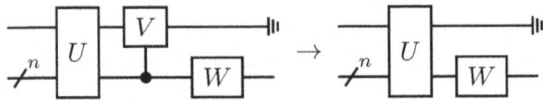

4   Evaluation

We conduct 3 sets of experiments to evaluate our method. In the first set, our method is applied to the quantum algorithm VQE. In the second set, an instance of QPE is optimized with our method. In the third set, our method is applied to random circuits. The demo implementation of optimization and experiments is accessible at https://github.com/i2-tum/demo-dead-gate-elimination.

VQE Algorithm. The VQE is a hybrid quantum-classical algorithm that finds the ground state energy of a quantum system, and it is widely used in areas like quantum chemistry and material science [12]. In each iteration of the VQE algorithm, an *Ansatz*, a parameterized circuit selected from a diverse range of designs, is executed, and measurements are performed on all output qubits. These measurement outcomes are then scheduled to an *optimizer*, a classical computing procedure. This procedure uses them to calculate expectation values of Hamiltonian terms, which are then used to update parameters in the Ansatz.

Due to the broad range of applications of VQE, it has been integrated into well-established toolchains such as Qiskit, allowing it to be utilized as a black-box subroutine [25]. While this facilitates the use of quantum computers for domain experts, it also introduces the risk of misalignment between quantum circuits and classical computing procedures, particularly because there are already numerous choices of Ansatz with different focuses, and many more are expected to be developed in the future [20,24,29].

Consider the instance of the VQE algorithm constructed in Fig. 4. In each iteration, the 4-qubit Ansatz A_1 is executed and measured. The resulting measurement outcomes, o_i ($i \in 0,\ldots,3$), are then sent to the optimizer. There, two expectation values, $\mathbb{E}_Z$ and $\mathbb{E}_X$, are computed and combined. The resulting values are used to update the parameters in the Ansatz–namely, $\vec{\theta_r}$ and θ_j ($j \in 1,\ldots,8$)–which are adapted for the next iteration.

Here, we observe that the calculation of $\mathbb{E}_Z$ and $\mathbb{E}_X$ only depends on the measurement outcomes o_2 and o_3, meaning q_0 and q_1 are dead qubits. By applying Algorithm 1 to the Ansatz A_1, a simplified Ansatz A_2 shown in Fig. 5 is obtained. Thus, we can optimize the VQE instance by replacing A_1 with A_2, reducing the number of parameterized gates by 4 and the number of two-qubit gates by 3 in each iteration. Considering that VQE requires many iterations to converge, the reduction in gate operations becomes even more significant.

QPE Algorithm. The QPE algorithm determines the phase associated with an eigenvalue of a given unitary operator [9,15,22]. QPE is employed as a subroutine in various quantum algorithms, among which one of the most famous examples is Shor's algorithm [31].

Consider an instance of QPE constructed in Fig. 6, where the QPE circuit is executed and its measurement outcomes constitute the estimated phase $\theta \in [0,1)$ received by the classical computing procedure $\mathbf{Proc}_c$. However, we can observe that the most significant bit of θ, namely θ_0, is always subtracted away in the expression $\lambda - \lfloor \lambda \rfloor$. Hence, the initial value of θ_0, i.e., o_0, is not contributory, and

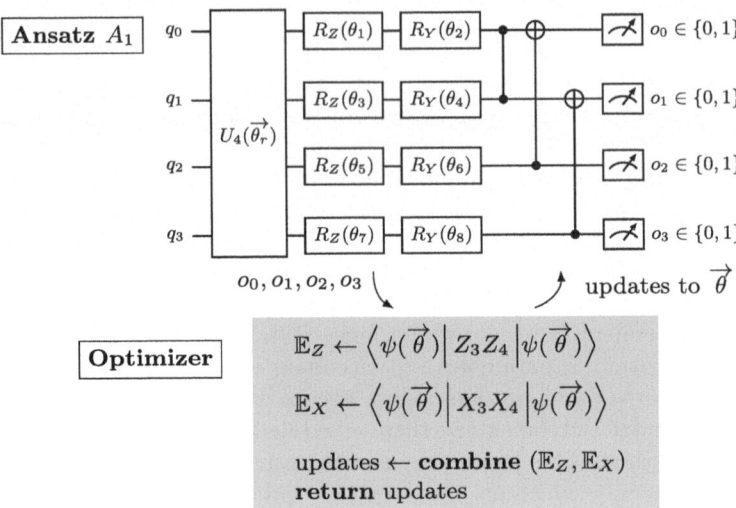

Fig. 4. An instance of VQE algorithm.

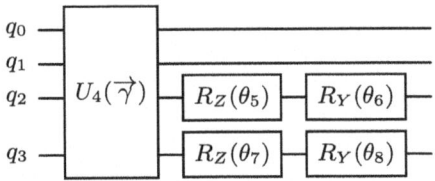

Fig. 5. A simplified Ansatz, A_2, that can replace A_1 in Fig. 4.

we identify q_0 as a dead qubit. Then, by running our optimization on the QPE circuit, we remove m two-qubit gates and a Hadamard gate and get a simplified circuit in Fig. 7. Thus we can optimize the QPE instance by replacing the circuit in Fig. 6 by the circuit Fig. 7.

Random Circuits. As an effort to ensure a broad and unbiased evaluation of our optimization algorithm, we also conduct experiments where Algorithm 1 is performed on randomly generated circuits. We consider hybrid programs consisting of alternating quantum and classical segments, illustrated as follows:

$$QC_0 \longrightarrow C_0 \longrightarrow \cdots \longrightarrow QC_{b-1} \longrightarrow C_{b-1}$$

Each hybrid program consists of a sequence of b quantum-classical blocks, where each block comprises a quantum circuit QC_i followed by a classical computation C_i. The parameter b controls the number of such blocks in the hybrid program. In our evaluation, we set $b = 60$ and generate random circuits of *circuit width* ranging from 2 to 100 qubits, where the circuit width is defined as the number of

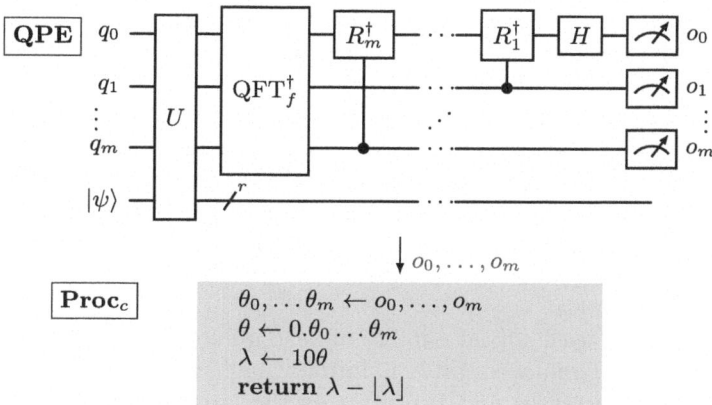

Fig. 6. An instance of QPE, where U consists of Hadamard gates and a sequence of controlled oracles to prepare the state ready for the inverse quantum Fourier transform (QFT [7,22]), $\text{QFT}_f^\dagger$ is the front part of the inverse QFT, measurement outcomes $o_i \in \{0, 1\}$ for all i, and $\lfloor \cdot \rfloor$ is the floor function that maps a real value to the greatest integer less than or equal to it.

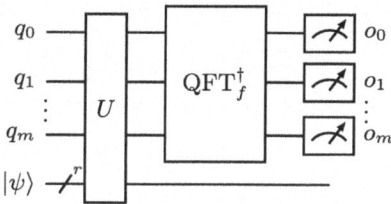

Fig. 7. A simplified QPE circuit to replace the circuit in Fig. 6.

qubits in the circuit. For each circuit, the total gate count is 100 times the circuit width. The circuits are constructed using the universal Clifford + T gate set with single-qubit gates comprising 10% of all generated gates, ensuring a reasonable balance between single- and multi-qubit operations. For each circuit width w, we generate 1000 hybrid programs. In each program, every quantum block QC_i is instantiated with a random circuit generated as described above. All gates are placed uniformly at random across the circuit, ensuring no positional bias in their distribution. These circuits are assumed to be first pre-optimized using circuit transpilers, such as Qiskit and t|ket⟩, ensuring that the input circuits are highly optimized by the state-of-the-art compilation toolchain.

We then apply our optimization algorithm to every quantum circuit QC_i within each hybrid program under various settings of dead qubit constraints. Specifically, we consider five settings: 1, 2, or 3 dead qubits, as well as when the number of dead qubits is set to 10% or 20% of the circuit width. For each circuit width and each dead qubit setting, we evaluate the gate count reduction achieved by our algorithm across all circuits in all generated hybrid programs.

The mean gate reduction serves as our primary performance metric, providing a robust and comprehensive assessment of the algorithm's effectiveness in realistic hybrid execution scenarios.

The result of our experiments is shown in Fig. 8. The experimental results show that our method consistently removes a non-trivial number of gates across settings. This is particularly notable given that our optimization is applied to circuits that are assumed to have already been pre-optimized using circuit transpilers. This demonstrates that our approach achieves further optimization beyond what is achievable with current state-of-the-art quantum compilation tools.

For the settings with a fixed number of dead qubits (1, 2, and 3), we observe that the number of removed gates is initially high when the circuit width is small, and then decreases and stabilizes as the circuit width increases. This trend reflects a key observation: in small circuits, a fixed number of dead qubits represents a large proportion of the total qubit count (e.g., 1 dead qubit out of 2 or 3 is 50%–33%), which creates substantial optimization opportunities. As the circuit grows wider, however, the proportion of dead qubits diminishes, leading to less pronounced impact from the optimization.

The early-stage behavior of settings with a fixed number of dead qubits directly parallels the trends observed in the percentage-based dead qubit settings (10% and 20%). In these settings, the number of removed gates grows stepwise with the circuit width, as the absolute number of dead qubits increases discretely with circuit size. These steps correspond to the increase in dead qubit count, and each jump leads to a corresponding spike in optimization gain. This confirms that our optimization method's effectiveness is primarily driven by the proportion of dead qubits rather than their absolute number alone.

To assess the practical efficiency of our method, we evaluate its runtime as a function of circuit width, as shown in Fig. 8. The results indicate that the execution time increases approximately linearly with the circuit width, even in cases with a significant proportion of dead qubits. This empirical observation suggests that our approach remains efficient in practice. While the asymptotic complexity analysis in Theorem 4 establishes a worst-case quadratic dependence on the number of gates, our experimental results demonstrate that, for realistic circuits, the algorithm exhibits near-linear scaling with circuit width. This suggests that our method is practically efficient and scalable.

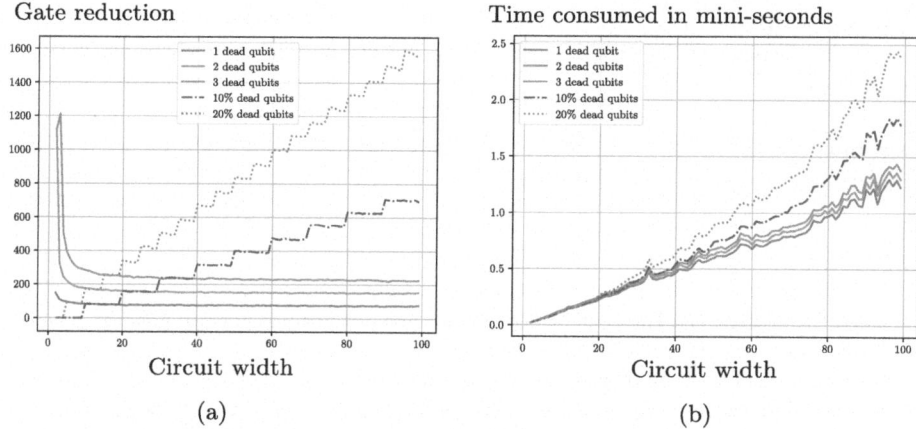

Fig. 8. (a) Gate reduction obtained by performing our optimization Algorithm 1 with different dead-qubit settings on hybrid programs embedded with random circuits of circuit width ranging from 2 to 100. (b) The corresponding time consumed in mini seconds.

5   Related Works

The concept of dead gates is inspired by the concept of dead variables in liveness analysis in compiler constructions of classical programming languages [21,30]. Static analysis stands among the most potential approaches to automate the detection of dead qubits from the context of circuits, and they have proven to be effective in bug detection, program analysis, and circuit optimization [5,6,23, 33,34].

6   Conclusion and Future Works

Our method demonstrates a practical approach to optimizing quantum circuits in hybrid programs by taking into account context from the classical host. By identifying dead gates and simplifying circuits accordingly, we achieve significant reductions in gate count when quantum-classical integration is suboptimal. The evaluation of our method on the VQE and QPE instances and on random circuits confirms its potential to improve the quality of circuits. It would be an interesting future work to investigate how to construct the dead gate analysis on dynamic circuits. We think one of the challenges there is that mid-circuit measurements come with side effect even if they are performed on dead qubits.

Acknowledgments. This work is supported by the Bavarian state government through the project Munich Quantum Valley with funds from the Hightech Agenda Bayern Plus.

Disclosure of Interests. The authors have no competing interests to declare that are relevant to the content of this article.

References

1. Abedi, E., Beigi, S., Taghavi, L.: Quantum lazy training. Quantum **7**, 989 (2023). https://doi.org/10.22331/q-2023-04-27-989

2. Bermot, E., et al.: Quantum generative adversarial networks for anomaly detection in high energy physics . In: 2023 IEEE International Conference on Quantum Computing and Engineering (QCE), pp. 331–341. IEEE Computer Society, Los Alamitos (2023). https://doi.org/10.1109/QCE57702.2023.00045. https://doi.ieeecomputersociety.org/10.1109/QCE57702.2023.00045

3. Bouland, A., van Dam, W., Joorati, H., Kerenidis, I., Prakash, A.: Prospects and challenges of quantum finance (2020). https://arxiv.org/abs/2011.06492

4. Chen, T.F., Jiang, J.H.R., Hsieh, M.H.: Partial equivalence checking of quantum circuits. In: 2022 IEEE International Conference on Quantum Computing and Engineering (QCE), pp. 594–604. IEEE (2022).https://doi.org/10.1109/qce53715.2022.00082

5. Chen, Y., Fulginiti, I., Mendl, C.B.: Probabilistic circuit model. In: 2024 International Conference on Quantum Computing and Engineering (2024). https://doi.org/10.1109/QCE60285.2024.10379

6. Chen, Y., Stade, Y.: Quantum constant propagation. In: Hermenegildo, M.V., Morales, J.F. (eds.) Static Analysis, pp. 164–189. Springer, Cham (2023). https://doi.org/10.1007/978-3-031-44245-2_9

7. Coppersmith, D.: An approximate fourier transform useful in quantum factoring (2002). https://arxiv.org/abs/quant-ph/0201067

8. De Luca, G.: A survey of nisq era hybrid quantum-classical machine learning research. J. Artif. Intell. Technol. **2**(1), 9–15 (2021). https://doi.org/10.37965/jait.2021.12002. https://ojs.istp-press.com/jait/article/view/60

9. Dobšíček, M.: Quantum computing, phase estimation and applications. arXiv preprint arXiv:0803.0909 (2008)

10. Elsharkawy, A., et al.: Integration of quantum accelerators with high performance computing – a review of quantum programming tools (2023). https://arxiv.org/abs/2309.06167

11. Elsharkawy, A., et al.: Challenges in hpcqc integration. In: 2023 IEEE International Conference on Quantum Computing and Engineering (QCE), vol. 02, pp. 405–406 (2023). https://doi.org/10.1109/QCE57702.2023.10304

12. Fedorov, D.A., Peng, B., Govind, N., Alexeev, Y.: VQE method: a short survey and recent developments. Mater. Theory **6**(1), 2 (2022)

13. Jojo, J., Khandelwal, A., Chandra, M.G.: Quantum algorithms for tensor-svd (2024). https://arxiv.org/abs/2405.19485

14. Kaye, P., Laflamme, R., Mosca, M.: An introduction to quantum computing. OUP Oxford (2006)

15. Kitaev, A.Y.: Quantum measurements and the abelian stabilizer problem. arXiv preprint quant-ph/9511026 (1995)

16. Li, P., Liu, J., Gonzales, A., Saleem, Z.H., Zhou, H., Hovland, P.: Qutracer: mitigating quantum gate and measurement errors by tracing subsets of qubits (2024). https://arxiv.org/abs/2404.19712

17. Matondo-Mvula, N., Elleithy, K.: Advances in quantum medical image analysis using machine learning: current status and future directions. In: 2023 IEEE International Conference on Quantum Computing and Engineering (QCE), vol. 01, pp. 367–377 (2023). https://doi.org/10.1109/QCE57702.2023.00049

18. McCaskey, A., Dumitrescu, E., Liakh, D., Humble, T.: Hybrid programming for near-term quantum computing systems. In: 2018 IEEE International Conference on Rebooting Computing (ICRC), pp. 1–12 (2018). https://doi.org/10.1109/ICRC.2018.8638598

19. McCaskey, A., Dumitrescu, E., Liakh, D., Humble, T.: Hybrid programming for near-term quantum computing systems. In: 2018 IEEE International Conference on Rebooting Computing (ICRC), pp. 1–12. IEEE (2018)

20. McClean, J.R., Romero, J., Babbush, R., Aspuru-Guzik, A.: The theory of variational hybrid quantum-classical algorithms. New J. Phys. **18**(2), 023023 (2016). https://doi.org/10.1088/1367-2630/18/2/023023

21. Muchnick, S.: Advanced Compiler Design Implementation. Morgan kaufmann (1997)

22. Nielsen, M.A., Chuang, I.L.: Quantum Computation and Quantum Information: 10th Anniversary Edition, 1 edn. Cambridge University Press, Cambridge. (2012). https://doi.org/10.1017/CBO9780511976667

23. Paltenghi, M., Pradel, M.: Analyzing quantum programs with lintq: a static analysis framework for qiskit. Proc. ACM Softw. Eng. **1**(FSE), 2144–2166 (2024)

24. Peruzzo, A., et al.: A variational eigenvalue solver on a photonic quantum processor. Nat. Commun. **5**(1), 4213 (2014)

25. Qiskit contributors: Qiskit: an open-source framework for quantum computing (2023). https://doi.org/10.5281/zenodo.2573505

26. Quetschlich, N., Forster, T., Osterwind, A., Helms, D., Wille, R.: Towards equivalence checking of classical circuits using quantum computing (2024). https://arxiv.org/abs/2408.14539

27. Rieffel, E., Polak, W.: An introduction to quantum computing for non-physicists. ACM Comput. Surv. (CSUR) **32**(3), 300–335 (2000)

28. Rohe, T., Grätz, S., Kölle, M., Zielinski, S., Stein, J., Linnhoff-Popien, C.: From problem to solution: a general pipeline to solve optimisation problems on quantum hardware (2024). https://arxiv.org/abs/2406.19876

29. Romero, J., Babbush, R., McClean, J.R., Hempel, C., Love, P.J., Aspuru-Guzik, A.: Strategies for quantum computing molecular energies using the unitary coupled cluster ansatz. Quant. Sci. Technol. **4**(1), 014008 (2018). https://doi.org/10.1088/2058-9565/aad3e4

30. Seidl, H., Wilhelm, R., Hack, S.: Compiler Design: Analysis and Transformation. Springer, Heidelberg (2012)

31. Shor, P.W.: Polynomial-time algorithms for prime factorization and discrete logarithms on a quantum computer. SIAM J. Comput. **26**(5), 1484–1509 (1997). https://doi.org/10.1137/S0097539795293172

32. Veshchezerova, M., et al.: A hybrid quantum-classical approach to the electric mobility problem . In: 2023 IEEE International Conference on Quantum Computing and Engineering (QCE), pp. 636–641. IEEE Computer Society, Los Alamitos (2023). https://doi.org/10.1109/QCE57702.2023.00078. https://doi.ieeecomputersociety.org/10.1109/QCE57702.2023.00078

33. Xia, S., Zhao, J.: Static entanglement analysis of quantum programs. In: 2023 IEEE/ACM 4th International Workshop on Quantum Software Engineering (Q-SE), pp. 42–49 (2023). https://doi.org/10.1109/Q-SE59154.2023.00013
34. Zhao, P., Wu, X., Li, Z., Zhao, J.: Qchecker: detecting bugs in quantum programs via static analysis. In: 2023 IEEE/ACM 4th International Workshop on Quantum Software Engineering (Q-SE), pp. 50–57. IEEE (2023)

Advances in Adapting Memory-Bound CFD Computations to RISC-V Multicore Architecture

Tomasz Olas[1]([⊠])[iD], Lukasz Szustak[1][iD], Roman Wyrzykowski[1][iD], Mateusz Olas[1], and Marco Lapegna[1,2][iD]

[1] Department of Computer Science, Częstochowa University of Technology, Częstochowa, Poland
{olas,szustak,roman}@icis.pcz.pl
[2] University of Naples Federico II, Naples, Italy
marco.lapegna@unina.it

Abstract. This paper tackles the challenge of adapting HPC codes to RISC-V architecture for real-world applications with memory-bound numerical codes. The Multidimensional Positive Definite Advection Transport Algorithm (MPDATA) application is the code we study as a use case. This work explores whether the methodology developed in our previous works for Intel and AMD x86 architectures can address performance trade-offs and bottlenecks of multicore RISC-V computing platforms while executing the memory-bound MPDATA code. The explored platforms include: (i) Banana Pi BPI-F3 low-power platform, and (ii) Milk-V system with the 64-core Sophon SG2042 processor. Special emphasis is given to efficient vectorization and using lower-precision computations. Besides performance, energy consumption is studied as well.

Keywords: RISC-V · SG2042 · SpacemiT K1 · CFD · MPDATA application · memory-bound codes · porting applications

1 Introduction

RISC-V is an open standard Instruction Set Architecture (ISA) that enables the royalty-free development of CPUs and a common software stack [6]. Following this community-driven ISA standard, a very diverse set of CPUs suited to a range of workloads have been, and continue to be, developed. While RISC-V has already become popular in some fields, it has yet to gain traction in general-purpose computing, including HPC and AI/ML. In particular, recent advances in RISC-V make it a more realistic proposition for HPC workloads than ever before. An example is the vectorization extension, which gives essential performance advantages for HPC codes but was only standardized in early 2022 as RVV 1.0, so we are only now seeing CPU designs fully implementing this extension [3,6].

© The Author(s), under exclusive license to Springer Nature Switzerland AG 2025
M. H. Lees et al. (Eds.): ICCS 2025, LNCS 15905, pp. 151–166, 2025.
https://doi.org/10.1007/978-3-031-97632-2_11

At the same time, the performance of publicly available RISC-V CPUs is still behind even mobile x86 and ARM CPUs, but developments in this area are progressing rapidly. Since the RISC-V software stack includes all necessary tools for application development, it is of considerable interest to study porting real-life codes to computing platforms based on RISC-V. Knowledge gained in this way will allow application programmers to identify bottlenecks in existing approaches to mapping and optimizing codes for HPC architectures, considering the characteristics of available computing platforms and the software stack supporting them. Furthermore, the lessons learned in this way can provide helpful feedback for future hardware and software solutions developers.

This work tackles the challenge of adapting HPC applications to RISC-V platforms for real-life problems with memory-bound codes, for which memory performance is the main factor affecting computation time [17]. The code we study as a use case implements the Multidimensional Positive Definite Advection Transport Algorithm (MPDATA) - a CFD (computational fluid dynamics) algorithm that allows numerical modeling of advection transport phenomena [10].

The paper is organized as follows. Related works are discussed in Sect. 2. Sections 3 and 4 outline the studied RISC-V platforms and MPDATA application, respectively. Section 5 introduces the parallelization methodology for MPDATA on RISC-V platforms, while the vectorization of codes is described in Sect. 6. The performance evaluation of MPDATA on RISC-V is presented in Sect. 7. Section 8 deals with using single precision and evaluating energy efficiency, while Sect. 9 concludes the paper.

2   Related Works

In recent years, the research community has been actively investigating the capabilities of the RISC-V architecture. Most of the papers focus on the development of the ISA, and applications of RISC-V in some areas like embedded and edge computing, with less attention paid to optimizing the performance of parallel codes on RISC-V CPUs. However, with the development of high-performance RISC-V platforms, bridging the gap between the HPC community and RISC-V technology has become increasingly relevant.

At the moment, not many studies have been published regarding performance analysis and optimization on RISC-V CPUs. In particular, an overview of RISC-V vector extensions and the corresponding computing platforms (at the end of 2022) is given in [6]. In [11], the authors present benchmarking results of OpenFOAM, one of the most widely used frameworks for scientific simulations, comparing the performance and power consumption across devices with ARM and RISC-V architectures. However, only a single-core RISC-V processor is tested, and the analyzed CFD code is not subject to optimization. Paper [19] explores an important computing kernel, the Fast Fourier Transform (FFT), demonstrating that RISC-V-specific optimizations can significantly speed up calculations. In [3], the authors optimized a production CFD code kernel to run efficiently on the FPGA emulator of a RISC-V CPU with long-vector capabilities. Thus, the studies of various authors demonstrate the significant potential of RISC-V technology for HPC while emphasizing the need for new developments in hardware and software optimization methods, using real-world applications.

This work studies MPDATA, a CFD algorithm that represents a general approach to modeling complex geophysical flows from micro to planetary scales and one of the main parts of the EULAG multiscale fluid model [9]. In our previous papers [12,13,15], we proposed an adaptation methodology that allowed us to develop the automatic transformation of the memory-bound MPDATA code. The resulting MPDATA code has been carefully optimized for achieving scalable, high performance on ccNUMA multicore platforms with Intel processors of various generations [13,14] and AMD EPYC Rome architecture [16].

However, there is still a lack of RISC-V-specific optimizations of real-life stencil-based parallel codes, including MPDATA. This research has been conducted in this direction, exploring whether the proposed methodology can address performance trade-offs and bottlenecks of resource-constrained multicore RISC-V platforms while executing the memory-bound MPDATA code. The studied platforms are based on two state-of-the-art commodity CPUs with opposing characteristics in terms of performance and power requirements; they also differ in the vector extension version. While the first CPU has not been covered in the literature so far, the second one has only been studied in papers on optimizing FFT [19] and in works published by Nick Brown et al. on benchmarking the SG2042 CPU using RAJA and NPB suites [5].

3   RISC-V Computing Platforms

Banana Pi BPI-F3 Low-Power Platform
Banana Pi BPI-F3 is an industrial-grade RISC-V development board powered by the SpacemiT K1 RISC-V CPU [2], including eight 64-bit cores operating at a frequency of 1.05 GHz and providing an eight-stage in-order dual-issue pipeline execution. This CPU, launched in late 2023, adheres to the RISC-V 64GCVB architecture and RVA22 standard. Despite a relatively low performance, the attractiveness of this board for the HPC area is that SpacemiT K1 is the world's first commodity processor supporting the vector extension RVV 1.0. SpacemiT K1 provides a 256-bit vector length VLEN with a 128-bit x 2 execution width.

Apart from the in-order execution instead of the out-of-order one adopted in x86 CPUs, the main distinctive feature of the RISC-V platform is a much simpler on-chip memory hierarchy with only L1 and L2 caches, without an L3 cache. Every core has L1 instruction and data caches, each with 32KB. The shared 1MB L2 cache is divided into two 512KB banks. The board integrates 4GB of LPDDR4-2666 memory (up to 16GB) with a single memory controller, providing a modest bandwidth of up 10.6GB/s. At the same time, the important advantage of the platform is the use of the low-power CPU with a TDP of ~3-5W [2].

Milk-V Pionier Platform with 64-core Sophon SG2042 Processor
Milk-V Pioneer is a developer motherboard based on the 64-core Sophon SG2042
RISC-V CPU in a standard microATX form factor [1]. SG2042 is the first mass-
produced, commodity-available, high-core count RISC-V CPU designed for HPC
workloads [5,18]. It runs at 2GHz and is organized in 16 clusters of four XuanTie
C920 cores. Clusters are connected through the network-on-chip (NoC) with a
2D mesh topology. Every 64-bit core, designed by T-Head, adopts a 12-stage out-
of-order multiple-issue superscalar pipeline execution, supporting the RV64GCV
instruction set.

Each C920 core has L1 instruction and data caches, each with 64KB, while
1MB of L2 cache is shared across a cluster of four cores. All cores in the CPU
share 64 MB of the system-level L3 cache, composed of 16 slices connected
through the NoC. SG2042 integrates four DDR4-3200 memory controllers and
32 lanes of PCIe Gen4. The board has 128GB of DDR4-3200 memory, providing
a maximum bandwidth of 102.4 GB/s - ten times higher than the previous
one. At the same time, the board consumes much more energy since the typical
power demand of SG2042 is 120W [1]. Finally, unlike the SpacemiT K1 CPU,
the SG2042 C920 core provides only version 0.7.1 of the vectorization extension
with a vector width of 128 bits. What is important is that, opposite to version
1.0, mainline compilers like gcc and Clang do not support RVV 0.7.1.

4   Overview of MPDATA

MPDATA corresponds to the second-order accurate nonoscillatory iterative algorithms and is defined using a finite-difference scheme over structured rectilinear
grids. MPDATA solves the advection of a non-diffusive quantity Ψ in a flow field:

$$\partial\Psi/\partial t + \mathrm{div}(V\Psi) = 0, \tag{1}$$

where V is the velocity vector [8]. In this work, we focus on modeling 3D advection problems, when MPDATA is defined in a 3D domain of sizes $n \times m \times l$
according to $i-$, $j-$, and $k-$dimensions, respectively.

In general, MPDATA is intended to run long simulations that engage even
many thousands of time steps. Each step takes five 3D arrays as input and
returns a single 3D array reused in the next step. Each step performs a series of
17 kernels, depending on each other [13]. Every kernel is a 3D stencil code that
updates all elements of its output array using a particular pattern.

5   Parallelization of MPDATA on RISC-V Platforms

Methodology for Adapting MPDATA to Multicore Architectures
MPDATA executes a set of stencil kernels with heterogeneous patterns. In the
basic version of the parallel code, kernels are executed sequentially, and each
kernel is processed in parallel using OpenMP. Data parallelism and vectorization

are employed to distribute kernels across cores and vector units. Particularly, `#pragma omp for` directive across outer-most loop ($i-$dimension) is applied to split loop iterations among cores, and `#pragma omp for simd` directive allows us to incorporate vectorization along inner-most loop ($k-$dimension).

Since the basic version is not optimized for cache reusing, its performance is limited by the main memory bandwidth. Consequently, the low operational intensity of each kernel does not allow us to utilize modern CPUs well. In our works [12,15], suitable optimizations were proposed to exploit multicore ccNUMA platforms more efficiently. The resulting methodology for adapting MPDATA to such platforms consists of the following optimizations [13]:

1. *(3+1)D decomposition* explores spatial blocking across kernels, using overlapped tiling with redundant computations. Moreover, loop fusion is used to group all kernels into five packages of kernels. Besides increasing the computational intensity, this approach reduces the main memory traffic and efficiently utilizes L3 and L2 levels of the cache hierarchy.
2. *Partitioning cores into work teams* relieves the overhead of data traffic within the cache hierarchy of the ccNUMA system by setting groups of cores – MPDATA work teams, also called *islands of cores*. The price for mitigating this overhead is the replication of some computations.
3. *Data-flow synchronization* – the aim is to reduce the cost of synchronization by synchronizing only interdependent threads following the data dependencies between the kernels instead of using the barrier approach.
4. *Vectorization of MPDATA kernels* – possible approaches include automatic vectorization, using intrinsics, or even assembly.

In order to parallelize the MPDATA workload across computing resources, the MPDATA domain is evenly split into S sub-domains of size $\frac{m}{S} \times n \times l$, processed in parallel by S hardware teams of cores (islands of cores) available in a given ccNUMA platform. Every team processes a given sub-domain following the (3+1)D decomposition. Each sub-domain is partitioned into blocks with a size that enables efficient utilization of L3 and L2 caches. The successive blocks are processed sequentially, one by one. Each block exploits data parallelism across $i-$ and $j-$dimensions to distribute workload among C_T cores of a given work team. As a result, each MPDATA block is partitioned into a set of C_T sub-blocks. Finally, the vectorization is performed along $k-$dimension for appropriate chunks of data arrays corresponding to the sub-blocks.

Transferring the Methodology to RISC-V Platforms

The expected use of the presented methodology depends on the platform features related primarily to the processor architecture. The lack of ccNUMA domains in the SpacemiT K1 CPU and its simple memory hierarchy with only two-level caches do not justify the usage of the islands-of-cores partitioning. At the same time, an important advantage of this CPU is the possibility of using a compiler-supported vectorization, either automatic or with intrinsics.

Unlike SpacemiT K1, the SG2042 CPU has the three-level cache hierarchy with a reasonably large L3 cache, and what is important is that this last-level cache is divided into slices distributed among four-core clusters connected through the network-on-chip with 4×4 mesh topology. The network is also used to connect four DD4 memory channels. These architecture features justify the need to consider all four optimizations the adaptation methodology provides, including the islands-of-cores partitioning. Moreover, it is advisable to study the scalability of MPDATA execution depending on how the MPDATA workload is distributed across 16 clusters of SG 2042. Finally, the lack of compiler support for version 0.7.1 of the vector extension forces us to incorporate manual vectorization in assembly language to utilize the resources of vector hardware.

6   Vectorization Using Various RVV Extensions

Using vector (or SIMD) unit has yielded notable performance improvements for Intel and AMD processors [13]. In RISC-V processors, the vector (or"V") extension of ISA is dedicated to supporting vectorization. The studied platforms implement different versions of this extension - either the obsolete 0.7.1 version for the SG2042 CPU or the up-to-date RVV 1.0 extension for SpacemiT K1, which forces us to use different approaches. In both cases, the vectorization is carried out for all 17 MPDATA kernels grouped into five packages [12,13].

SG2042 with RVV 0.7.1 Extension

Due to difficulties in finding a compiler that supports at least vector intrinsics for this CPU, we use a manual vectorization of code fragments written in assembly. For this aim, the gcc compiler in version 9.2.0 is used, allowing us to compile the assembly code corresponding to the `xtheadvector` extension adopted by this processor. Here, this gcc compiler is used exclusively for compiling assembly code fragments, while the remaining C++ code is compiled by the gcc 13.2.0 compiler, which also handles the linking stage for the entire code.

Figure 1 presents an example of vectorization in assembly language for a function corresponding to a fragment of kernel 4. The code includes three stages: (i) initialization, (ii) processing loop, and (iii) updating pointers and loop control. First, the processing range for a given thread is initialized with `lCoreStart` passed to `t0` register. The loop begins with calculating the remaining range to be processed (`t1 = lCoreEnd - t0`) and setting the vector length VL in `t2` register. Next, the kernel operations are performed: loading data from memory (`tmp_A[k]` and `tmp_A[k+1]`) into vector registers v1 and v2, performing subtraction v3 = `tmp_A[k+1]` - `tmp_A[k]`, and then writing the result to `tmp_f3` array in memory. In the third stage, the data pointers (`tmp_A` and `tmp_f3`) are advanced by the number of processed elements, while index `t0` is incremented by VL. The loop is repeated until the entire range has been processed. The other functions being vectorized have a similar structure, but typically, they handle more arguments (up to 24) and perform more complex operations.

A major challenge for assembly vectorization is optimally using registers to avoid accessing memory. To this end, unused registers such as `s0` and `gp` are leveraged. Their states are saved on the stack before the function begins and restored upon completion.

SpacemiT K1 with RVV 1.0 Extension

Since the state-of-the-art versions of Clang and gcc compilers incorporate vectorization support for the RVV 1.0 extension, it becomes possible to burden a compiler with automatic code vectorization. The codes of MPDATA kernels must be suitably modified for this aim - in a way similar to x86 [13]. Clang directives, including #pragma clang loop vectorize(enable), are used to enforce vectorization and pass additional configuration parameters. We also leverage compiler-supported vectorization with intrinsics to improve the performance obtained by auto-vectorization. This solution makes programming vector operations much more productive than using assembly.

```
        # Arguments:
        # a0: pointer to tmp_A (input)
        # a1: pointer to tmp_f3 (output)
        # a2: lCoreStart (starting index)
        # a3: lCoreEnd (end index)
        # Initialization
        mv t0, a2            # t0 <- lCoreStart (current index)
        # Processing loop
loop:
        sub t1, a3, t0      # t1 <- lCoreEnd - t0
        vsetvli t2, t1, e64 # Set VL (Vector Length) for 64-bit elements
        # Load input data
        vle.v v1, (a0)      # v1 <- tmp_A[k]
        addi t3, a0, 8
        vle.v v2, (t3)      # v2 <- tmp_A[k+1]
        vfsub.vv v3, v2, v1 # v3 <- tmp_A[k+1] - tmp_A[k] = v2 - v1
        vse.v v3, (a1)      # Store the result into tmp_f3
        # Updating pointers
        slli t4, t2, 3      # t4 <- VL * 8 (data size in bytes)
        add a0, a0, t4      # Advance tmp_A pointer by VL * 8 bytes
        add a1, a1, t4      # Advance tmp_f3 pointer by VL * 8 bytes
        add t0, t0, t2      # Increment t0 index by VL
        # Loop control
        blt t0, a3, loop    # If t0 < lCoreEnd, repeat the loop
        ret                 # Return from the function
```

Fig. 1. Vectorization of a function implementing vector subtraction.

7 Performance Evaluation of MPDATA on RISC-V

7.1 Evaluation Metodology

For Banana Pi BPI-F3, the experiments presented in this section focus on evaluating the influence of optimization steps on the execution time of MPDATA.

Besides the basic, non-optimized code, the studied versions of MPDATA embrace the (3+1)D decomposition, data-flow synchronization, and three variants of vectorization: (i) automatic, (ii) using intrinsic, and (iii) in assembly. The measured execution times are obtained for the Clang compiler in version 20.0.0git. They correspond to 100 time steps and 3D grid of size 512 ÃŮ 480 ÃŮ 64.

For the second platform, the range of experiments is much broader. First, the memory throughput is tested, followed by benchmarking the platform performance and scalability with the NPB (NAS Parallel Benchmark) test suite. Then, the impact of optimizations on performance is tested together with evaluating the scalability of codes, assuming allocation of 1, 2, 3, or 4 threads per each four-core cluster of SG2042. While gcc 9.2.0 is used only for compiling assembly parts, the gcc 13.2.0 compiler handles compiling and linking for the entire code. For both platforms, computations are performed in double precision.

Table 1. Execution time for running different versions of MPDATA code on Banana Pi BPI-F3, where Std denotes standard deviation.

Code version	T_{mean} [s]	T_{med} [s]	Std	T_{Min} [s]	T_{Max} [s]
Basic	361.9	383.7	37.3	316.5	397.6
(3+1)D	235.2	236.1	4.15	229.7	241.3
(3+1)D + auto-vec	203.5	205.1	4.2	197.4	210.1
(3+1)D + intr vec	174.1	172.9	4.5	168.9	182.2
(3+1)D + asm vec	165.5	165.0	4.3	158.8	171.9
(3+1)D + df synchr	216.7	217.5	3.4	211.4	220.0
(3+1)D +df synchr +auto-vec	184.4	183.7	3.9	179.4	189.9
(3+1)D +df synchr +intr vec	160.4	161.4	3.5	154.7	166.4
(3+1)D +df synchr +asm vec	157.1	157.6	6.1	146.3	167.2

7.2 Evaluation of Banana Pi BPI-F3 Platform

Table 1 presents execution times for nine versions of MPDATA, starting with the non-optimized code and ending with three codes implementing (3+1)D decomposition together with the data-flow synchronization (denoted as df synchr) and vectorization using either auto-vectorization (auto-vec), intrinsic (intr vec) or assembly (asm vec). For each version, we provide the mean T_{mean} and median T_{med} values of the execution time obtained for 10 repeated measurements, giving results in the range from T_{min} to T_{max}.

The results in Table 1 prove the effectiveness of the proposed optimizations. The (3+1)D decomposition achieves the most significant effect, which decreases the median execution time by 1.63 times. Subsequently, even auto-vectorization is more productive than data-flow synchronization (DFS). At the same time, leveraging assembly-based vectorization yields considerably better results than auto-vectorization, almost avoiding the usage of DFS. In fact, vectorization in assembly without DFS permits decreasing the median execution time 2.33 times against the basic version, while mixing DFS with assembly-based vectorization speeds up MPDATA 2.43 times.

7.3   Evaluation of Milk-V Pionier Platform

Benchmarking the Platform
Measurements of memory throughput (in MB/s) for various thread (core) numbers (Table 2) show that the total throughput rises only slightly with increasing the thread number, which radically reduces the throughput per thread - e.g., from 2591.9 MB/s (16 threads) to 738.8 MB/s (64 threads) for `daxpy`, or 3.5 times. This decrease in throughput per thread with scaling thread number correlates with the scalability of codes from the NPB suite (Fig. 2). Only embarrassingly parallel computations are scalable up to 64 cores. Some codes feature good scalability up to 32 cores, with a slight growth for 48 cores (LU) or a fall after 32 cores (CG). Finally, benchmarks such as MG provide only a slight performance growth for 32 cores compared to 16, with a fall afterward.

Table 2. Total throughput B(P) and per thread B1(P) in MB/s for P = 16 and P = 64 threads and various functions measured on SG2042.

Function	B(16)	B1(16)	B(64)	B1(64)
Copy	36215.8	2263.5	44940.7	702.2
Scale	36438.0	2277.4	44874.6	701.2
Add	40847.4	2553.0	47626.5	743.2
Triad	38731.5	2420.7	45744.6	714.8
Daxpy	41470.4	2591.9	47285.8	738.8

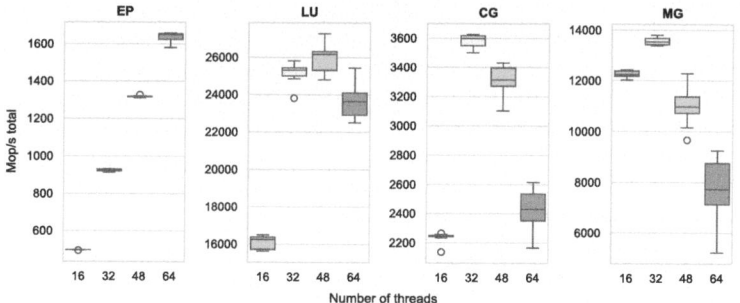

Fig. 2. Scalability of selected NPB codes (class C) on SG2042, where EP, LU, CG and MG are for embarrassingly parallel, LU decomposition, conjugate gradient and multigrid codes, respectively.

Table 3. Execution times for running various MPDATA versions on SG2042.

Code version	P	T_{mean} [s]	T_{med} [s]	Std	T_{min}[s]	T_{max}[s]
Basic	16	51.85	51.38	1.31	49.86	54.39
Basic	32	47.33	47.12	1.26	45.75	49.19
Basic	48	94.85	90.97	14.45	79.42	116.76
Basic	64	128.59	123.56	20.15	104.89	168.96
(3+1)D	16	28.20	28.34	0.33	27.52	28.50
(3+1)D	32	20.04	20.33	1.53	16.94	21.94
(3+1)D	48	20.80	21.53	1.56	17.76	22.17
(3+1)D	64	23.70	16.24	16.36	9.18	61.30
(3+1)D + DFS	16	25.49	25.55	0.18	25.23	25.71
(3+1)D + DFS	32	15.22	15.22	0.23	14.85	15.62
(3+1)D + DFS	48	13.54	13.50	0.55	12.88	14.47
(3+1)D + DFS	64	9.18	7.92	2.78	7.87	14.82
(3+1)D + DFS + vec	16	15.7	15.7	0.2	15.5	16.0
(3+1)D + DFS + vec	32	10.5	10.5	0.2	10.3	10.9
(3+1)D + DFS + vec	48	10.4	9.9	1.4	8.9	13.2
(3+1)D + DFS + vec	64	10.55	6.60	6.63	6.50	22.90

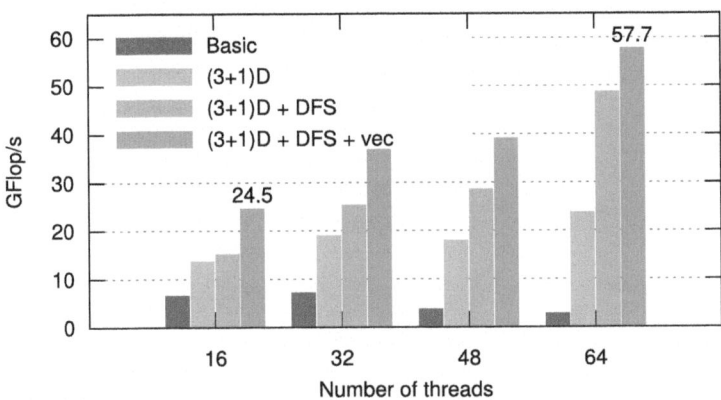

Fig. 3. Performance (in Gflop/s) of different versions of MPDATA on SG2042.

Evaluating Scalability and Efficiency of Optimization Steps

Table 3 shows execution times for various core numbers and four versions of MPDATA: the basic, non-optimized code, and three codes with growing optimization levels - from using only the (3+1)D decomposition, through leveraging also the DFS step, to the code including vectorization as well. We have not yet been able to leverage the islands-of-cores optimization step to speed up the computation. Consequently, each work team includes only a single core. Taking advantage of this optimization will be the subject of our future work.

The median values T_{med} from Table 3 are the basis of our further analysis. They are also used to calculate the performance of MPDATA codes (in Gflop/s) presented in Fig. 3. The obtained results once again confirm the effectiveness of the proposed optimizations. While the basic code is not scalable, yielding only a slight performance gain when going from 16 to 32 cores, with a radical slowdown afterward, the resulting optimized code provides quite good scalability, reducing the median execution time by 1.5 and 2.34 times for 32 and 64 cores, respectively, when compared to 16 cores. Even more appealing is the final speedup S_F yielded by the fully optimized code against the basic one: $S_F = 47.12/6.60 = 7.14$ times.

The analysis of the impact of various optimizations on performance is of considerable interest. This impact, measured by the speedup achieved by a given code against the code with a lower level of optimization, is changing with increasing the core number. Thus, the impact of the (3+1)D decomposition is increasing from 1.9 times on 16 cores, through 2.32 times on 32 cores, to 7.6 times on 64 cores. The impacts of two subsequent optimizations are interrelated. While on 16 cores, switching on DFS speeds up MPDATA only $S_{DFS} = 1.11$ times, and adding vectorization allows us to shorten the execution time by $S_{vec} =1.63$ times, for high numbers of cores, these speedups are as follows: $S_{DFS} = 1.34$, $S_{vec} = 1.45$ on 32 cores, and $S_{DFS} = 2.05$, $S_{vec} = 1.2$ for 64 cores. The increased impact of DFS can be explained by the performance bottlenecks of NoC which heavily favor increasing computing locality achieved by DFS. At the same time, the reason for the decreased impact of vectorization lies in the limited memory bandwidth.

Performance Analysis Based on the Roofline Model

Figure 4 presents the preliminary Roofline model [16] built to analyze MPDATA performance on the Milk-V platform. This model expresses the attainable performance AP (in Gflop/s) as a function of the operational intensity O (in flop/B). Figure 4 shows that the proposed optimizations allow us to increase the intensity significantly. While for 17 kernels of the basic code, $0.14 \le O \le 0.54$, we have $0.72 \le O \le 1.04$ for five packages of the optimized code. Multiplying the intensity O by the measured memory bandwidth $B_{DRAM} = 47.6$ GB/s permits us to estimate the range of AP from 34.6 to 58.1 Gflop/s. Here AP = 39.5 Gflop/s for P_1 package, which takes $\sim$40% of the total execution time, and AP = 58.1 Gflop/s for P_3 package, which takes only $\sim$15% of the total time.

Comparing these estimations of AP with the performance MP = 57.7 Gflop/s measured on 64 cores (Fig. 3), we conclude that, like x86 architectures [13,16], the packages $P_2 - P_5$ leverage the L3 cache. Its bandwidth is significantly high, leading to an adequate increase in the estimation of the attainable performance AP. Our future work will focus on reliably incorporating the impact of L3/L2 cache performance characteristics into the model.

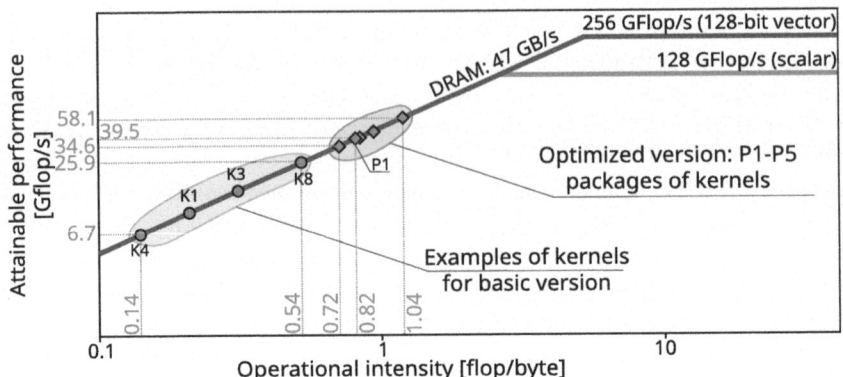

Fig. 4. The Roofline model for the double-precision MPDATA on Milk-V.

8   Using Single Precision for Improving Performance and Evaluation of Energy Efficiency

Using Single Precision

Paper [7] showed that in some application domains, using the single precision format of data instead of double precision is enough to provide the required accuracy. This transition to lower precision increases the performance of computing units and allows more efficient utilization of the available memory bandwidth.

Figure 5 compares the median execution time of the basic and fully optimized single-precision codes on SG2042. One can conclude that while using single precision for the basic code allows us only to slightly shorten the execution time - from $T_b^d(32) = 47.12$ s for double precision to $T_b^s(32) = 40.0$ s (both values achieved on 32 cores), the optimized code is quite scalable, accelerating the execution by 1.67 and 2.54 times for respectively 32 and 64 cores when compared to 16 cores. Interestingly, unlike the double-precision code, the contribution of DFS to this acceleration decreases with scaling the core number while the contribution of vectorization grows. Consequently, the final speedup S_F achieved by the fully optimized code versus the basic one is given by $S_F = T_b^s(32)/T_o^s(64) = 40.0/3.65 = 10.96$. An even more practically interesting result is that the transition from double to single precision in the optimized code permits improving the performance by the ratio of $T_o^d(64)/T_o^s(64) = 6.60/3.65 = 1.80$, i.e., slightly less than twice.

Evaluation of Energy Efficiency

By reducing the execution time, the proposed optimizations also allow us to decrease the energy consumed by MPDATA [14,16]. We employ the Yokogawa WT310 digital power meter to obtain accurate and reliable measurements of the energy and power consumed by the tested platforms. The power meter passes the power to the platform under the load and implements measurements in real time. The USB interface and YokoTool software allow us to collect data without a noticeable influence on energy/power measurements.

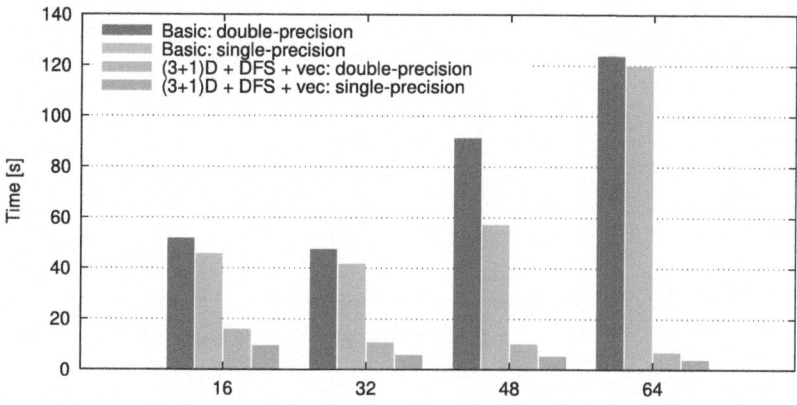

Fig. 5. Performance of single-precision codes on SG2042 versus double-precision.

Figure 6 summarizes the measurement results for double precision. For both platforms, the proposed optimization improves computations' energy efficiency significantly. These improvements are in line with the reduction in execution time. At the same time, energy gains can be noticeably higher than time reductions. For example, for 64 cores running on Milk-V Pionier, the energy consumption is decreased by more than 11.5 times compared to the basic code, while the code is executed only 7.14 times shorter. What is also interesting is that the energy consumed by Milk-V is decreasing with the increasing number of cores. In particular, the energy consumed for 64 cores is 1.79 and 1.36 times lower than the energy required for 16 and 32 cores, respectively.

We also compare the energy efficiency (in MFlop/s/W) of MPDATA codes for both platforms. While Banana BPI-F3 beats the second platform for eight cores, already by using 32 cores Milk-V Pionier catches up with BPI-F3, and by exploiting 64 cores Milk-V beats the opponent by 36%. The energy efficiency evaluation is finished by analyzing the average power consumption of MPDATA versions on both platforms. This analysis shows that BPI-F3 consumes practically the same power for all versions - about 8.3 W. The power consumed by Milk-V is much higher (Fig. 7). It starts with about 85 W for all versions running on eight cores, while already for 32 cores, we can observe some increase in average power consumed by more optimized versions of MPDATA compared to the basic one. This increase is particularly visible for 64 cores when the average power consumption increases from about 100 W for the basic version to about 140 W for the most optimized code.

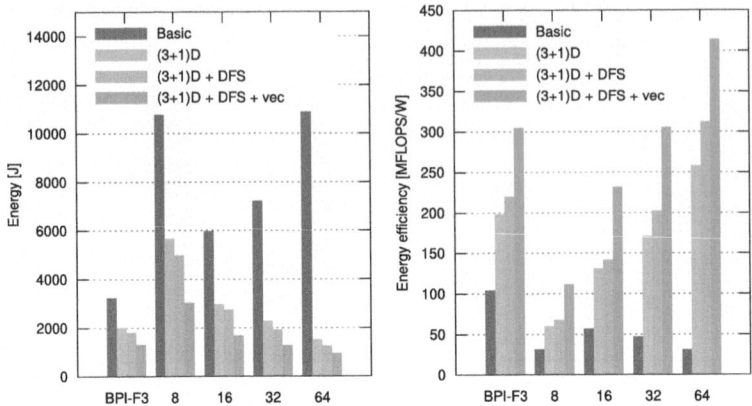

Fig. 6. Energy consumption in joules (left) and energy efficiency in MFlop/s/W (right) for double-precision MPDATA on two tested platforms.

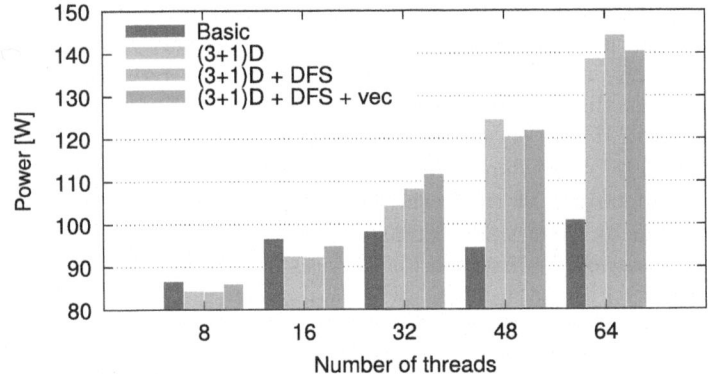

Fig. 7. Average power consumption in watts for double-precision versions of MPDATA on the Milk-V platform.

9   Conclusions and Future Work

The RISC-V architecture is rapidly expanding. The current level of infrastructure development allows porting state-of-the-art software onto existing RISC-V platforms and identifying the most promising RISC-V-specific approaches to improving performance. In this paper, we tackle the challenge of adapting HPC codes to RISC-V architecture for real-world applications with memory-bound numerical codes such as MPDATA CFD application. Our findings can be summarized as follows:

1. We demonstrate that our optimization methodology developed previously for Intel and AMD x86 architectures can efficiently address performance trade-offs and bottlenecks of two resource-constrained, multicore RISC-V platforms while executing the memory-bound MPDATA code.

2. To efficiently utilize the vector hardware of the considered CPUs, besides using the auto-vectorization for the SpacemiT K1 CPU, we develop a manual vectorization of MPDATA codes for both versions of the RVV extension and different numerical precisions.
3. The experimental evaluation of MPDATA codes on the Sophgon SG2042 CPU shows that, unlike most tests from the NPB test suite and the basic MPDATA code, our optimized code is scalable up to all 64 cores of this CPU.
4. In double precision, the code optimizations allow us to speed up computation more than 7 times compared to the original code. For single precision, this speedup is even higher, exceeding 10 times.
5. The experimental evaluation of energy consumption demonstrates convincingly the energy savings achieved by the code optimizations performed for both platforms. In particular, on the Milk-V Pioneer platform, energy consumption is reduced radically - by more than 11 times.
6. The evaluation of energy efficiency shows that while for eight cores, the Banana BPI-F3 low-power platform beats the Milk-V Pioneer platform by more than two times, already by using 32 cores, Milk-V catches up with BPI-F3, and by exploiting 64 cores, Milk-V beats the opponent by 36%.

Below, we outline three possible directions for future work. The first one concerns using the islands-of-cores step in the optimization methodology, as well as including the L3/L2 cache performance characteristics in the performance analysis of Sect. 7. The second direction involves porting other real-life applications, including ML/AI workloads such as Bayesian network learning considered in paper [4]. The third direction concerns exploiting long vectors [3] offered by upcoming RISC-V platforms.

References

1. Milk-V Pioneer. https://milkv.io/docs/pioneer/ (2024)
2. Banana Pi BPI-F3. https://wiki.banana-pi.org/Banana_Pi_BPI-F3 (2024)
3. Blancafort, M., et al.: Exploiting long vectors with a CFD code: a co-design show case. In: 2024 IEEE Int. Conf. Parallel and Distributed Processing Symposium (IPDPS). pp. 453–464 (2024)
4. Bratek, P., Szustak, L., Zola, J.: Parallel auto-scheduling of counting queries in machine learning applications on HPC systems. In: Euro-Par 2023: Parallel Processing Workshops. vol. 14325, pp. 327–333. Lect. Notes Comp. Sci. (2024)
5. Brown, N., Jamieson, M.: Performance characterisation of the 64-core SG2042 RISC-V CPU for HPC. In: ISC High Performance 2024 Int. Workshops. vol. 15058, pp. 354–377. Lect. Notes Comp. Sci. (2024)
6. Lee, J., Jamieson, M., Brown, N., Jesus, R.: Test-driving RISC-V vector hardware for HPC. In: ISC High Performance 2023 Int. Workshops. vol. 13999, pp. 419–432. Lect. Notes Comp. Sci. (2023)
7. Rojek, K., Halbiniak, K., Kuczynski, L.: CFD code adaptation to the FPGA architecture. Int. J. High Performance Comput. Appl. **35**(1), 33–46 (2015)
8. Rosa, B., et al.: Adaptation of multidimensional positive definite advection transport algorithm to modern high-performance computing platforms. Int. J. Model. Optimization **5**(3), 171–176 (2015)

9. Smolarkiewicz, P., Charbonneau, P.: EULAG, a computational model for multi-scale flows: an MHD extension. J. Comput. Phys. **236**, 608–623 (2013)
10. Smolarkiewicz, P., Margolin, L.: MPDATA: a finite-difference solver for geophysical flows. J. Comput. Phys. **140**(2), 459–480 (1998)
11. Suarez, D., Almeida, F., Blanco, V.: Comprehensive analysis of energy efficiency and performance of ARM and RISCâĂŚV SoCs. J. Supercomput. **80**(9) (2024)
12. Szustak, L.: Strategy for data-flow synchronizations in stencil parallel computations on multi-/manycore systems. J. Supercomput. **74**(4) (2018)
13. Szustak, L., Bratek, P.: Performance portable parallel programming of heterogeneous stencils across shared-memory platforms with modern intel processors. Int. J. High Performance Comput. Appl. **33**(3), 507–526 (2019)
14. Szustak, L., Wyrzykowski, R., T., O., Mele, V.: Correlation of performance optimizations and energy consumption for stencil-based application on intel xeon scalable processors. IEEE Trans. Parallel Distrib. Syst. **31**(11) (2020)
15. Szustak, L., et al.: Adaptation of MPDATA heterogeneous stencil computation to Intel Xeon Phi coprocessor. Sci. Program. **2015** (2015)
16. Szustak, L., et al.: Architectural adaptation and performance-energy optimization for CFD application on AMD EPYC rome. IEEE Trans. Parallel Distrib. Syst. **32**(12), 2852–2866 (2021)
17. Volokitin, V., et al.: Case study for running memory-bound kernels on RISC-V CPUs. In: Paralel Computing Technologies (PaCT 2023). vol. 14098, pp. 51–65. Lect. Notes Comp. Sci. (2023)
18. Wei, C.: SG2042 technical reference manual. https://github.com/milkv-pioneer/pioneer-files/blob/main/hardware/SG2042-TRM.pdf (2023)
19. Zhao, X., Zhang, X., Zhang, Y.: Optimization of the FFT Algorithm on RISC-V CPUs. In: ISC High Performance 2023 Int. Workshops. vol. 13999, pp. 515–525. Lect. Notes Comp. Sci. (2023)

Detecting Potential HIV Inhibitors Using the Cross Siamese Network

Konrad Witkowski[(✉)] [iD], Agnieszka Duraj [iD], and Piotr S. Szczepaniak [iD]

Institute of Information Technology Lodz University of Technology, al. Politechniki 8, 93-590 Lodz, Poland
konrad.witkowski@dokt.p.lodz.pl

Abstract. The issue of in silico analysis plays a crucial role in designing new medicines in modern day pharmaceutical industry. Selecting the best candidate for a new drug among countless molecules is a challenge which can be facilitated by machine learning methods. Following article addresses the problem of computational prediction of Human Immunodeficiency Virus (HIV) inhibition level among molecules. We introduced the cross siamese network (CSN) - a novel architecture based on siamese neural network - generating an embedding aiming to enhance the prediction process. The proposed neural net is a hybrid type model which combines embeddings generated from several subnetwork trained in estimating HIV inhibition level and other chemical properties like: solubility, lipophilicity or toxicological effects. To verify the efficiency of the proposed solution we trained a set of k-nearest neighbors classifiers on starting molecules' fingerprints and embeddings outputted by the experimental models. The results from this test showed that some versions of model enhanced the molecular embeddings, improving their utility for predicting HIV inhibition.

Keywords: In silico drug design · HIV inhibition prediction · siamese neural network

1 Introduction

It is estimated that at the end of 2023 there were 39.9 millions people with HIV, 65% of them living in the WHO African Region [25]. The virus targets the human immune system making it easier for various infections to attack the transmitter. The late HIV infection stage is called acquired immunodeficiency syndrome (AIDS). The disease is not curable, however with proper treatment it is considered to be a manageable health condition. Patients who undergo antiretroviral therapy may in fact lower the level of HIV in their blood [11]. We hope that our work may help in finding better medicines for HIV or any other disease.

In this paper, we propose a novel architecture that aims to embed an Extended-Connectivity Fingerprint (ECFP) [16] molecule representation in such a way that the subsequent classification process may be improved. The main idea consists in combining outputs of several submodels, each trained to recognize other chemical features (e.g. toxicologial effects), in one common embedding. We hypothesize that this procedure may not only lead to more information-rich vector representations but also spare resources for the laboratory experiments necessary to indicate exact measures of additional chemical features.

The main objective of this experiment is to verify whether our main model is able to make use of auxiliary information gathered from the submodels and, in effect, improve the classification of potential molecules-HIV inhibitors. For this task, we collected data available through MoleculeNet [24] - a publicly available repository of datasets containing molecules and their explored chemical parameters.

This paper is organized as follows. At the beginning of the article we present some of the solutions inspired by siamese neural network used to predict the potential inhibition of HIV and other bioactivities among molecules. In the next part we describe the siamese neural network and explain how it contributed to the architecture of CSN. The end of the article is dedicated to the experiment itself, its results and conclusions.

2   Related Works

Similarly to other publications [3,27] our network uses similarity metric to learn molecule embedding for better prediction. Attempts to improve the siamese network in cheminformatics can be performed in various ways, which means that the novelties may be differentiated for example by such factors as training data selection [27], model architecture [2] or combining multiple predictors' outputs creating this way a hybrid model [3]. In our case CSN fulfills the last 2 criteria.

Zhang et al. [27] introduced similarity based pairing of molecules dedicated for training the siamese network for regression tasks. The selection of training pairs was orchestrated by Tanimoto similarity calculated on ECFP fingerprints of molecules - each molecule from training set was assigned with its most similar counter part. For testing the new method the authors used 3 physiochemical datasets: lipophilicity [7], freesolv [17] and ESOL [8].

In [3] Altalib MK et al. attempted to improve the retrieval recall by connecting models from their earlier publications [2] using different variants of decision fusion layers and features fusion layers. Each of 4 hybrid constructions consists of 2 versions of authorial siamese neural net. To create an enhanced molecule representation a molecular fingerprint is processed parally by 2 submodels ending up with 2 feature vectors which are merged by a features fusion layer. The role of calculating the similarity measurement lies in a decision fusion layer. The effectiveness of constructed variants was tested on MDL Drug Data Report [1] and Maximum Unbiased Validation MUV [21].

Paykan et al. [10] approached the challenge of predicting the bioactivity level of previously unseen molecules in a slightly different manner. Instead of training a model to specify similarity between 2 molecules they proposed an authorial solution called BioAct-Het which calculates the likelihood of association between the molecules structure and bioactivity class. BioAct-Het is a heterogenous siamese neural net which merges molecules' embeddings generated by preatrained graph convolutional models (loaded from DGL-LifeSci [14]) with a vector representing predicted bioactivity class. To come out with an embedding for a single bioactivity class the authors introduced Bio-Prof - a novel method mapping Morgan Fingerprints of molecules inside of the training dataset into a single vector whose indices are indicators of how significant a given substructure for occurrence of an explored bioactivity is. The following datasets were used to test the efficacy of the solution: Sider [13], Tox21 [19] and MUV [21].

Li TH et al. [15] proposed a siamese neural net inspired model SNRMPACDC to estimate the synergy value of drug combination on different tissue types cell lines. Data for training and testing the model was downloaded from large-scale cancer screeing uploaded by Merck & Co [20]. SNRMPACDC is a 2-piece model from which each unit is responsible for something else. The first part is a siamese neural net which transforms ECFP fingerprint, physicochemical properties and binary toxicophores representation of a molecule into a drug feature vector. Additionally, vectors from both branches of siamese neural net are mixed by Random Matrix Projection (original solution which enabled the model to calculate the interaction potential between 2 molecules) and merged into a single vector representing drug combination features. The second part focuses on utilizing the matrices depicting cell line genomic features and mutation features. These 2 matrices are preprocessed and run through convolutional neural network module, resulting in cell line features. The outputs of 2 part of SNRMPACDC are combined together by Hadamard product and processed by multi-layer network to obtain the synergy value.

3   Model Architecture

Cross siamese network is based on the idea that the distance between molecules with similar properties should be smaller than the distance between molecules with radically different characteristics. For creating such representations on hyperplane we have inspired ourselves by the siamese network designed by Bromley et al. [6] and its further applications in chemistry [2,3,27].

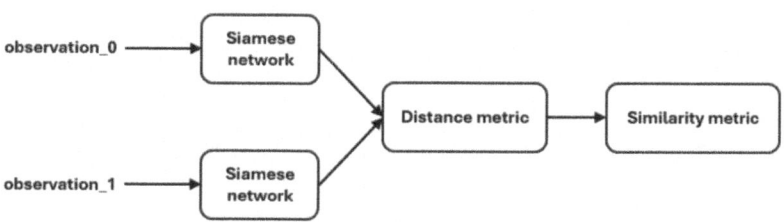

Fig. 1. Simplified architecture of siamese neural network

Siamese neural network consists of 2 parallel neural networks that share the same weights. Calculating the similarity measurement between 2 observations starts with inputting them into separate parts of siamese neural network to generate their embeddings whose remoteness is calculated by distance metric like 12-norm. The task of turning the distance between 2 samples into their similarity level belongs to the similarity metric which is performed at the end of the whole procedure. The simplified architecture of siamese neural network was shown in Fig. 1.

3.1   Cross Siamese Network

Cross siamese network is a hybrid model which merges output embeddings of each of n auxiliary siamese neural networks into 1 feature vector $\boldsymbol{f}_{merged}$ at fusion layer phase. The feature vector $\boldsymbol{f}_{merged}$ is calculated as follows:

$$\boldsymbol{f}_{merged} = \sum_{i=1}^{n} \boldsymbol{f}_i \boldsymbol{w}_i, \tag{1}$$

where $\boldsymbol{f}_i$ are feature embeddings generated by submodels and $\boldsymbol{w}_i$ are learnable weights. The main goal of the feature fusion layer consists in generating an enriched vector which in further steps should facilitate the predictions. Nevertheless, such an uncomplex way of combining multiple outputs of submodels may not be sufficient for preserving all of the gained information and result in a too mixed up embeddings.

Before the final output is created the consolidated vector generated by feature fusion layer is processed by convolutional blocks and linear block.

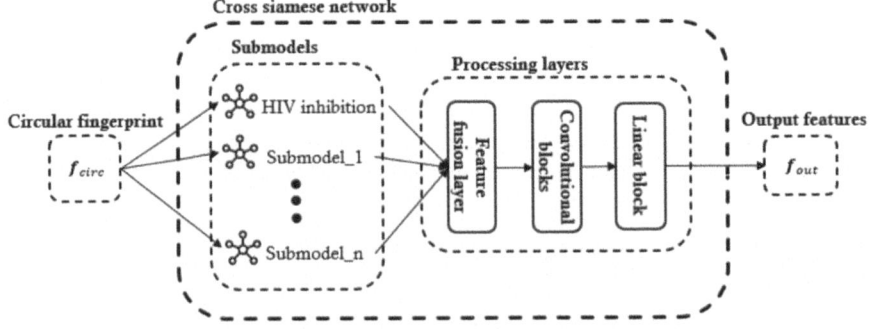

Fig. 2. Architecture of the cross siamese network

The Fig. 2 shows the architecture of a network based on cross siamese network approach. This network is designed to analyze chemical and biological data, such as biological activity related to HIV inhibition, which is why the "circular fingerprint" is defined at the top of the diagram. Circular fingerprint (CF), which was more extensively described in Sect. 4.2, is a way of representing a molecule in a numerical form, based on local molecular structures. Submodels, created using CFs, analyze various aspects of molecular features. The HIV inhibition submodel is a siamese neural net that was trained using molecules tagged with their HIV inhibition level. The remaining submodels analyze other chemical or biological properties. Before the final output is generated the data is processed by: feature fusion layer, convolutional blocks and a linear block. The 2 last structures were presented in Fig. 3. To introduce the nonlinearity in the convolutional and linear blocks we used the ReLU activation function which is described by the following equation:

$$f(x) = max(0, x). \tag{2}$$

The convolutional phase is responsible for reorganizing the merged vector and extracting more features from it. It consists of 5 convolutional blocks. Each of these blocks is constructed upon convolutional layer 1d with kernel of size 1, ReLU activation function and batch normalization.

Additionally, 2 residual connections are applied. The input and output number of channels for each convolutional block is 32 except for the last one where the is 32 input channels and 1 output channel.

The linear block, whose role consists in transforming the data into a form suitable for output predictions, is composed of linear layer, ReLU activation function and batch normalization.

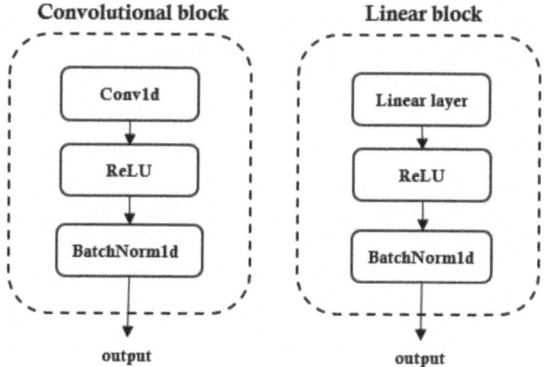

Fig. 3. Structure of convolutional and linear block

Both parts of the CSN serve the purpose of creating a goal-oriented molecular fingerprint whose form in the experiment should be designed to enhance the detection of HIV inhibitors.

3.2   Submodels

All of the auxiliary models, which we called siamese mol nets (SMNs), have the same kern substructure responsible for producing the feature vectors. The kern substructure was presented in Fig. 4. It is based on 3 pairs of linear and batch normalization layers. The input number of features of the first linear layer is 2048 which corresponds to the length of fingerprints. However, its number of output features is 4096 and the size of the processed vector stays this way until the end, which means that the output feature vector also has 4096 elements.

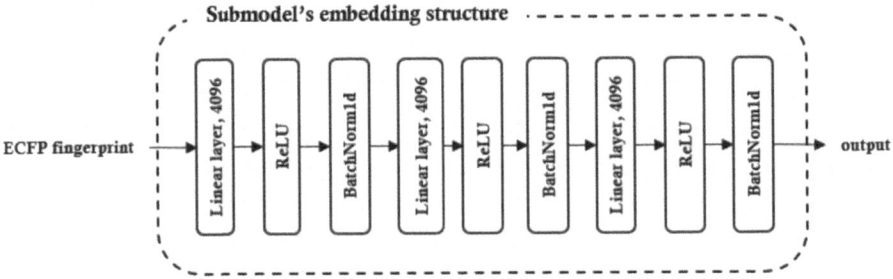

Fig. 4. Schema of kern substructures of the auxiliary models

In the case of the classification task, the structure in Fig. 4 suffices for the entire model. On the other hand, the regression task requires additional layers whose goal is to stack feature vectors coming from 2 molecules, calculate their averages for each index, and using several linear layers generate the predicted real value output. The first linear layer reduces the 4096 element vectors to their equivalents of length 2048. This number is further condensed to 64 by the second linear layer. The output receives its final shape by the third and final layer, which generates a single-element vector. The described processing flow was shown in Fig. 5.

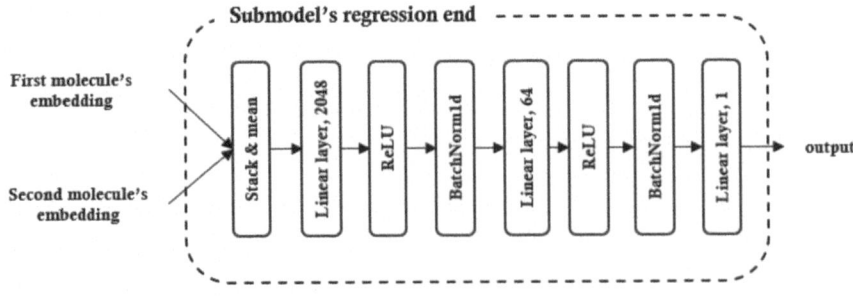

Fig. 5. Schema of submodel's regression end

In both scenarios, the shape of the output has a significant value for the task the model was trained for. In the case of classification, a large number of output vector elements allows to embed it in the molecular representation space more accurately. In contrast, the regression end outputs a single-element vector which is the estimated distance between two molecules' embeddings.

4   Experiment

The goal of the experiment consisted in testing the new architecture in several configurations and verifying whether its embeddings can enhance the classification of potential HIV inhibitor. To perform the experiment, we decided to train the models on publicly available molecular datasets. Finally, we gathered the results from the whole experiment and described them in the next chapter.

4.1   Datasets

To compose the CSNs we chose to use the datasets from MoleculeNet. In order to elastically download the molecules and their ECFP fingerprints we utilized the DeepChem [23] library written in Python.

The starting point for our experiment was the HIV dataset provided by Drug Therapeutics Program which tested 40 000 molecules for inhibition of HIV replication and assigned to each of the explored compound a label describing their level of activeness: confirmed active, confirmed moderately active and confirmed inactive. The authors of DeepChem chose to merge the first 2 labels into a single category i.e. 1 [26], while treating all remaining samples as 0. For training and testing purposes, we used the scaffold splitter - a dedicated cheminformatics tool that identifies the core substructures (scaffolds) of the molecules. Although the scaffold splitter was designed to resemble real-world conditions in cheminformatics, it has certain limitations. This method divides the dataset by identifying ring structures to ensure that the training and testing sets remain distinct. However, this approach may disadvantage molecules that lack such structural features.

In order to build the CSNs we selected 3 auxiliary groups of molecules:

- LIPO is a part of data provided by AstraZeneca and stored by ChEMBL [7]. It contains 4200 molecules with their lipophilicity scores measured by octanol/water distribution. The train and test parts of the dataset were divided using scaffold splitter.
- Delaney contains information about solubility of 1128 molecules. The train and test parts of the dataset were divided using scaffold splitter.
- TOX21 [19] dataset consists of molecules and their toxicity measurement understood as compound activity in all nuclear receptor signaling pathways. For the purpose of experiment we choose following categories: androgen receptor (NR_AR), androgen receptor ligand binding (NR_AR_LBD), androgen receptor aryl hydrocarbon receptor (NR_AR_AHR) and aromatase receptor (NR_AROMAT). The train and test parts of the dataset were divided using scaffold splitter.

4.2 Extended-Connectivity Circular Fingerprint

The ECFP is a molecular fingerprint created using a refinement of Morgan Algorithm [18]. The final fingerprint f is a vector of a predefined length S. Its form is a product of R layers that during processing of a single molecule update the information assigned to single atoms. The process of generating an ECFP fingerprint f (of length S) of a molecule operates as follows:

1. Initialize the final fingerprint with zeros: $f \leftarrow \mathbf{0}_S$
2. For each layer L loop over each atom a and perform following operations:
    (a) concatenate atom feature vectors of neighbors of atom a to an auxiliary vector v: $v \leftarrow [r_a, r_0, ..., r_n]$
    (b) update atom's a feature vector: $r_a \leftarrow hash(v)$
    (c) retrieve an index i of fingerprint f to update its value to 1: $i \leftarrow mod(r_a, S)$
3. Output fingerprint f

The final fingerprint is a binary vector whose final form is independent of the task for which it was created.

4.3 Training Process

Depending on the predicted value the training process required different loss function which we describe in further details in the following subsection.

For the LIPO and Delaney datasets, where the labels have real values, we decided to measure the quality of the models with the mean square error:

$$L(y_i, \hat{y}_i) = \sum_{i=1}^{n} (y_i - \hat{y}_i)^2, \tag{3}$$

where y_i stands for the absolute value of difference between labels of 2 selected molecules, $\hat{y}_i$ is the output of model and n is the number of pairs in mini batch.

For other datasets whose task consisted in binary classification we used triplet margin loss [4] which can be formulated in the following way:

$$L(a_i, p_i, n_i) = max\{\|a_i - p_i\|_2 - \|a_i - n_i\|_2 + margin, 0\}, \tag{4}$$

where a_i is the i-th anchor, p_i is the i-th positive sample, n_i is the negative sample. In case of the experiment the margin was set to 1. To force the models to focus on more challenging examples we decided to use the hard batch-hard mining which for a given observation looks inside of a batch for the nearest negative sample and for farthest positive sample. Additionally, we made sure that the samples from the minority class were equally split across batches. We hoped that this way we could achieve a more stable training process.

To mitigate the risk of overtraining the models in classification task we came up with an idea of boosting the loss of those triplets where the anchor came from the minority class as follows:

$$
w_i = \begin{cases} \frac{\text{number of negative samples}}{\text{number of positive samples}}, & a_i \text{ is a positive sample.} \\ 1, & a_i \text{ is a negative sample} \end{cases} \tag{5}
$$

We designed the experiment so that one part of the classification models was trained with stable weights and the other part with boosted weights. The training process using these 2 weighting strategies required different number of batch samples to obtain a controllable loss decrease flow. Specifically, for models dedicated for HIV data, we set the batch size to 32 in the boosted weights scenario and 128 in the stable weights scenario. All training processes were performed by Adam optimizer [12], with the learning rate set to 1e-5. The initial weights of the convolutional and linear layers in all models were generated using the Xavier uniform distribution [5], with the gain of 1.0. The starting biases of all layers were manually set to 0.01.

4.4 Evaluation Methods

Our primary goal was to determine whether the models could effectively map ECFP fingerprints to equivalent vectors, ensuring that vectors from the same class were closer together, while those from opposite classes were farther apart. To evaluate the quality of the embeddings generated by the models, we used the k-nearest neighbors algorithm for each version of both SMN and CSN. This procedure consisted in training the classifier on embeddings outputted by an explored type of siamese neural network and testing it on the equivalent embeddings derived from the test dataset.

The k-nearest algorithm [22] is a nonparametric classification method. Given a training set $D = \{(x_n, y_n)\}_{n=1}^{N}$ an observation x_i, where $i \in [1, N]$, will be assigned with the most frequent class among k nearest neighbors. For calculating the distance we chose the l2 metric. In the experiment we made use of 2 variants of k-nearest algorithm, both trained on the embeddings from the training set. The first type was set to 4 and the second one to 3 nearest neighbors. This approach ensured that an observation was assigned to the class with the majority representation among its 3 nearest neighbors.

In order to estimate the performance of the classifiers, we used metrics such as accuracy, precision and recall [9].

The accuracy, assessing the overall quality of the classification process, is given as follows:

$$
accuracy = \frac{TP + TN}{TP + TN + FP + FN}, \tag{6}
$$

where TP is the number of true positives, TN is the number of true negatives, FP is the number of false positives and FN is the number of false negatives.

For measuring how well the algorithm works strictly on the embeddings from the minority class (in this case the positive one) we used:

$$precision = \frac{TP}{TP + FP} \tag{7}$$

and recall formulated in the following way:

$$recall = \frac{TP}{TP + FN}. \tag{8}$$

The calculated metrics belong to the standard procedures for measuring the classification process. Thanks to them, it is possible not only to assess the prediction process but also to notice the potential threat of overemphasizing the majority class.

5   Results

We divided the description of the results obtained from the experiment into 2 subsections addressing the weighting strategies. To measure how well the SMN and CSN performed, we used the results of the k-nearest neighbors classifier (Table 1) trained on the initial ECFP fingerprints from the HIV dataset as a reference point.

Table 1. Performance metrics of k-nearest neighbor trained on starting ECFP fingerprints of HIV dataset.

Phase	Accuracy	Precision	Recall
Train	0,9683	0,6262	0,3807
Test	0,9694	0,5588	0,1462

The abbreviations in the tables that represent model types combine the underlying structure with the task for which the model was trained. The full list of created models looks in the following way:

- SMN_HIV - SMN trained on HIV inhibition data
- CSN_HIV - CSN composed of a SMN trained on HIV inhibition data
- CSN_HIV_LIPO - CSN composed of a SMN trained on HIV inhibition data and a SMN trained on lipophilicity data
- CSN_HIV_TOX_NR_AR - CSN composed of a SMN trained on HIV inhibition data and a SMN trained on compound activity in androgen receptor data
- CSN_HIV_TOX_NR_AROMAT - CSN composed of a SMN trained on HIV inhibition data and a SMN trained on compound activity in aromatase receptor data

- CSN_HIV_TOX_NR_AR_LBD - CSN composed of a SMN trained on HIV inhibition data and a SMN trained on compound activity in androgen receptor ligand binding data
- CSN_HIV_TOX_NR_AR_AHR - CSN composed of a SMN trained on HIV inhibition data and a SMN trained on compound activity in androgen receptor aryl hydrocarbon receptor data
- CSN_HIV_DELANEY - CSN composed of a SMN trained on HIV inhibition data and a SMN trained on solubility data

5.1   Stable Weights

The training process based on constant weights, whose loss scores were shown in Table 2, led to CSN_HIV_TOX_NR_AR achieving the smallest loss i.e. 0.86563. However, this version of siamese network did not produce the most effective embeddings for the k-nearest neighbor classifier. As presented in Table 3 the CSN_HIV_TOX_NR_AR was among the weakest models in terms of precision, achieving only 0.6123 on the training dataset and 0.5161 on the testing dataset. These results indicate that the embeddings generated by CSN_HIV_TOX_NR_AR perform even worse for classification purposes than the original molecular fingerprints.

Table 2. Loss results for the normal weights variant of training process

Model type	Loss
SMN_HIV	0.99962
CSN_HIV	0.89432
CSN_HIV_LIPO	1.04699
CSN_HIV_TOX_NR_AR	**0.86563**
CSN_HIV_TOX_NR_AROMAT	1.03654
CSN_HIV_TOX_NR_AR_LBD	0.93201
CSN_HIV_TOX_NR_AHR	0.97224
CSN_HIV_TOX_NR_ER	1.00304
CSN_HIV_TOX_NR_ER_LBD	1.26426
CSN_HIV_DELANEY	0.96148

The most effective molecular fingerprints were undeniably generated by SMN_HIV, enabling the k-nearest neighbor classifier to achieve a precision of 0.8668 on the training set and 0.8571 on the test set. It should be noted that in this case the recall levels obtained at the training phase (0.3011) and the testing phase (0.0462) were noticeably different. This difference can be likely attributed to the scaffold splitter, which divided the main dataset in a such way that the molecules of the same class from the training and testing phase were as structurally various

Table 3. Training and Testing performance metrics (accuracy, precision, recall) for different model variants.

Model type	Training			Testing		
	Acc.	Prec.	Rec.	Acc.	Prec.	Rec.
SMN_HIV	0.9721	0.8668	0.3011	0.9696	0.8571	0.0462
CSN_HIV	0.9673	0.6137	0.3417	0.9686	0.5110	0.1769
CSN_HIV_LIPO	0.9666	0.6255	0.2670	0.9691	0.6000	0.0922
CSN_HIV_TOX_NR_AR	0.9664	0.6123	0.3289	0.9688	0.4857	0.1632
CSN_HIV_TOX_NR_AROMAT	0.9666	0.6140	0.3433	0.9686	0.4857	0.1323
CSN_HIV_TOX_NR_AR_LBD	0.9678	0.6310	0.3117	0.9686	0.5926	0.1231
CSN_HIV_TOX_NR_AR_AHR	0.9670	0.6755	0.2281	0.9686	0.5568	0.0769
CSN_HIV_DELANEY	0.9683	0.6250	0.3856	0.9696	0.5439	0.2385

as possible. The fingerprints generated by SMN_HIV significantly enhanced the starting ECFP versions in terms of precision.

Among the CSNs, the best performance was delivered by the embeddings of CSN_LIPO and CSN_HIV_TOX_NR_AR_LBD, achieving precision values of 0.6 and 0.5926 on the test set, respectively. However, these scores are comparable with those achieved on the initial ECFP fingerprints.

In the normal weighting scenario the loss value did not align with the performance of the k-nearest classifier. The most effective neural net, SMN_HIV, reached a loss of 0.9962, which was not indicative of its ability to generate high-quality embeddings. This may suggest that the CSNs may have focused on optimizing different components of the triplet loss compared to SMN_HIV. The CSNs may owe the biggest decrease in their loss values to the minimized distance between samples from the same class. While this ensures that observations within the same class are closely grouped, it does not necessarily establish a clear distinction between different classes.

5.2   Boosted Weights

In the boosted weighting variant of loss calculation (Table 4), CSN_HIV_TOX_NR_AROMAT emerged as the most promising model, achieving a loss of 1.56074. It is worth noticing that the CSN_HIV_TOX_NR_AR, which had the lowest level under the normal weighting strategy, also performed well with a loss of 1.67667.

Nevertheless, the CSN_HIV_TOX_NR_AROMAT did not reach the best results for k-nearest classifier among models trained with boosted weights. As presented in Table 5 the CSN_HIV_TOX_NR_AROMAT's precision levels (0.6498 on the training dataset and 0.6 on the testing dataset) were surpassed by SMN_HIV's which reached 0.8788 on the training dataset and 0.7143 on the testing dataset.

Table 4. Loss results for the boosted weights of training process

Model type	Loss
SMN_HIV	2.0035
CSN_HIV	1.69165
CSN_HIV_LIPO	2.09365
CSN_HIV_TOX_NR_AR	1.67667
CSN_HIV_TOX_NR_AROMAT	**1.56074**
CSN_HIV_TOX_NR_AR_LBD	1.81694
CSN_HIV_TOX_NR_AHR	2.06221
CSN_HIV_TOX_NR_ER	1.72272
CSN_HIV_TOX_NR_ER_LBD	1.93214
CSN_HIV_DELANEY	1.68214

The best CSN (in terms of precision) turned out to be the CSN_HIV_TOX _NR_AHR which scored 0.6708 on the training dataset and 0.8235 on the testing dataset.

However, this successful score did not align with the equivalent from the scenario with stable weights. The experiment revealed that for CSNs there is no definitive pattern indicating how the neural net will perform under different weighting strategies. This suggests that, in more complex models, the observations themselves may play a more critical role in training than the weighting strategies. Additionally, the inconsistent results could also be explained by the datasets consisting of different molecules for each task.

Table 5. Training and Testing performance metrics (accuracy, precision, recall) for different model variants.

Model type	Training			Testing		
	Acc.	Prec.	Rec.	Acc.	Prec.	Rec.
SMN_HIV	0.9829	0.8788	0.6299	0.9713	0.7143	0.1538
CSN_HIV	0.9672	0.6423	0.2784	0.9677	0.4000	0.0462
CSN_HIV_LIPO	0.9671	0.6118	0.3287	0.9684	0.5000	0.1538
CSN_HIV_TOX_NR_AR	0.9685	0.6471	0.3482	0.9703	0.6333	0.1462
CSN_HIV_TOX_NR_AROMAT	0.9672	0.6498	0.2711	0.9691	0.6000	0.0692
CSN_HIV_TOX_NR_AR_LBD	0.9676	0.6267	0.3312	0.9699	0.5938	0.1462
CSN_HIV_TOX_NR_AR_AHR	0.9693	0.6708	0.3523	0.9711	0.8235	0.1077
CSN_HIV_DELANEY	0.9675	0.6153	0.3531	0.9684	0.5000	0.1231

Similarly as under the normal weighting strategy, the achieved loss did not translate into the most effective model. Moreover, none of the evaluated models from both strategies scored an accuracy below 96%, suggesting that the majority class was not overlooked. The key distinction between the variants of SMN_HIV was the relatively high recall level scored by the boosted weighting version: 0.6299 on the training dataset and 0.1538 on the testing dataset. This improvement indicates that the SMN trained using boosted weights works better at detecting rare classes.

6   Conclusions

The main conclusion is that models based on siamese neural nets are able to enhance the classification of potential HIV inhibitors. The embeddings generated by the SMN_HIVs managed to outperform the starting fingerprints leading to a significant improvement of the precision of the k-nearest neighbors obtained on the test dataset. Specifically, the classifier achieved a precision of 0.5588 using the initial fingerprints, 0.8571 with embeddings generated by the SMN_HIV trained with the stable weights, and 0.7143 with embeddings produced by the SMN_HIV trained with the boosted weights. Furthermore, the boosted weighting strategy allowed to increase the control over triplet loss function enabling a precision-recall trade-off. However, the embeddings generated by the CSNs were comparable in their classification quality to that of the initial fingerprints, raising questions about the underlying reasons for this outcome. In summary, the proposed SMN_HIV architecture, along with the weighting strategy, may find application in the search for potential drugs beyond HIV. Nevertheless, the perspective of CSN remains an important topic for further discussion.

References

1. Accelrys: Mdl drug data report (mddr). http://www.accelrys.com, accelrys Inc.: San Diego, CA, USA. Accessed 31 Oct 2021
2. Altalib, M.K., Salim, N.: Similarity-based virtual screen using enhanced siamese multi-layer perceptron. Molecules **26**, 6669 (2021). https://doi.org/10.3390/molecules26216669
3. Altalib, M.K., Salim, N.: Hybrid-enhanced siamese similarity models in ligand-based virtual screen. Biomolecules **12**(11), 1719 (2022). https://doi.org/10.3390/biom12111719
4. Balntas, V., Riba, E., Ponsa, D., Mikolajczyk, K.: Learning local feature descriptors with triplets and shallow convolutional neural networks. BMVC Proceedings pp. 119.1–119.11 (2016). https://doi.org/10.5244/C.30.119
5. Bengio, Y., Glorot, X.: Understanding the difficulty of training deep feed forward neural networks. In: International Conference on Artificial Intelligence and Statistics pp. 249–256 (2010)
6. Bromley, J., Guyon, I., LeCun, Y., Sickinger, E., Shah, R.: Signature verification using a "siamese" time delay neural network. In: Advances in Neural Information Processing Systems. vol. 6, pp. 737–744 (1993)
7. Chemdbl: Chembl3301361. https://www.ebi.ac.uk/chembl/document_report_card/CHEMBL3301361/ Accessed 30 Sept 2024

8. Delaney, J.S.: Esol: estimating aqueous solubility directly from molecular structure. J. Chem. Inf. Comput. Sci. **44**(3), 1000–1005 (2004). https://doi.org/10.1021/
   ci034243x
9. Google: classification: accuracy, recall, precision, and related metrics. https://
   developers.google.com/machine-learning/crash-course/classification/accuracy-
   precision-recall Accessed 5 Dec 2024
10. Heyrati, M.P., Ghorbanali, Z., Akbari, M., Pishgahi, G., Zare-Mirakabad, F.:
    Bioact-het: a heterogeneous siamese neural network for bioactivity prediction using
    novel bioactivity representation. ACS Omega **8**(47), 44757–44772 (2023). https://
    doi.org/10.1021/acsomega.3c05778
11. HIV.gov: What are hiv and aids. https://www.hiv.gov/hiv-basics/overview/about-
    hiv-and-aids/what-are-hiv-and-aids Accessed 26 Sept 2024
12. Kingma, D., Ba, J.: Adam: a method for stochastic optimization. In: International
    Conference on Learning Representations (2014)
13. Kuhn, M., Letunic, I., Jensen, L.J., Bork, P.: The sider database of drugs and
    side effects. Nucleic Acids Res. **44**, D1075–D1079 (2016). https://doi.org/10.1093/
    NAR/GKV1075
14. Li, M., Zhou, J., Hu, J., Fan, W., Zhang, Y., Gu, Y., Karypis, G.: Dgl-lifesci: an
    open-source toolkit for deep learning on graphs in life science. ACS Omega **6**(41),
    27233–27238 (2021). https://doi.org/10.1021/acsomega.1c04017
15. Li, T.H., Wang, C.C., Zhang, L., Chen, X.: Snrmpacdc: computational model
    focused on siamese network and random matrix projection for anticancer syn-
    ergistic drug combination prediction. Briefings Bioinform. **24**(1), bbac503 (2023).
    https://doi.org/10.1093/bib/bbac503
16. Micheli, A.: Neural network for graphs: a contextual constructive approach. IEEE
    Trans. Neural Netw. **20**(3), 498–511 (2009). https://doi.org/10.1109/TNN.2008.
    2010350
17. Mobley, D.L., Guthrie, J.P.: FreeSolv: a database of experimental and calculated
    hydration free energies, with input files. J. Comput. Aided Mol. Des. **28**(7), 711–
    720 (2014). https://doi.org/10.1007/s10822-014-9747-x
18. Morgan, H.L.: The generation of a unique machine description for chemical
    structures-a technique developed at chemical abstracts service. J. Chem. Doc. **5**(2),
    107–113 (1965)
19. National center for advancing translational studies: Tox21 data challenge (2014).
    https://tripod.nih.gov/tox21/challenge/data.jsp Accessed 1 Oct 2024
20. O'Neil, J., et al.: An unbiased oncology compound screen to identify novel combi-
    nation strategies. Mol. Cancer Ther. **15**, 1155–1162 (2016)
21. Rohrer, S.G., Baumann, K.: Maximum unbiased validation (muv) data sets for
    virtual screening based on pubchem bioactivity data. J. Chem. Inf. Model. **49**(2),
    169–184 (2009). https://doi.org/10.1021/ci8002649
22. Shi, Y., Yang, K., Yang, Z., Zhou, Y.: Chapter two - primer on artificial intel-
    ligence. In: Shi, Y., Yang, K., Yang, Z., Zhou, Y. (eds.) Mobile Edge Artificial
    Intelligence, pp. 7–36. Academic Press (2022). https://doi.org/10.1016/B978-0-
    12-823817-2.00011-5
23. The DeepChem Project: Model classes. https://deepchem.readthedocs.io/en/
    latest/api_reference/models.html Accessed 21 Mar 2024
24. The deepchem project: moleculenet. https://deepchem.readthedocs.io/en/latest/
    api_reference/moleculenet.html Accessed 29 Sept 2024
25. WHO: Hiv and aids. https://www.who.int/news-room/fact-sheets/detail/hiv-aids
    Accessed 26 Sept 2024

26. Wu, Z., Ramsundar, B., Feinberg, E.N., Gomes, J., Geniesse, C., Pappu, A.S., Leswing, K., Pande, V.: Moleculenet: a benchmark for molecular machine learning. Chem. Sci. **9**(2), 513–530 (2017). https://doi.org/10.1039/c7sc02664a
27. Zhang, Y., Menke, J., He, J., Nittinger, E., Tyrchan, C., Koch, O., Zhao, H.: Similarity-based pairing improves efficiency of siamese neural networks for regression tasks and uncertainty quantification. J. Cheminform. **15**, 75 (2023). https://doi.org/10.1186/s13321-023-00744-6

Generation of Quality Green's Function Libraries in Complex Three-Dimensional Crustal Structures by Adaptive Mesh Refinement

Kai Nakao[1](✉), Hideaki Ito[1](✉), Tsuyoshi Ichimura[1], Kohei Fujita[1], Lalith Wijerathne[1], and Muneo Hori[2]

[1] Earthquake Research Institute and Department of Civil Engineering,
The University of Tokyo, Tokyo, Japan
{k-nakao,hideakiito,ichimura,fujita,lalith}@eri.u-tokyo.ac.jp
[2] Research Institute for Value-Added-Information Generation,
Japan Agency for Marine-Earth Science and Technology,
Yokohama, Kanagawa, Japan
horimune@jamstec.go.jp

Abstract. Fault slip estimation from crustal deformation is essential for understanding earthquake mechanisms, and three-dimensional subsurface models have become incorporated in the estimation to account for the geometric and material heterogeneity in the crust. In the estimation process, a precomputed Green's function library (GFL), which represents the displacement field due to unit point loads at each receiver on ground surface, can be employed for computing arbitrary source responses through convolution of the source terms and the GFL. However, challenges remain in generating meshes adapted to the singularity of the point loads and in assessing the accuracy of the GFL. We developed a GFL computation method, Adaptive Finite Element Method for Green's Function Library (AFEM-GFL). By combining initial meshes with good element quality and a mesh refinement algorithm that is resistant to element quality degradation, our method can generate meshes well adapted to computing GFLs. The accuracy of the results is assessed through convergence of the solution and comparisons between the convergent solutions from different initial meshes. In numerical experiments on a two-layered half-space and a realistic crustal structure of Japan, accurate and convergent GFLs were obtained with moderate amount of computational resources in the settings where it was difficult to achieve with uniform meshes even using a massively parallel supercomputer. These quality GFLs will serve as a robust foundation for reliable fault slip estimation.

Keywords: Green's function library · Adaptive finite element method · Crustal structure

---

K. Nakao and H. Ito—Equally contributed to this work.

M. H. Lees et al. (Eds.): ICCS 2025, LNCS 15905, pp. 183–197, 2025.
https://doi.org/10.1007/978-3-031-97632-2_13

1   Introduction

Estimation of coseismic fault slip based on crustal deformation is crucial for deepening our understanding of earthquake mechanisms and assessing disaster risks. In the estimation, it is common to model the crust as a flat and homogeneous semi-infinite elastic medium [12] and estimate parameters of the fault model that can reproduce the observed crustal deformation. On the other hand, geometric and material heterogeneity in the crust should be considered in cases where the subsurface structure is complex. Attempts have been made to construct a three-dimensional subsurface structure model that can properly represent these features and employ numerical crustal deformation simulations by the finite element method or the spectral element method as the forward analysis in the estimation process [6]. When there is a nonlinear relationship between the fault model parameters and crustal deformation, the computational cost for the estimation can be huge performing costly numerical simulations every time the parameters are updated in the optimization process. In such cases, methods that reduce the computational costs utilizing the reciprocity theorem [5] are effective. According to the reciprocity theorem, the displacement observed at a receiver point on the ground surface due to a load applied at a subsurface source point is equal to the displacement observed at the source point due to a load applied at the receiver point. A fault slip can be approximated as a set of forces at several points in numerical simulations. If we precompute a set of the subsurface displacement field due to unit point loads applied at each receiver point on the ground surface, which we call a Green's Function Library (GFL), the displacement on the ground surface due to arbitrary fault slips can be obtained by convolution of the source terms and the GFL without performing simulations.

Challenges in constructing GFL for 3D subsurface structures include the quantitative evaluation of the accuracy of the GFL and the generation of high-quality meshes. In finite element analysis of crustal deformation due to fault slips, it is known that the size of the elements near the fault affects the accuracy of the solution [9]. Similarly, the accuracy of the GFL is expected to be affected by the elements near the loading points on the surface. Therefore, it is desirable to generate GFL using methods that allow verification of sufficiency of the mesh resolution based on the quantitative evaluation. Moreover, realistic crustal structures often have thin layers with complex geometry near the ground surface. When controlling the mesh size to refine the elements around the loading points, it is challenging to generate meshes that accurately represent the geometry while maintaining good element quality (i.e., avoiding elements with high aspect ratio) when using general-purpose mesh generators.

To address the aforementioned challenges, we developed a method to generate GFL, which we call Adaptive Finite Element Method for Green's Function Library (AFEM-GFL). Adaptive finite element method (AFEM) is a technique that iteratively conducts finite element analysis, estimates the error, and refines the mesh based on the error to accurately compute fields with significant variations [1]. In AFEM-GFL, algorithms that can protect the layers geometry and are resistant to element quality degradation are employed for initial mesh generation and mesh refinement. This allows for the generation of high-quality meshes well adapted for the computation of GFLs. Regarding the accuracy evaluation for the GFLs, the convergence of the solution can be assessed from the difference of the solutions between subsequent iterations. Furthermore, inadequate mesh quality could lead to convergent but inaccurate solutions. By comparing solutions obtained from runs with different initial meshes, it is possible to detect if any significant issues that affect the accuracy of the GFLs arise during the mesh refinement. The GFL, whose quality is assured from these evaluations, can be a robust foundation for the reliable fault slip estimation.

This paper describes the details of the AFEM-GFL method in Sect. 2 and demonstrates its effectiveness through numerical experiments computing GFL for a simple two-layered half-space and a realistic crustal structure of Japan in Sect. 3. A summary of this study is provided in Sect. 4.

2   Method: AFEM-GFL

This section describes the details of the developed GFL computation method, AFEM-GFL. As shown in Algorithm 1, AFEM-GFL generates n different initial meshes, and GFLs are obtained by adaptive mesh refinement starting from each of them, then accuracy of them is evaluated. In this study, we assume $n = 2$ for the number of initial meshes, but it can be increased to three or more to ensure robustness. Within this framework, by employing initial meshes with fine resolution and good element quality, along with a mesh refinement strategy that is resistant to element quality degradation, high-quality adaptive meshes can be generated even for crustal structures with complex geometries. The degrees of freedom tend to be large from the early stages in this approach, and that large cost is handled by parallelizing the computation by MPI. Below, we explain the details of each step in Algorithm 1.

Algorithm 1. AFEM-GFL

1: Generate n initial meshes (2.1).
2: **for** mesh in (mesh1, mesh2, $\cdots$, mesh_n) **do** *
3:      **for** $i = 1, \cdots, i_{\max}$ **do**
4:           Compute GFL by finite element method (2.2).
5:           Estimate posterior error for each element (2.3).
6:           Refine mesh as indicated by error estimator (2.4).
7:      **end for**
8: **end for**
9: Evaluate accuracy of GFLs   (2.5).
10: **if** the accuracy is not sufficient **then**
11:      Improve initial mesh quality and go to *
12: **end if**

2.1   Generation of Initial Meshes

When targeting models with complex layer structures, generating the initial meshes is not straightforward. In this study, we utilize the mesh generation algorithm by Ichimura et al. (2009) [7], which can produce meshes that can properly represent complex layer structures with good element quality. This algorithm covers the entire domain with cubic background cells, and approximate the geometry of the layers for each cell to avoid generating elements with high aspect ratios, then decompose the domain into tetrahedral elements. Furthermore, this approach can form an octree structure of background cells, allowing it to hierarchically merge cells, thus reducing the number of elements in regions distant from the layer boundaries.

This method takes two parameters: ds, which is the side length of the background cells, and nk, which controls the largest merged cell size, ds × 2^{nk-1}. If ds is too large, the approximation of the geometry of the layers may become too coarse. In AFEM-GFL, whether ds is sufficiently small that this approximation does not significantly affect the GFLs can be checked based on the accuracy evaluation described in Sect. 2.5. Also, excessive merging of cells may result in leaving inadequately refined regions in the mesh. nk is set to one or two in this study.

2.2   Computation of GFL

To calculate the displacement fields by loading in x, y, and z directions at each loading point on the top surface of the model, the following governing equation is discretized and solved by the finite element method in the semi-infinite domain:

$$\nabla \cdot \sigma + \boldsymbol{f}_c = \boldsymbol{0}. \tag{1}$$

Here, σ is the stress tensor, and $\boldsymbol{f}_c$ is the body force equivalent to a point load:

$$\boldsymbol{f}_c = \boldsymbol{s}\delta(\boldsymbol{x} - \boldsymbol{\xi}), \tag{2}$$

where $\boldsymbol{s}$ is the load vector and $\boldsymbol{\xi}$ is the loading point. Each component of stress tensor can be expressed in terms of the displacement as

$$\sigma_{ij} = \lambda \delta_{ij} \frac{\partial u_k}{\partial x_k} + \mu \left(\frac{\partial u_i}{\partial x_j} + \frac{\partial u_j}{\partial x_i} \right) \tag{3}$$

where λ and μ are Lamé parameters, and δ_{ij} is the Kronecker delta. Discretizing the equation using a set of basis functions, solving Eq. (1) reduces to solving the following linear equation:

$$\boldsymbol{K}\boldsymbol{u} = \boldsymbol{f}. \tag{4}$$

Here, $\boldsymbol{K}$ is the stiffness matrix, and $\boldsymbol{u}$ and $\boldsymbol{f}$ are nodal displacement and force vectors, respectively. Specifically, $\boldsymbol{f}$ has non-zero values only for components corresponding to nodes in elements that contain $\boldsymbol{\xi}$. To better represent the infinite conditions with a smaller computational domain, infinite elements are implemented on the sides and bottom of the model [8]. The adaptive conjugate gradient method [3] is used to solve Eq. (4). When there are M loading patterns, we obtain the GFL $\{u_i \mid 1 \le i \le M\}$ by solving Eq. (4) for each corresponding nodal forces f_i $(i = 1, \cdots, M)$.

2.3   Posterior Error Estimation

Based on the results of finite element analysis, posterior error estimation is performed for each element, and those predicted to have large errors are targeted for refinement. In this study, we employ the posterior error estimator for elasticity problems proposed by Vefürth (1999) [14]. Once the displacement field u_h is determined, the posterior error estimator η_K for element K is given by:

$$\eta_K = \left\{ h_K^2 \, \|R_K(u_h)\|_K^2 + \sum_{E \in \mathcal{E}_K} \frac{1}{2} h_E \, \|R_E(u_h)\|_E^2 \right\}^{\frac{1}{2}}. \tag{5}$$

Here, h_K and h_E are the diameters of element K and face E, respectively, and $\mathcal{E}_K$ is the set of faces comprising K. $R_K(u_h)$ and $R_E(u_h)$ denote the residuals at K and E, respectively. These residuals are defined as:

$$R_K(u_h) = \int_K \|\nabla \cdot \sigma(u_h) + f_c\|^2 \mathrm{d}V \tag{6}$$

$$R_E(u_h) = \begin{cases} \int_E \|J_E(n_E \cdot \sigma(u_h))\|^2 \mathrm{d}S & \text{if } E \in \mathcal{E}_\Omega \\ 0 & \text{otherwise} \end{cases}, \tag{7}$$

where J_E and n_E are the jump operator and normal vector on E, respectively. $\mathcal{E}_\Omega$ is the set of faces which are not on the boundary of the model. However, for elements containing the loading points, the singularity of f_c makes it difficult to compute $R_K(u_h)$. In such cases, η_K is not computed, and those elements are always selected for refinement. In the computation of GFL, M posterior error estimators η_K^i $(i = 1, \cdots, M)$ are obtained corresponding to the M loading patterns. From these, we define the following metric: $\tilde{\eta}_K = \max_i \eta_K^i / \bar{u}_K^i$. Here, $\bar{u}_K^i$ is the average displacement norm for loading pattern i at nodes in the element K. Normalizing the error estimator by the displacement aims to mitigate locally large errors near the loading points. The elements satisfying $\tilde{\eta}_K > \theta \tilde{\eta}_{\max}$ are selected for refinement, where $\tilde{\eta}_{\max}$ is the maximum value of $\tilde{\eta}_K$ among elements not containing the loading points, and θ is a threshold value set to 0.01 in this study.

2.4   Mesh Refinement

In mesh refinement, the selected elements are subdivided into smaller elements. In refining the unstructured meshes for complex geometries, a highly robust algorithm must be used, capable of preserving the representation of the layer structures while maintaining good element quality. In this study, we refine the mesh by tetrahedral bisection. Tetrahedral bisection involves adding a new node at the midpoint of an edge of the target tetrahedron to divide it into two tetrahedra, and dividing tetrahedra containing hanging nodes into multiple tetrahedra. By choosing bisecting edges according to the algorithm proposed by Arnold & Mukherjee (1999) [2], it is analytically guaranteed that this process yields a conforming mesh when repeated a finite number of times, providing high robustness for the complex geometries. Furthermore, it is ensured that the shapes of the generated tetrahedra are closed to a finite set of shape patterns determined by the original tetrahedron, supporting that this method is resistant to element quality degradation. Protection of the layer structures is achieved by inheriting the properties of each original tetrahedron to the ones generated by bisection.

Additionally, smoothing of the mesh is performed to locally improve the quality of the elements generated by bisection. The nodes of the newly generated elements and their neighbors are moved according to

$$\boldsymbol{P}_i^{\text{new}} = \boldsymbol{P}_i^{\text{old}} + \frac{\lambda}{|N_i|} \sum_{j \in N_i} \left(\boldsymbol{P}_{i,j} - \boldsymbol{P}_i^{\text{old}} \right) \tag{8}$$

until the maximum aspect ratio of those elements stops decreasing. In Eq. (8), $\boldsymbol{P}_i^{\text{old}}$ and $\boldsymbol{P}_i^{\text{new}}$ denote the positions of node i before and after the move, respectively, while $\boldsymbol{P}_{i,j}$ is the position of node j which is adjacent to node i. N_i represents the set of nodes adjacent to node i, and λ is a small constant. The nodes are moved in the direction of the average position of their neighbors to mitigate the degradation of element quality by bisection. To prevent destruction of the layer structures, the smoothing is performed only for nodes not on the layer boundaries.

2.5 Accuracy Evaluation of GFLs

Although the accuracy of the solutions can be predicted by the error estimator η_K, it cannot be calculated for elements containing the loading points. Consequently, mesh refinement does not necessarily lead to a reduction in η_K for all the elements. For a more direct accuracy evaluation, displacements are sampled at several points in the model, and used to calculate accuracy metrics.

As an indicator of convergence of the GFL, the following d_j^i is defined:

$$d_j^i = \frac{1}{M} \sum_{l=1}^{M} \frac{\sqrt{\sum_{k=1}^{3} \left(u_{j,k,l}^i - u_{j,k,l}^{i-1} \right)^2}}{\sqrt{\sum_{k=1}^{3} \left(u_{j,k,l}^{i-1} \right)^2}}. \tag{9}$$

Here, $u_{j,k,l}^i$ is the k-th component of the displacement at the sampling point j for loading pattern l obtained on the mesh after i-th refinement. The convergence of d_j^i for all j as i increases indicates that the sampled displacements are converging. d_j^i can be calculated for each of GFL obtained on mesh1 and mesh2.

In cases where inadequately refined regions remain in the mesh, extremely poor quality elements are generated, or the approximation of the layer structures is too coarse, d_j^i approaching zeros might indicate convergence of the GFL to incorrect solutions. To detect such situations, the following e_j^i is defined:

$$e_j^i = \frac{1}{M} \sum_{l=1}^{M} \frac{\sqrt{\sum_{k=1}^{3} \left(u_{j,k,l}^{i1} - u_{j,k,l}^{i2} \right)^2}}{\sqrt{\sum_{k=1}^{3} \left(u_{j,k,l}^{i1} \right)^2}}. \tag{10}$$

Here, $u_{j,k,l}^{i1}$ and $u_{j,k,l}^{i2}$ are the displacements sampled from GFLs on mesh1 and mesh2, respectively. The convergence of e_j^i for all j as i increases indicates that the convergent GFLs are consistent between mesh1 and mesh2. In such a situation, it is evaluated that no issues that distort the sampled displacements have occurred during the mesh refinement process, and high-accuracy GFLs have been obtained. If e_j^i remains large and does not decrease in the iterative mesh refinement process, it is necessary to improve the quality of the initial meshes and rerun AFEM-GFL.

3   Numerical Experiments

In this section, we describe the numerical experiments targeting a two-layered half-space and a realistic crustal structure in Japan. Section 3.1 discusses a numerical experiment where the response due to point loads at a single point on the top surface of the two-layered half-space is computed by AFEM-GFL. Through this experiment, we demonstrate AFEM-GFL's capability to compute displacements in high accuracy in terms of the metrics d_j^i and e_j^i, and examine how the mesh refinement progresses and how the response converges. Section 3.2 discusses a numerical experiment for a realistic crustal structure in Japan, where GFL is obtained by computing responses due to loads at multiple points and crustal deformation at the loading points is computed using the obtained GFL. Through this experiment, we demonstrate that AFEM-GFL is also effective for analysis targeting models with complex geometries and that high-accuracy crustal deformation can be computed from the obtained GFL.

3.1   Two-Layered Half-Space

For a model consisting of two horizontal layers with different material properties, we computed the response due to loads at a single point on the top surface of the model. The layer structure and point load settings are shown in Table 1, with an overview of the model shown in Fig. 1. We denote the settings where the initial mesh was generated with (ds, nk) = (1.4 km, 2) and (2.8 km, 1) as AFEM-GFL1 and AFEM-GFL2, respectively. This numerical experiment involved 15 iterations of mesh refinement for both AFEM-GFL1 and AFEM-GFL2. For calculating the

Table 1. Settings for the two-layered half-space model

Domain size	$0\,\mathrm{km} \leq x, y \leq 403.2\,\mathrm{km}$
Layer 1	$(E, \nu) = (2.89\,\mathrm{GPa}, 0.44)$, $250.176\,\mathrm{km} \leq z \leq 260.176\,\mathrm{km}$
Layer 2	$(E, \nu) = (23.6\,\mathrm{GPa}, 0.26)$, $0\,\mathrm{km} \leq z \leq 250.176\,\mathrm{km}$
Loading Point	$(203.338\,\mathrm{km},\ 203.448\,\mathrm{km},\ 260.176\,\mathrm{km})$
Loads	$(1\,\mathrm{N}, 0\,\mathrm{N}, 0\,\mathrm{N}), (0\,\mathrm{N}, 1\,\mathrm{N}, 0\,\mathrm{N}), (0\,\mathrm{N}, 0\,\mathrm{N}, 1\,\mathrm{N})$

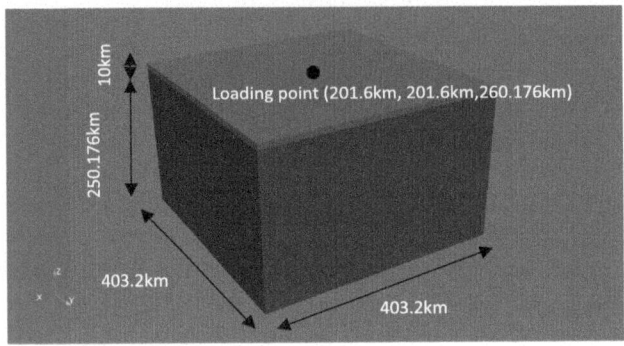

Fig. 1. Overview of the two-layered half-space model and the loading point.

metrics d_j^i and e_j^i, displacements were sampled at 1575 points within the range $22.4\,\mathrm{km} \leq x, y \leq 380.8\,\mathrm{km}$, $179.2\,\mathrm{km} \leq z \leq 254.8\,\mathrm{km}$.

Figure 2 shows the mesh refinement process. Figure 2(a) displays only the elements subjected to tetrahedral bisection, illustrating that refinement initially occurred around the loading point and spread over the domain after the mesh around the loading point was refined. Figure 2(b) shows the mesh after 15 iterations of refinement.

Fig. 2. (a) Elements subjected to refinement during the 3rd and 8th iterations. (b) Mesh after 15 refinements for AFEM-GFL2 model, cut at $x = 201.6\,\mathrm{km}$, showing the side containing $x = 0\,\mathrm{km}$. The left panel shows the overview and the right panel shows a close-up view around the loading point.

d_j^i was computed from the sampled displacements and its distribution is presented in Fig. 3(a). The maximum value of d_j^i at $i = 15$ is 0.014% and 0.028% for AFEM-GFL1 and AFEM-GFL2, respectively, indicating that displacements were converged through mesh refinement. Similarly, e_j^i was computed and its distribution is shown in Fig. 3(b). At $i = 15$, the value of e_j^i is confined to 0.050%. It implies that convergent displacements for AFEM-GFL1,2 are consistent, suggesting that high accuracy of the solutions obtained by AFEM-GFL.

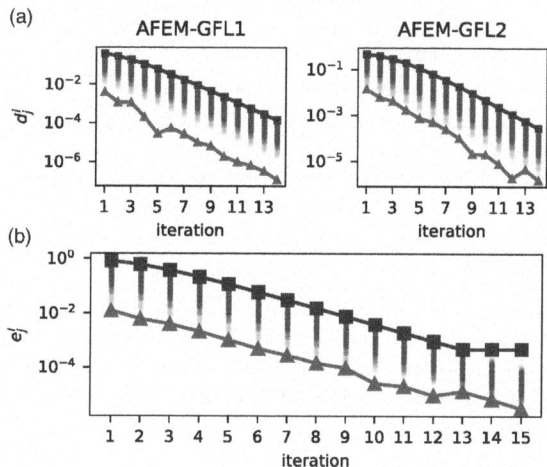

Fig. 3. (a) Distribution of d_j^i for AFEM-GFL1 and AFEM-GFL2. (b) Distribution of e_j^i. In both figures, the value of each metric at each sample point is shown as a dot, with square and triangle markers representing maximum and minimum values, respectively.

Mainly two patterns for convergence of the displacements were observed. Figure 4 presents examples of each pattern, compared with displacements computed using three meshes with uniform element sizes of 2.8 km , 1.4 km, and 0.7 km. At sample points far from the loading point, as shown in Fig. 4(a), the AFEM-GFL1 and AFEM-GFL2 results converged to the same value and the size of the element containing the loading point has a dominant effect on the convergence. For sample points near the loading point, as shown in Fig. 4(b), the two solutions converged to slightly different values. However, as indicated by the metric e_j^i, these differences are small.

We compare the computational cost of analysis using the uniform mesh 0.7 km and AFEM-GFL1 in Table 2. The degrees of freedom for the AFEM-GFL1's final mesh is no more than 1/100 of that for the uniform mesh 0.7 km. Further significant refinement would be required to achieve convergence with uniform meshes and such analysis would necessitate even larger computational resources, while high-accuracy convergent solutions were achieved with moderate-scale computer cluster (320 CPU cores) by AFEM-GFL. This result highlights the attractiveness of the method in terms of capability computing, although the total elapsed time for it was large due to the repetitive finite element analyses and the mesh refinement process.

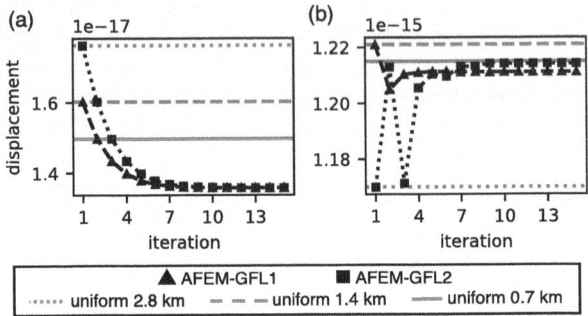

Fig. 4. y components of the displacement due to a load in x direction at sample points are shown. (a) Displacement at $(89.6\,\text{km}, 89.6\,\text{km}, 201.6\,\text{km})$. (b) Displacement at $(201.6\,\text{km}, 201.6\,\text{km}, 254.8\,\text{km})$.

Table 2. Comparison of the uniform mesh 0.7 km and AFEM-GFL1. Supercomputer Fugaku [13] and a Xeon Gold 6230-based CPU cluster were used for uniform 0.7 km and AFEM-GFL1, respectively. The elapsed time for the uniform mesh includes the time for mesh generation, partitioning, and finite element analysis. One for AFEM-GFL1 includes, in addition to the above, the time for error estimation and mesh refinement.

	Processor	# of CPU cores	Elapsed time(hour)	DoF
Uniform 0.7 km	A64FX	38400	2.6	9.9×10^8
AFEM-GFL1	Xeon Gold 6230	320	22.3	7.3×10^6

3.2   Crustal Structure in Japan

Numerical experiments were conducted for a model of the crustal structure in the Tohoku region of Japan. The Japan Integrated Velocity Structure Model (JIVSM) [10], a unified underground structure model of the Japanese islands, provides a Digital Elevation Model (DEM) representing the layer structure in the form of (latitude, longitude, elevation) and material properties of each layer. We utilized these data to generate the crustal structure model. DEM data for the longitude range $139.7\,°\text{E} \sim 142.7\,°\text{E}$ and latitude range $37.4\,°\text{N} \sim 39.4\,°\text{N}$ were extracted from the JIVSM, and too thin layers near the ground surface was deleted [11], then converted to a Cartesian coordinate system [6] to obtain a DEM in the range of $0 \leq x \leq 240\,\text{km}, 0 \leq y \leq 200\,\text{km}, 0 \leq z \leq 206\,\text{km}$. Using this data, meshes generated with the settings $(\text{ds}, \text{nk}) = (0.625\,\text{km}, 2)$ and $(1.25\,\text{km}, 1)$ were used as initial meshes for AFEM-GFL1 and AFEM-GFL2 settings, respectively, and 15 iterations of mesh refinement were performed for both settings. Three loading patterns of $(1\,\text{N}, 0\,\text{N}, 0\,\text{N})$, $(0\,\text{N}, 1\,\text{N}, 0\,\text{N})$, $(0\,\text{N}, 0\,\text{N}, 1\,\text{N})$ were applied to each of the 16 points evenly distributed on the surface within the range $93\,\mathrm{km} \leq x \leq 172\,\mathrm{km}, 78\,\mathrm{km} \leq y \leq 144\,\mathrm{km}$. Displacements sampled at 350 points within the range of $60\,\mathrm{km} \leq x \leq 180\,\mathrm{km}, 60\,\mathrm{km} \leq y \leq 140\,\mathrm{km}, 40\,\mathrm{km} \leq z \leq 195\,\mathrm{km}$ were used to evaluate d_j^i and e_j^i. Figure 5 presents an overview of the model and loading points.

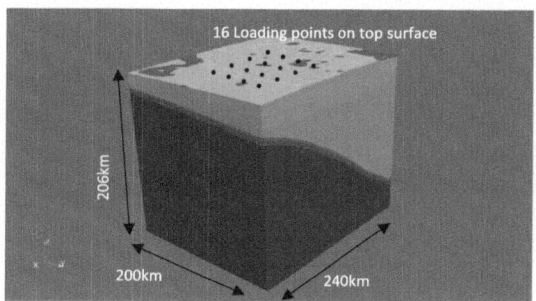

Fig. 5. Overview of the crustal structure model and the loading points.

Figure 6 shows the mesh refinement process. In Fig. 6(a), only the elements subjected to tetrahedral bisection are displayed, showing a similar trend to the setting in Sect. 3.1 where refinement initially focuses on the loading points and then spreads over the domain. Figure 6(b) shows the final mesh for AFEM-GFL2. The maximum aspect ratios in the final meshes for AFEM-GFL1 and AFEM-GFL2 are 36.6 and 36.2, respectively, while those in the initial meshes are 8.85 and 8.64. Although the maximum aspect ratio increased around 4 times during mesh refinement, no degenerate elements (e.g., aspect ratio greater than 1000) were generated even for the crustal structure model.

Regarding the accuracy of the computed GFLs, Figs. 7(a) and (b) show the distribution of d_j^i and e_j^i, respectively. For d_j^i, the maximum values at $i = 15$ are 0.02% and 0.04% for AFEM-GFL1 and AFEM-GFL2, respectively. As for e_j^i, the maximum value was 0.60% at $i = 15$. This confirms that even when targeting complex structure models, accurate GFL can be computed by AFEM-GFL from the perspective of d_j^i and e_j^i. However, compared to the setting in Sect. 3.1, the reduction of e_j^i stopped earlier, and the residual of e_j^i after 15 iterations was larger. One reason might be that while the geometry of the layer structure was accurately represented by the mesh in the two-layered half-space model, it was approximated when generating the initial mesh for the crustal structure model. Although the algorithm used in this study cannot improve geometry representation through mesh refinement, using such methods that correct geometries referring the original DEM when adding nodes during tetrahedral bisection may potentially reduce e_j^i further.

Fig. 6. (a) Elements subjected to refinement during the 3rd and 8th iterations. (b) Mesh after 15 refinements for AFEM-GFL2 model, cut at $y = 120\,\text{km}$, showing the side containing $y = 0\,\text{km}$. The left panel shows the overview and the right panel shows a close-up view around the loading point at $x = 93.4\,\text{km}$, $y = 100.0\,\text{km}$.

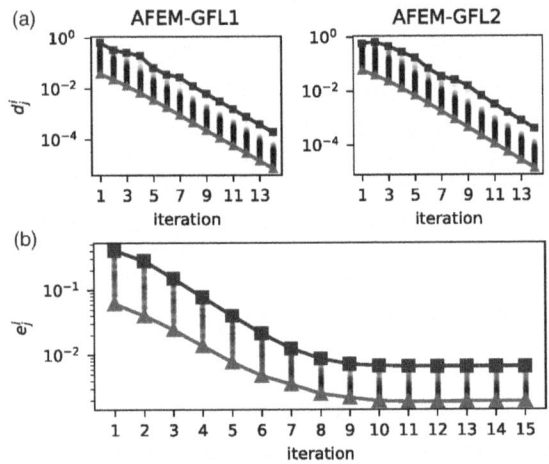

Fig. 7. (a) Distribution of d_j^i for AFEM-GFL1 and AFEM-GFL2. (b) Distribution of e_j^i. In both figures, the value of each metric at each sample point is shown as a dot, with square and triangle markers representing maximum and minimum values, respectively.

Furthermore, the accuracy of crustal deformation computed from the obtained GFLs was also evaluated. Here, we calculated the displacements at the 16 loading points due to a point source at position $(62\,\text{km}, 102\,\text{km}, 162\,\text{km})$ with a moment tensor of $m_{xx} = m_{yy} = m_{zz} = m_{xy} = m_{yz} = 0, m_{zx} = 1.0 \times 10^{19}\,\text{N} \cdot \text{m}$. The point source was approximated as a set of forces acting on vertices of a cube with a side length of $5\,\text{km}$ [4], and displacement was obtained by calculat-

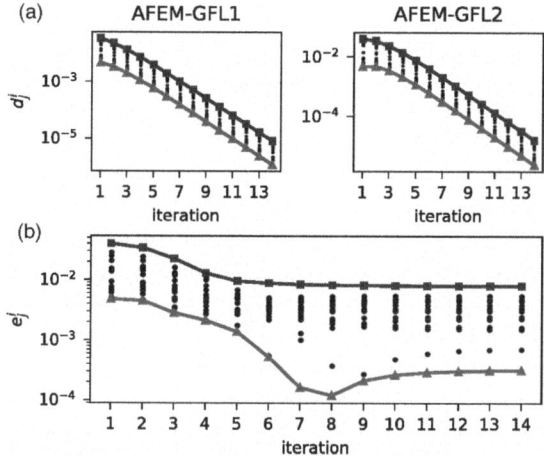

Fig. 8. (a) Distribution of $\tilde{d}_j^i$ for AFEM-GFL1 and AFEM-GFL2. (b) Distribution of $\tilde{e}_j^i$. In both figures, the value of each metric at each sample point is shown as a dot, with square and triangle markers representing maximum and minimum values, respectively.

ing the convolution with the GFLs. Similar to the metrics in Eqs. (9) and (10), the accuracy of point source response was evaluated using metrics defined as follows:

$$\tilde{d}_j^i = \frac{\sqrt{\sum_{k=1}^3 \left(u_{j,k}^i - u_{j,k}^{i-1} \right)^2}}{\sqrt{\sum_{k=1}^3 \left(u_{j,k}^{i-1} \right)^2}} \qquad (11)$$

$$\tilde{e}_j^i = \frac{\sqrt{\sum_{k=1}^3 \left(u_{j,k}^{i1} - u_{j,k}^{i2} \right)^2}}{\sqrt{\sum_{k=1}^3 \left(u_{j,k}^{i1} \right)^2}}. \qquad (12)$$

Here, in Eq. (11), $u_{j,k}^i$ represents the k-component of the point source response at the loading point j obtained using the GFL after i-th refinement. In Eq. (12), $u_{j,k}^{i1}$ and $u_{j,k}^{i2}$ represent the responses obtained using the GFL in AFEM-GFL1 and AFEM-GFL2 settings, respectively. The distribution of $\tilde{d}_j^i$ and $\tilde{e}_j^i$ are shown in Figs. 8(a) and (b), respectively. The maximum values of $\tilde{d}_j^i$ at $i = 15$ is 0.0008% for AFEM-GFL1 and 0.001% for AFEM-GFL2, confirming that the point source response was converged as well as the GFL. As for $\tilde{e}_j^i$, the maximum value is 0.79% at $i = 15$, confirming that the two responses from AFEM-GFL1 and AFEM-GFL2 are aligned. Evaluating the accuracy of the GFLs at the source point, the maximum value of e_j^{15} is 0.2%. The order of magnitude of the maximum value of e_j^{15} and $\tilde{e}_j^{15}$ is consistent, suggesting that the point source responses were computed with a similar level of accuracy as the

GFLs. In this study, the accuracy of the source response is evaluated with the source representation fixed, but future works should consider the impact of the source representation (e.g., the size of the cube for the point source).

4   Conclusion

In this study, we presented a method called AFEM-GFL for calculating GFLs using adaptive mesh refinement with the quantitative accuracy evaluation of solutions. This method generates adaptive meshes suitable for GFL calculations even for models with complex layer structures and allows the quantitative evaluation of accuracy from the perspectives of solution convergence and consistency of solutions obtained from different initial meshes.

Numerical experiments on a two-layered half-space confirmed that high-accuracy convergent solutions can be obtained using this method, as both metrics d_j^i and e_j^i became small. While a convergent solution could not be obtained even with 9.9×10^8 degrees of freedom using a uniform mesh, the degrees of freedom in the final mesh of AFEM-GFL were 7.3×10^6, enabling execution on a moderate scale computer cluster. This feature shows the superiority of the proposed method over analysis using uniform meshes, although the elapsed time for AFEM-GFL was longer than analysis using uniform meshes. In numerical experiments targeting the realistic crustal structure model, it was confirmed that this method is robust, assuring a certain level of accuracy even for models with the complex geometry. It was also verified that point source responses can be calculated with similar accuracy to that of the obtained GFLs.

We can obtain quality GFLs using the proposed method. Those will be utilized for reliable source estimation and enhancing our understanding of earthquakes.

CRediT Authorship Contribution Statement. Kai Nakao: Investigation, Methodology, Software, Visualization, Writing - original draft **Hideaki Ito**: Investigation, Methodology, Software, Visualization, Writing - original draft **Tsuyoshi Ichimura**: Conceptualization, Funding acquisition, Methodology, Project administration, Supervision, Writing - original draft, Writing - review & editing **Kohei Fujita**: Conceptualization, Supervision, Writing - review & editing. **Lalith Wijerathne**: Conceptualization, Supervision, Writing - review & editing. **Muneo Hori**: Conceptualization, Writing - review & editing.

Acknowledgments. This work was supported by MEXT, Japan as "Program for Promoting Researches on the Supercomputer Fugaku" (Large-scale numerical simulation of earthquakes by Fugaku supercomputer, JPMXP1020230213). This work was supported by JSPS KAKENHI, Japan Grant Numbers 23H00213, 22K18823. This work used computational resources of Fugaku provided by the RIKEN Center for Computational Science (Project ID: hp230203, hp240217). This work was supported by JST SPRING, Grant Number JPMJSP2108.

Disclosure of Interests. The authors have no competing interests to declare that are relevant to the content of this article.

References

1. Adjerid, S., Flaherty, J.E.: A local refinement finite-element method for two-dimensional parabolic systems. SIAM J. Sci. Stat. Comput. **9**(5), 792–811 (1988). https://doi.org/10.1137/0909053
2. Arnold, D.N., Mukherjee, A.: Tetrahedral Bisection and Adaptive Finite Elements, pp. 29–42. Springer, New York (1999). https://doi.org/10.1007/978-1-4612-1556-1_2
3. Golub, G.H., Ye, Q.: Inexact preconditioned conjugate gradient method with inner-outer iteration. SIAM J. Sci. Comput. **21**(4), 1305–1320 (1999). https://doi.org/10.1137/S1064827597323415
4. Graves, R.W.: Simulating seismic wave propagation in 3D elastic media using staggered-grid finite differences. Bull. Seismol. Soc. Am. **86**(4), 1091–1106 (1996). https://doi.org/10.1785/BSSA0860041091
5. Graves, R.W., Wald, D.J.: Resolution analysis of finite fault source inversion using one- and three-dimensional green's functions: 1. strong motions. Jo. Geophys. Res. Solid Earth **106**(B5), 8745–8766 (2001). https://doi.org/10.1029/2000JB900436
6. Hori, T., Agata, R., Ichimura, T., Fujita, K., Yamaguchi, T., Iinuma, T.: High-fidelity elastic Green's functions for subduction zone models consistent with the global standard geodetic reference system. Earth Planets Space **73**(1), 1–12 (2021). https://doi.org/10.1186/s40623-021-01370-y
7. Ichimura, T., Hori, M., Bielak, J.: A hybrid multiresolution meshing technique for finite element three-dimensional earthquake ground motion modelling in basins including topography. Geophys. J. Int. **177**(3), 1221–1232 (2009). https://doi.org/10.1111/j.1365-246X.2009.04154.x
8. Ichimura, T., Agata, R., Hori, T., Hirahara, K., Hori, M.: Fast numerical simulation of crustal deformation using a three-dimensional high-fidelity model. Geophys. J. Int. **195**(3), 1730–1744 (2013). https://doi.org/10.1093/gji/ggt320
9. Kim, M., So, B.D., Kim, S., Jo, T., Chang, S.J.: Mesh size effect on finite source inversion with 3-D finite-element modelling. Geophy. J. Int. **237**(2), 716–728 (2024). https://doi.org/10.1093/gji/ggae060
10. Koketsu, K., Miyake, H., Suzuki, H.: Japan integrated velocity structure model version 1. In: Proceedings of the 15th World Conference on Earthquake Engineering, p. 1773 (2012)
11. Murakami, S., Hashima, A., Iinuma, T., Fujita, K., Ichimura, T., Hori, T.: Detectability of low-viscosity zone along lithosphere–asthenosphere boundary beneath the nankai trough, japan, based on high-fidelity viscoelastic simulation. Earth Planets Space **76** (2024). https://doi.org/10.1186/s40623-024-02008-5
12. Okada, Y.: Surface deformation due to shear and tensile faults in a half-space. Bull. Seismol. Soc. Am. **75**(4), 1135–1154 (1985). https://doi.org/10.1785/BSSA0750041135
13. RIKEN Center for Computational Science: Supercomputer fugaku. https://www.r-ccs.riken.jp/en/fugaku/
14. Verfürth, R.: A review of a posteriori error estimation techniques for elasticity problems. Comput. Methods Appl. Mech. Eng. **176**(1), 419–440 (1999). https://doi.org/10.1016/S0045-7825(98)00347-8

An Iterative Scheme for the Solidification Benchmark Modeling

Xiaoyu Feng[1,2]($\boxtimes$)(ID), Huangxin Chen[3], Bo Yu[4], and Shuyu Sun[1]($\boxtimes$)(ID)

[1] Computational Mathematics, School of Mathematical Sciences, Tongi University, Shanghai 200092, China
[2] Computational Transport Phenomena Laboratory, Division of Physical Science and Engineering, King Abdullah University of Science and Technology, Thuwal 23955–6900, Saudi Arabia
xiaoyu.feng@kaust.edu.sa, suns@tongji.edu.cn
[3] School of Mathematical Sciences and Fujian Provincial Key Laboratory on Mathematical Modeling and High Performance Scientific Computing, Xiamen University, Fujian 361005, China
[4] School of Petroleum Engineering, Yangtze University, Wuhan 430100, China

Abstract. The processes of solidification and macro-segregation involve intricate interactions across multiple physical, phase, and compositional fields, including mass, momentum, energy, and material transfer. Accurate prediction of phase transitions, chemical heterogeneities, and compositional flows is crucial in fields such as materials science, energy science, and planetary science. Numerical benchmark studies provide an effective means to explore these phenomena. This paper presents an iterative scheme based on operator splitting and evaluates its accuracy, stability, and implementation through a relevant benchmark problem. The results demonstrate strong performance of the scheme, particularly in capturing key physical phenomena such as channel segregation, freckle formation, and edge effects.

Keywords: Solidification · Multi-phase · Operator-splitting · Iterative Scheme · Benchmark Modeling

1 Introduction

Solidification is a complex process involving the transfer of mass, momentum, energy, and species, with multi-phase and multi-component interactions. Key phenomena such as chemical heterogeneity, macro-segregation, and phase transitions between solid, mushy, and liquid regions are essential to understanding material behaviors in fields like material science [20], energy storage [11], magma ocean evolution [12], safe operation of pipelines [19] and the high-efficiency recovery of natural gas hydrate (NGH) from the subsurface [16].

Experimental studies using both opaque alloys (e.g., Al−Cu, Sn−Pb) and transparent analogs (e.g., NH_4Cl, Na_2CO_3) have provided valuable data and made progress for deep understanding of this physical process [9,10]. Numerical simulation [15,22] is also a vital tool for understanding these complex processes, especially when time and space constraints limit experimental methods. These simulations are commonly based on mesh-free methods like SPH [7,8], moving-grid methods like ALE [2], and fixed-mesh methods such as VOF, level-set [17] and LBM [4]. Among these, the enthalpy-porosity methods [14,18] and the phase field methods [21] are widely used for solidification simulations due to their simplicity in dealing with phase transitions on a macroscale.

This paper introduces a novel iterative scheme based on operator-splitting and matrix-based methods, designed to improve convergence rates and computational efficiency. The scheme is validated through benchmark tests, demonstrating its capability to capture important physical phenomena such as macrosegregation and phase transitions [3,18]. Additionally, the integration of vectorization and forward matrix assembly techniques enhances the scalability and efficiency of the method, making it suitable for extending to 3D simulations.

The structure of the paper is as follows: Sect. 2 provides a concise summary of the classical models and their assumptions. Section 3 introduces and derives our proposed numerical scheme. Section 4 presents a classic benchmark example and discusses the validation results. Finally, Sect. 5 concludes the study with final remarks.

2   Mathematical Model

The most fundamental and universal single-domain continuum mixture model is derived from mass averaging and classical mixture theory. This model suggests that the properties of the mixture are the result of the individual components, with its governing equations resembling those of the individual phases. It leverages the continuous phase transition of the mixture, represented by phase fractions, while ensuring the conservation of mass, momentum, energy, and components within the system. Since all governing equations follow a similar convection-diffusion form, by introducing a general scalar quantity ϕ_ω associated with phase ω in a multiphase system, the general physical relationship can be established. We define the partial volume density of phase for phase ω, namely $\bar{\rho}_\omega = \frac{V_\omega}{V} \frac{m_\omega}{V_\omega} = g_\omega \rho_\omega$ and $\bar{\rho}_\omega^\theta = \frac{V_\omega^\theta}{V_\omega} \frac{m_\omega^\theta}{V_\omega^\theta} = g_\omega^\theta \rho_\omega^\theta$ as the partial volume density of the component θ within phase ω.

The we have the mass fraction of one phase $f_\omega = \frac{\bar{\rho}_\omega}{\sum_\omega \bar{\rho}_\omega}$. Then obviously the relations for the volume fraction and mass fraction as follows: $\sum_\omega f_\omega = 1; \sum_\omega g_\omega = 1$. Based on the mass-averaged velocities and general variables, the equations are as follows:

$$\mathbf{u} = \frac{1}{\rho} \sum_\omega \bar{\rho}_\omega \mathbf{u}_\omega = \sum_\omega f_\omega \mathbf{u}_\omega, \tag{1}$$

$$\phi = \frac{1}{\rho} \sum_\omega \bar{\rho}_\omega \phi_\omega = \sum_\omega f_\omega \phi_\omega, \tag{2}$$

where the density of one mixture is $\rho = \sum_\omega \bar{\rho}_\omega$. The conservation law for general scalar quantities can be written as:

$$\frac{\partial}{\partial t} \int_V [\rho_\omega \phi_\omega] \, d\bar{V}_\omega + \int_S [\rho_\omega \mathbf{u}_\omega \phi_\omega] \cdot \mathbf{n} dS_\omega = \int_S \mathbf{J}_\omega \cdot \mathbf{n} dS_\omega + \int_V Sr_\omega d\bar{V}_\omega. \quad (3)$$

The terms correspond to the dynamic term, convection flux, diffusion flux, and the source term. Assuming smoothness and differentiability of the arguments under the surface integrals, the divergence theorem yields:

$$\frac{\partial}{\partial t} (\bar{\rho}_\omega \phi_\omega) + \nabla \cdot (\bar{\rho}_\omega \mathbf{u}_\omega \phi_\omega) = \nabla \cdot (g_\omega \mathbf{J}_\omega) + g_\omega Sr_\omega. \quad (4)$$

The conservation of mass, momentum, energy and component relations can be referred by assigning the general quantities in the governing equation (4) with different variables and doing summation for all phases ω.

Generally, the above conservation relations are applicable for multiple phases and components. Here, we only consider the solid-liquid phase change ($\omega = l, s$) of the binary mixture ($\theta = A, B$). Based on the above identity, the model for solidification of this problem can be presented as:

Mass conservation:

$$\frac{\partial \rho}{\partial t} + \nabla \cdot (\rho \mathbf{u}) = 0, \quad (5)$$

where $\rho = g_s \rho_s + g_l \rho_l$, $\mathbf{u} = f_s \mathbf{u}_s + f_l \mathbf{u}_l$.

Momentum equation:

$$\frac{\partial (\rho \mathbf{u})}{\partial t} + \nabla \cdot (\rho \mathbf{u} \otimes \mathbf{u}) = -\nabla p + \nabla \cdot (\mu_l \frac{\rho}{\rho_l} \nabla \mathbf{u}) + \rho \mathbf{g} - \frac{\mu_l}{K} \frac{\rho}{\rho_l} (\mathbf{u} - \mathbf{u}_s), \quad (6)$$

where the body force is $\boldsymbol{g}$ and the diffusion term can be derived with some trivial simplification, and the details can be referred to [3].

The flow in the mushy zone is considered laminar and Newtonian with constant viscosity, treated as an isotropic porous medium without direction coupling effects. Thus, the off-diagonal elements of the permeability tensor are zero. The phase interaction term, based on the Kozeny-Carman formula with relative phase velocity, acts as a damping term in equation (6). The permeability for flow in the mushy region is given by:

$$K = \frac{\lambda_2^2 f_l^3}{180 (1 - f_l)^2}, \quad (7)$$

where λ_2 represents the secondary dendrite arm spacing, a key microstructural feature that influences inter-dendritic flow and is commonly used to determine permeability.

Energy equation:

$$\frac{\partial (\rho h)}{\partial t} + \nabla \cdot (\rho \mathbf{u} h) = \nabla \cdot (k \nabla T) - \nabla \cdot (\rho (h_l - h)(\mathbf{u} - \mathbf{u}_s)), \quad (8)$$

where enthalpy $h = f_s h_s + f_l h_l$ and the conductivity uses volume average instead of mass average, $k = g_s k_s + g_l k_l$, because it is related to the size of the material itself.

Defining average heat capacities $\bar{c}_{pl} = \frac{1}{T} \int_0^T c_{pl} dT$ and $\bar{c}_{ps} = \frac{1}{T} \int_0^T c_{ps} dT$, the solid and liquid phase enthalpies are

$$h_s = \int_0^T c_{ps} dT + h_s^0 = \bar{c}_{ps} T, \quad h_l = \int_0^T c_{pl} + h_l^0 = \bar{c}_{pl} T + L,$$

where we set $h_s^0 = 0$ and $h_l^0 = L$, L is the latent heat of phase change. Based on the mass average of heat capacity $\bar{c}_p = f_s \bar{c}_{ps} + f_l \bar{c}_{pl}$, then we can easily arrive at

$$h = \bar{c}_p T + f_l L, \tag{9}$$

Transport equation:

$$\frac{\partial (\rho C)}{\partial t} + \nabla \cdot (\rho \mathbf{u} C) = \nabla \cdot (\rho f_l D_l \nabla C) + \nabla \cdot (\rho f_l D_l \nabla (C_l - C)) - \nabla \cdot (\rho (C_l - C)(\mathbf{u} - \mathbf{u}_s)). \tag{10}$$

Considering the binary solid-liquid system, for the sake of conciseness of the notation, C represents the concentration (or mass fraction) of the primary solute of the binary mixture system. The diffusion in the solid phase can be ignored compared with the liquid phase $(D_l^A \gg D_s^A)$. With this assumption and the identity $\nabla C_l = \nabla C + \nabla (C_l - C)$, (10) can be deduced.

Lever rule:

The component conservation and the constraint on mass fraction for the binary solid-liquid system indicate:

$$C = f_l C_l + f_s C_s. \tag{11}$$

$$f_l + f_s = 1. \tag{12}$$

Then, the "lever rule" can be deduced as:

$$\begin{cases} f_s = \frac{C_l - C}{C_l - C_s}, \\ f_l = \frac{C - C_s}{C_l - C_s}. \end{cases} \tag{13}$$

The implication of the "lever rule" can be easily appreciated from the above equations. They describe the relationship between the mass fraction of the component and the phase fraction in a binary system.

Phase equilibrium relation:

For this dynamic problem, even the local thermodynamic equilibrium has been assumed, but the phase equilibrium relations still need to be specified to ensure the closure of the PDE system. Usually, this relation is characterized by the equation of state (EoS). The van der Waals (VdW) EoS or Peng-Robinson (PR) EoS is widely recognized for gas-liquid problems [6]. Generally, the solute fraction and phase fraction can be determined based on the EoS and the constrains. While there is no widely accepted EoS model for the solid-liquid problem,

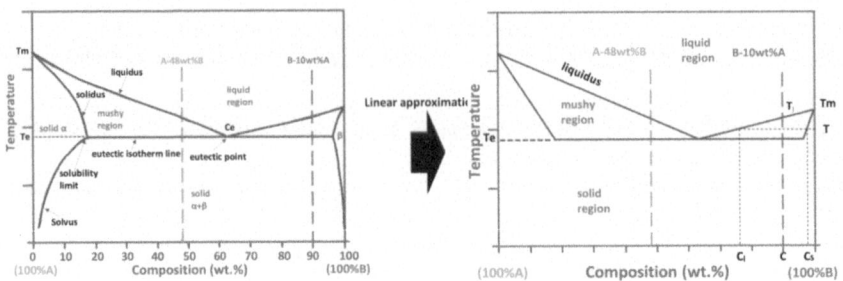

Fig. 1. Linear approximation of the phase diagram

the phase diagram from experiments with some approximations are still used as the basis for simulation. Combining the "lever rule" with the phase diagram, C_l, C_s and f_l, f_s can be determined with certain T and C.

Simplified model:

- The velocity of the solid phase is zero ($\mathbf{u}_s = \mathbf{0}$).
- Heat capacities of both phases are constant and equal ($c_{pl} = c_{ps} = c_p$).
- The densities of liquid and solid phases are equal and constant, thus $f_l = g_l$; $f_s = g_s$.
- The thermal and solutal driven buoyancy body forces are characterized by the Boussinesq approximation: $\Delta\rho = (-\beta_T (T - T_{ref}) - \beta_c (C_l - C_{ref})) \rho_0$.
- The linear approximation of the phase diagram is shown in Fig. 1 (the slopes for the liquidus and solidus lines are fixed, and they correspond to k_p and $\mathcal{L}$ being constants).

Then, the simplified and reduced model can be expressed as follows:

$$\nabla \cdot \mathbf{u} = 0, \quad (14a)$$

$$\frac{\partial \mathbf{u}}{\partial t} + (\mathbf{u} \cdot \nabla)\mathbf{u} = \nabla \cdot \left(\frac{\mu_l}{\rho_0}\nabla\mathbf{u}\right) - \frac{1}{\rho_0}\nabla p - \frac{\mu_l}{\rho_0}K^{-1}\mathbf{u} + \frac{(\rho_0 + \Delta\rho)}{\rho_0}\mathbf{g}, \quad (14b)$$

$$\frac{\partial T}{\partial t} + \nabla \cdot (\mathbf{u}T) = \nabla \cdot (\frac{k}{\rho_0 c_p}\nabla T) - \frac{L}{c_p}\frac{\partial g_l}{\partial t}, \quad (14c)$$

$$g_l = 1 - \frac{1}{1 - k_p}\frac{T - T_l}{T - T_m}, \text{(where } T_l = \mathcal{L}C + T_m), \quad (14d)$$

$$\frac{\partial C}{\partial t} + \nabla \cdot (\mathbf{u}C_l) = \nabla \cdot (g_l D_l \nabla C_l), \quad (14e)$$

$$C_l = \frac{C}{1 - (1 - k_p)(1 - g_l)}. \quad (14f)$$

Then unknowns become $\mathbf{u}, p, T, C, C_l, g_l$ and they can be determined by the above system with 6 equations.

3   Numerical Scheme

On the basis of the aforementioned mathematical model (14), an iterative numerical scheme are proposed and the classical fully decoupled scheme is also presented as comparison. In this section, operator-splitting and matrix-based techniques are employed for constructing the schemes. The sequence of numerical solutions follows the real physical process: the cooling is the fundamental startup source, then local phase equilibrium is assumed and T, g_l are strong couplings, the temperature is determined by the energy equation, then the thermal and solutal driven force induces the flow, and the flow will eventually cause the transport of the component. Thus, the temporal discretization is given in this order. The superscripts $n + 1$ and n represent implicit and explicit terms, respectively.

3.1   The Operator-Splitting Method for Energy Equation

In this part, schemes are designed to solve the energy equation and to handle the local phase equilibrium relation concurrently, namely the strong coupling effect between the temperature and phase fraction.

A fully decoupled scheme for the energy equation:

The solution procedure for the energy equation is split into two steps: first step takes into account heat convection and diffusion,

$$\frac{T^* - T^n}{\Delta t} + \mathbf{u}^n \cdot \nabla T^* = \nabla \cdot \left(\frac{k}{\rho_0 c_p} \nabla T^* \right). \tag{3.1}$$

The second step is responsible for the correction of latent heat when a phase transition occurs,

$$\frac{T^{n+1} - T^*}{\Delta t} = -\frac{L}{c_p} \frac{g_l^{n+1} - g_l^n}{\Delta t}. \tag{3.2}$$

The phase equilibrium relations yields:

$$g_l^{n+1} = 1 - \frac{1}{1 - k_p} \frac{T^{n+1} - T_l(C^n)}{T^{n+1} - T_m}, \tag{3.3}$$

and

$$T_l = T_m + \mathcal{L}C^n, \tag{3.4}$$

where T_l corresponds to the liquidus temperature at the current concentration C. Because of the segregation process, the deviation of concentration C^n from the uniform concentration C_0 will result in a range of $T_l(C^n)$ across the entire computational domain. Then, by substituting it in the second step of operator-splitting expressions, a quadratic equation of the liquid volume fraction g_l^{n+1} will be found:

$$a \cdot (g_l^{n+1})^2 + b \cdot g_l^{n+1} + c = 0, \tag{3.5}$$

where coefficients are

$$\begin{cases} a = 1 - k_p, \\ b = k_p - g_l^n (1 - k_p) + \frac{c_p}{L} (1 - k_p) (T_m - T^*), \\ c = -\left(g_l^n + \frac{c_p}{L} T^* \right) k_p + \frac{c_p}{L} (T_l - (1 - k_p) T_m). \end{cases} \tag{3.6}$$

By employing the bound $[0, 1]$ for the phase fraction and the truncation operation for the root, the unique solution for g_l will be obtained. It must be emphasized that the second step is crucial for accurately quantifying the change in temperature when it involves the phase change and the latent heat releases. The second step will also determine T^{n+1}. It is worth remarking that the cooling decreases temperature and cancels the latent heat at the initial stage, but when it comes to the eutectic temperature, phase change will occur isothermally; that is, the temperature at any point can only decrease once the material at that point has solidified completely.

An iterative scheme for the energy equation:

The fully decoupled scheme loses accuracy to some extent, while solving nonlinear systems directly is not expected. So, based on the operator-splitting idea, an iterative scheme is proposed here to enhance the accuracy at each time step. Given a tolerance ϵ and $T_0^{n+1} = T^n$. For $q \geq 0$, firstly we solve

$$\frac{\hat{T}_{q+1}^{n+1} - T^n}{\Delta t} + \nabla \cdot (\mathbf{u}^n T_q^{n+1}) = \nabla \cdot \left(\frac{k}{\rho_0 c_p} \nabla T_q^{n+1}\right) - \frac{L}{c_p} \frac{g_{l,q+1}^{n+1} - g_l^n}{\Delta t}, \quad (3.7)$$

where the subscript q represents the iteration number at the current time step. After we get the $\hat{T}_{q+1}$ from the first step, Secondly, we solve

$$\frac{T_{q+1}^{n+1} - \hat{T}_{q+1}^{n+1}}{\Delta t} + \nabla \cdot (\mathbf{u}^n T_{q+1}^{n+1} - \mathbf{u}^n T_q^{n+1}) = \nabla \cdot \left(\frac{k}{\rho_0 c_p} \nabla T_{q+1}^{n+1} - \frac{k}{\rho_0 c_p} \nabla T_q^{n+1}\right).$$
$$(3.8)$$

As the iteration progresses, if the difference between T_{q+1}^{n+1}, T_q^{n+1} and $\hat{T}_{q+1}$ becomes sufficiently small, then it is considered convergence. The convergence criteria is set: $\left\|T_{q+1}^{n+1} - T_q^{n+1}\right\|_{L^\infty} \leq \epsilon$ and $\left\|T_{q+1}^{n+1} - \hat{T}_{q+1}^{n+1}\right\|_{L^\infty} \leq \epsilon$ both hold. Otherwise, we continue the iteration by setting $q := q + 1$ and going back to compute a new T_{q+1}^{n+1}.

Remark 1. In the first step, solving the quadratic equation of the liquid fraction is still necessary. In contrast with a fully decoupled one, the coefficient b will be updated by $b := b - \frac{c_p}{L}(1 - k_p)\Delta t \gamma_q$ and c will be updated by $c := c - \frac{c_p}{L} k_p \Delta t \gamma_q$, where $\gamma_q = \nabla \cdot \left(\frac{k}{\rho c_p} \nabla T_q^{n+1}\right) - \nabla \cdot (\mathbf{u}^{n+1} T_q^{n+1})$.

3.2   The Semi-implicit Pressure Correction Method for Momentum Equation

$$\frac{\mathbf{u}^{n+1} - \mathbf{u}^n}{\Delta t} + \mathbf{u}^n \cdot \nabla \mathbf{u}^{n+1} = \nabla \cdot \left(\frac{\mu_l}{\rho_0} \nabla \mathbf{u}^{n+1}\right) - \frac{1}{\rho_0} \nabla p^{n+1} - \frac{\mu_l}{\rho_0} K^{-1} \mathbf{u}^{n+1} + \frac{(\rho_0 + \Delta \rho)}{\rho_0} \mathbf{g},$$
$$(3.9)$$

$$\nabla \cdot \mathbf{u}^{n+1} = 0. \quad (3.10)$$

The pressure correction scheme is used for solving velocity. Combining the divergence-free condition with the momentum equation gives

$$\frac{\mathbf{u}^* - \mathbf{u}^n}{\Delta t} + \mathbf{u}^n \cdot \nabla \mathbf{u}^* = \nabla \cdot \left(\frac{\mu_l}{\rho_0} \nabla \mathbf{u}^*\right) - \frac{1}{\rho_0} \nabla p^n - \frac{\mu_l}{\rho_0} K^{-1} \mathbf{u}^* + \frac{(\rho_0 + \Delta \rho)}{\rho_0} \mathbf{g}, \quad (3.11)$$

and the pressure correction Poisson equation

$$\nabla^2 \left(p^{n+1} - p^n\right) = \frac{\rho_0}{\Delta t} \nabla \cdot \mathbf{u}^*. \tag{3.12}$$

Then the velocity can be updated through pressure correction:

$$\frac{\mathbf{u}^{n+1} - \mathbf{u}^*}{\Delta t} = -\frac{1}{\rho_0} \nabla \left(p^{n+1} - p^n\right) \tag{3.13}$$

3.3   The Matrix-Based Technique for the Species Transport Equation

$$\frac{C^{n+1} - C^n}{\Delta t} + \nabla \cdot \left(\mathbf{u}^{n+1} C_l^{n+1}\right) = 0, \tag{3.14}$$

If we consider the diffusion effect in the liquid phase as well, we have

$$\frac{C^{n+1} - C^n}{\Delta t} + \nabla \cdot \left(\mathbf{u}^{n+1} C_l^{n+1}\right) = \nabla \cdot \left(g_l D_l \nabla C_l^{n+1}\right). \tag{3.15}$$

It must be noted that convection terms in the momentum equation, transport equation, and energy equation are all treated with an upwind implicit scheme. However, the main difficulty caused by the transport equation comes from C_l^{n+1} in the convection and diffusion terms, which differ from C^{n+1} in the temporal term. According to the explicit relationship between C_l and C:

$$C_l^{n+1} = \frac{C^{n+1}}{1 - (1 - k_p)(1 - g_l^{n+1})}, \tag{3.16}$$

in matrix form like:

$$C_l^{n+1} = A_g C^{n+1}, \tag{3.17}$$

where A_g is a diagonal matrix. Given the upwind coefficient matrix A_{con} and the Laplacian coefficient matrix A_{lap}, the temporal discrete transport equation in matrix and vector form can be expressed as:

$$\left(\frac{1}{\Delta t} I + A_{con} * A_g - A_{lap} * A_g\right) C^{n+1} = rhs, \tag{3.18}$$

where I is the identity matrix. The notation rhs denotes the right-hand side vector of the linear system.

This method with fully implicit scheme for C will enhance numerical stability and accuracy without introducing more explicit information of C^n in the discretization of the last two terms in the original expression used in traditional methods:

$$\frac{\partial C}{\partial t} + \nabla \cdot (\mathbf{u} C) = \nabla \cdot (g_l D_l \nabla C) + \nabla \cdot [g_l D_l \nabla (C_l - C)] - \nabla \cdot [(C_l - C) \mathbf{u}].$$

3.4   Updating the Concentration

With the determined g_l^{n+1} and C^{n+1}, the C_l^{n+1} and C_s^{n+1} can be updated by

$$C_l^{n+1} = \frac{C^{n+1}}{1 + \left(1 - g_l^{n+1}\right)(k_p - 1)}, \tag{3.19}$$

and

$$C_s^{n+1} = k_p C_l^{n+1}. \tag{3.20}$$

Now, the progress of single time step has been completed. The iterative scheme based on operator splitting and matrix-based techniques, enhances accuracy and numerical stability.

3.5   Spatial Discretization

The finite volume method based on staggered grids, as shown in Fig. 2 is applied. The computational domain $\Omega = [0, l_x] \times [0, l_y]$ includes a finite number of rectangular subdivisions. The mesh vertex points are located at:

$$x_i = i * h_x, \quad i = 0, 1, \cdots, n_x,$$

$$y_j = j * h_y, \quad j = 0, 1, \cdots, n_y,$$

where n_x and n_y are the number of meshes in each direction and $h_x = l_x/n_x$ and $h_y = l_y/n_y$ are mesh sizes. Four sets of mesh points (west-east edge points, south-north edge, cell-centered, vertex) are defined:

$$E_{we} = \left\{ \left(x_i, \frac{y_{j-1} + y_j}{2} \right) \mid i = 0, 1, \ldots, n_x; j = 1, 2, \ldots, n_y \right\},$$

$$E_{sn} = \left\{ \left(\frac{x_{i-1} + x_i}{2}, y_j \right) \mid i = 1, 2, \ldots, n_x; j = 0, 1, \ldots, n_y \right\},$$

$$E_c = \left\{ \left(\frac{x_{i-1} + x_i}{2}, \frac{y_{j-1} + y_j}{2} \right) \mid i = 1, 2, \ldots, n_x; j = 1, 2, \ldots, n_y \right\},$$

$$E_v = \{ (x_i, y_j) \mid i = 0, 1, \cdots, n_x; j = 0, 1, \cdots, n_y \}.$$

The east and north interfaces of one cell are considered the front interfaces. On the contrary, the west and south interfaces are the back interfaces. Next, we have the discretized spaces:

$$U_h = \{ u : E_{we} \to \mathbb{R} \}, \quad V_h = \{ v : E_{sn} \to \mathbb{R} \},$$

$$P_h = \{ P : E_c \to \mathbb{R} \}, \quad N_h = \{ interpolation : E_v \to \mathbb{R} \}.$$

U_h, V_h are for physical variables u, v and all other physical scalar quantities (pressure P, temperature T, concentration C, phase fraction g_l and g_s, etc.) correspond to P_h. The space N_h is for the interpolation of coefficients such as

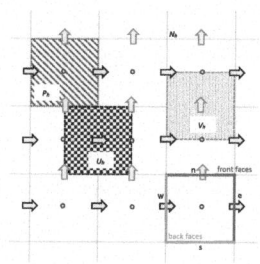

Fig. 2. Spatial discretization based on the staggered grid

μ. Multi-physical governing equations all obey the general convection-diffusion form:

$$\frac{\partial \phi}{\partial t} + \nabla \cdot (\mathbf{u}\phi) = \nabla \cdot (D\nabla\phi) + Sr(\phi), \qquad (3.21)$$

where ϕ denotes a general physical variable and D is the general diffusion coefficient; the convection velocity is denoted by $\mathbf{u}$ and $S_r(\phi)$ represents the source term. It is worth mentioning that, for the Darcy-like damping source term in the momentum equation, the fully implicit scheme for $\mathbf{u}^{n+1}$ is vital as well, which corresponds to $S_r(\phi_{i,j}^{n+1})$ in (3.21).

4   Numerical Simulation

This section presents simulations and analyses of a numerical benchmark case using the proposed methods. The solidification and macro-segregation processes of two common binary alloy systems, Sn-10%Pb and Pb-48%Sn, are examined. Data on the physical properties of these two kinds of alloys can be referred to [1,18]. The findings on physical phenomena, as well as the evolution of velocity, temperature, phase fraction, and chemical component distributions, are presented. All numerical simulations were conducted on a MacOS Mojave system equipped with a 2.5GHz quad-core Intel Core i7 processor. The codes were developed from scratch using Matlab.

Example: Solidification of Sn-10%Pb Alloy in a 2D Domain. This benchmark case examines the solidification of a Sn-10%Pb binary alloy within a rectangular cavity [1]. The computational domain, initial conditions, and boundary conditions are illustrated in Fig. 3. Initially, the cavity is filled with a still liquid alloy at a uniform temperature $T_0 = T_l$ and a consistent chemical concentration C_0. At $t = 0$, the solidification process is triggered by cooling the left and right walls of the cavity via natural convection, described by the heat transfer coefficient h_T:

$$q_T = h_T(T - T_{\text{ext}}).$$

The top and bottom walls of the cavity are thermally insulated, and it is assumed that the cavity walls are rigid and nonslip. Phase transition occurs

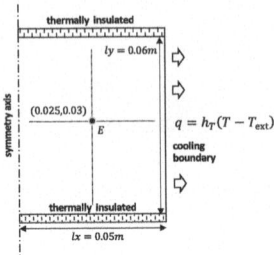

Fig. 3. Schematic of physical settings in Example 1

within the cavity, with inner flow driven by thermal and solutal buoyancy forces. A central observation point E is established to track the evolution of the flow and concentration, allowing us to validate the accuracy of the proposed scheme.

Numerical results were obtained using a 150×180 mesh with a time step of $\Delta t = 5 \times 10^{-3} s$. Figures show the solidification process at $t = 5s, 38s, 168s, 350s$, depicting temperature, concentration, liquid fraction, velocity, streamlines, and phase interfaces. The $g_l = 0.99$ and $g_l = 0.01$ contours represent the liquid/mush and mush/solid interfaces.

At the initial stage of solidification ($t = 5s$ and $t = 38s$), thermal buoyancy dominates, causing intense downward flow near the liquid/mush interface, with a clockwise circulation inside the liquid (see Figs. 4(a) and 5(a)). The Pb-enriched melt increases density, intensifying downward flow. At $t = 5s$, the liquid/mush interface follows the temperature contour, but by $t = 38s$, the interface deviates due to higher Pb concentration (see Fig. 4(a), 6(b)).

From $t = 38s$ to $168s$, the high Pb concentration lowers the liquidus temperature, causing the bottom section to reach mush last. Severe flow transports Pb, forming a platform-shaped Pb-enriched region and raising it along the cavity's centerline. The temperature gradient and component transport cause bending of liquid fraction contours, resulting in channel segregation (see Fig. 5(c)).

At $t = 168s$, the channels develop, and most of the domain is mush, except for the left-bottom corner (see Fig. 5(c)). A thin isothermal phase transition layer appears at the right-bottom corner signaling the approach of full solidification.

By $t = 350s$, the mush/solid interface reaches the center, and solidification completes by $t = 450s$, with the channel segregation pattern matching previous results [5,13] (see Fig. 5(d), 7(b)). Figure 7(a) shows the velocity magnitude change at point E, which aligns well with our predictions, exhibiting two peaks and one valley. The first peak corresponds to the most intense upward flow, and the second peak, higher than the first, corresponds to the severe downward flow near the liquid/mush interface. The minimum velocity occurs when point E is at the center of the vortex.

Figure 7(b) shows the liquid fraction profile at point E. The final segregation results are consistent with those of Combeau et al. [5] and Shen et al. [13]. At around $t = 350s$, the liquid fraction drops sharply, indicating the transition

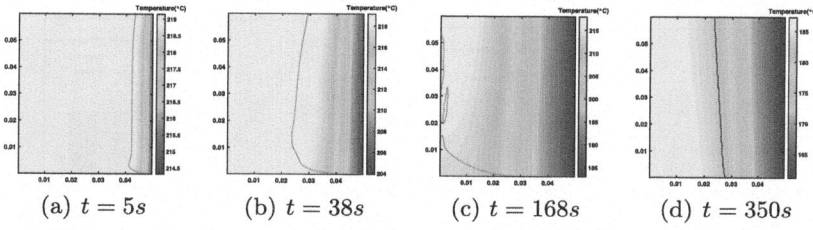

Fig. 4. Evolution of the temperature field with time; the liquid/mush interface (magenta dotted line); the mush/solid interface (black dotted line).

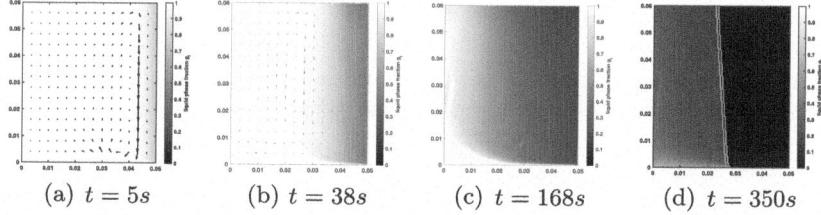

Fig. 5. Evolution of liquid fraction with time in Example 1; the iso-thermal phase transition layer (enclosed by green solid line); liquid fraction (colorbar). (Color figure online)

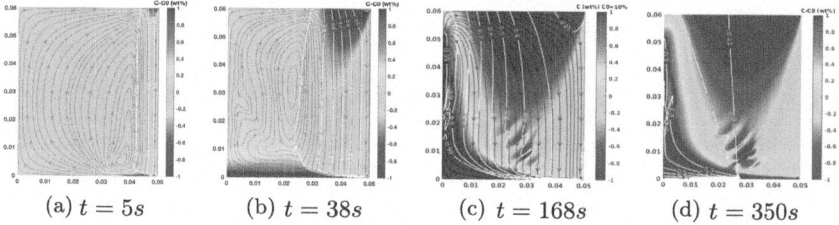

Fig. 6. Evolution of concentration variation $C - C_0$ with time; the liquid phase fraction contour line (white solid line); the streamline (black solid line with arrow); $C - C_0$ (colorbar)

from a non-isothermal to an isothermal process, with latent heat being released during phase change while the temperature remains constant until complete solidification. This confirms that our scheme effectively captures these critical properties and agrees with other benchmark results.

In Fig. 4, the L_∞ norms of the differences among intermediate temperatures and temperatures of neighboring iterations decrease continuously. The convergence criteria are set to be $\epsilon = 10^{-6}$ and these two kinds of norms will be less than the criteria within 38 iterations. The iterative method can obtain more accurate results for each time step and less numerical error accumulated through the entire dynamic physical process. To assess the accuracy, we can examine the final segregation maps depicted in Fig. 8. The patterns obtained from our fully

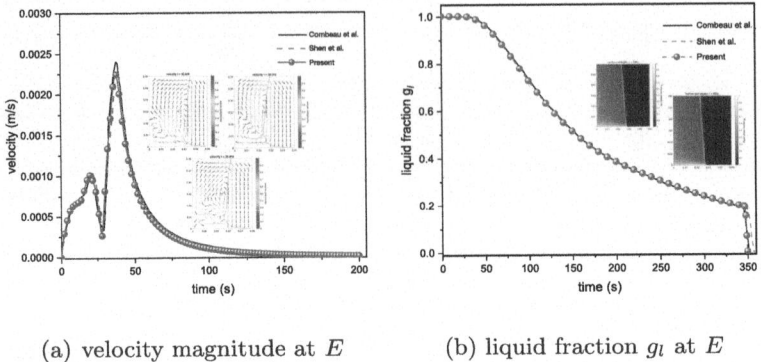

(a) velocity magnitude at E (b) liquid fraction g_l at E

Fig. 7. (a) Time-dependent profile of velocity magnitude at central sample point E; (b) time-dependent profile of liquid fraction g_l at central sample point E.

decoupled code are like the results of Shen's decoupled method [13]. The iterative scheme we employ yields a greater concentration of enrichment in the segregated channel layers, and the channel pattern is more agreeable to the results reported by Combeau et al. [5], which were simulated by high-order algorithms developed by the Institute Jean Lamour and the Fluent software.

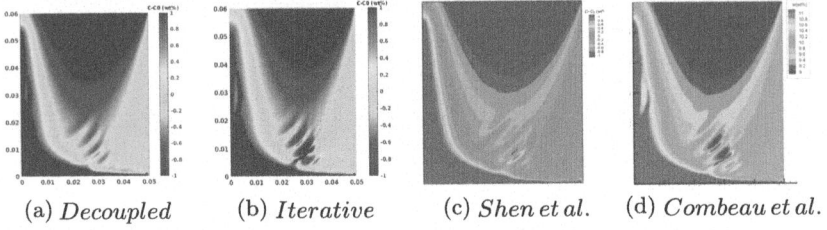

(a) *Decoupled*   (b) *Iterative*   (c) *Shen et al.*   (d) *Combeau et al.*

Fig. 8. The final segregation maps computed via (a) fully decoupled scheme; (b)iterative scheme; (c) and (d) are reference results from literature

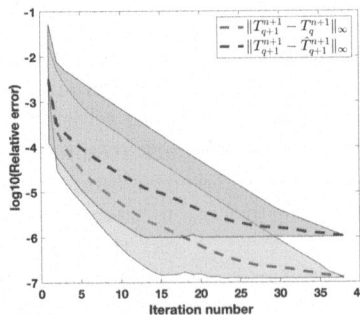

Fig. 9. Profile of the L_∞ norms with the number of iterations and their deviation(averaged by multiple time steps).

5   Conclusions

This work presents a complete mathematical model for solidification and macrosegregation, emphasizing their broad applicability. We propose an iterative scheme and matrix-based methods to address these problems. A classic 2D benchmark case is presented to validate the accuracy and computational efficiency of the approach. The results computed by this iterative method demonstrate better convergence and accuracy compared to traditional fully decoupled methods, with contours showing good agreement with numerical results from high-order algorithms and commercial software. Future work can enhance this framework by incorporating real-phase equilibrium relations like equations of state.

Acknowledgments. This work is partially supported by King Abdullah University of Science and Technology (KAUST) through the grants BAS/1/1351–01, URF/1/5028–01. The work is also supported by National Key Research and Development Project of China (Grant Number: 2023YFA1011702), National Natural Science Foundation of China (Grant Numbers: 12122115, 51936001).

Disclosure of Interests. The authors declare that they have no known competing financial interests to influence the work reported in this paper.

References

1. Bellet, M., et al.: Call for contributions to a numerical benchmark problem for 2d columnar solidification of binary alloys. Int. J. Therm. Sci. **48**(11), 2013–2016 (2009)
2. Bellet, M., Fachinotti, V.D.: ALE method for solidification modelling. Comput. Methods Appl. Mech. Eng. **193**(39–41), 4355–4381 (2004)
3. Bennon, W., Incropera, F.: A continuum model for momentum, heat and species transport in binary solid-liquid phase change systems—i. model formulation. Int. J. Heat Mass Transf. **30**(10), 2161–2170 (1987)
4. Chakraborty, S., Chatterjee, D.: An enthalpy-based hybrid lattice-boltzmann method for modelling solid-liquid phase transition in the presence of convective transport. J. Fluid Mech. **592**, 155–175 (2007)
5. Combeau, H., et al.: A numerical benchmark on the prediction of macrosegregation in binary alloys. Proceedings of frontiers in solidification science. Warrendale, USA: TMS2011 (2011)
6. Feng, X., Chen, M.H., Wu, Y., Sun, S.: A fully explicit and unconditionally energy-stable scheme for Peng-Robinson VT flash calculation based on dynamic modeling. J. Comput. Phys. **463**, 111275 (2022)
7. Feng, X., Qiao, Z., Sun, S., Wang, X.: An energy-stable smoothed particle hydrodynamics discretization of the Navier-Stokes-Cahn-Hilliard model for incompressible two-phase flows. J. Comput. Phys. **479**, 111997 (2023)
8. Ghoneim, A.Y.: A smoothed particle hydrodynamics-phase field method with radial basis functions and moving least squares for meshfree simulation of dendritic solidification. Appl. Math. Model. **77**, 1704–1741 (2020)

9. Hachani, L., Zaidat, K., Fautrelle, Y.: Experimental study of the solidification of sn–10 wt.% pb alloy under different forced convection in benchmark experiment. Int. J. Heat Mass Transf. **85**, 438–454 (2015)

10. Hebditch, D.J., Hunt, J.D.: Observations of ingot macrosegregation on model systems. Metall. Mater. Trans. B. **5**(7), 1557–1564 (1974). https://doi.org/10.1007/BF02646326

11. Hong, Y., Ye, W.B., Du, J., Huang, S.M.: Solid-liquid phase-change thermal storage and release behaviors in a rectangular cavity under the impacts of mushy region and low gravity. Int. J. Heat Mass Transf. **130**, 1120–1132 (2019)

12. Kuritani, T.: Magmatic differentiation examined with a numerical model considering multicomponent thermodynamics and momentum, energy and species transport. Lithos **74**(3–4), 117–130 (2004)

13. Li, W., Shen, H., Liu, B.: A response to numerical benchmark for solidification of binary alloys: macro-segregation with thermo-solutal convection. China Foundry **9**(2) (2012)

14. Pardeshi, R., Voller, V., Singh, A., Dutta, P.: An explicit-implicit time stepping scheme for solidification models. Int. J. Heat Mass Transf. **51**(13–14), 3399–3409 (2008)

15. Prakash, C., Voller, V.: On the numerical solution of continuum mixture model equations describing binary solid-liquid phase change. Numer. Heat Transf. Part B Fundam. **15**(2), 171–189 (1989)

16. Song, R., Feng, X., Wang, Y., Sun, S., Liu, J.: Dissociation and transport modeling of methane hydrate in core-scale sandy sediments: a comparative study. Energy **221**, 119890 (2021)

17. Sun, D., Tao, W.: A coupled volume-of-fluid and level set (VOSET) method for computing incompressible two-phase flows. Int. J. Heat Mass Transf. **53**(4), 645–655 (2010)

18. Voller, V.R., Cross, M., Markatos, N.: An enthalpy method for convection/diffusion phase change. Int. J. Numer. Meth. Eng. **24**(1), 271–284 (1987)

19. Wang, M., Yu, G., Zhang, X., Zhang, T., Yu, B., Sun, D.: Numerical investigation of melting of waxy crude oil in an oil tank. Appl. Therm. Eng. **115**, 81–90 (2017)

20. Wang, Z., Li, J., Liu, S., Zhao, J., Wang, X., Yang, W.: Investigation on freckle formation of nickel-based single crystal superalloy specimens with suddenly reduced cross section. J. Alloy. Compd. **918**, 165631 (2022)

21. Yang, X.: Efficient linear, stabilized, second-order time marching schemes for an anisotropic phase field dendritic crystal growth model. Comput. Methods Appl. Mech. Eng. **347**, 316–339 (2019)

22. Zaeem, M.A., Yin, H., Felicelli, S.D.: Modeling dendritic solidification of al-3% cu using cellular automaton and phase-field methods. Appl. Math. Model. **37**(5), 3495–3503 (2013)

Robust, Efficient, and Long-Time Accurate Schemes to Simulate Gas Storage in Geological Formation

Huangxin Chen[1,2], Yuxiang Chen[2], Jisheng Kou[3], Shuyu Sun[1(✉)], Dunhui Xiao[1], Xuejun Xu[1], Haitao Yu[4], Tao Zhang[5], and Xiaoying Zhuang[6]

[1] School of Mathematical Sciences, Tongji University, Shanghai 200092, China
`{suns,xiaodunhui,xuxj}@tongji.edu.cn`
[2] School of Mathematical Sciences and Fujian Provincial Key Laboratory on Mathematical Modeling and High Performance Scientific Computing, Xiamen University, Fujian 361005, China
`chx@xmu.edu.cn, chenyuxiang@stu.xmu.edu.cn`
[3] Key Laboratory of Rock Mechanics and Geohazards of Zhejiang Province, School of Civil Engineering, Shaoxing University, Shaoxing, Zhejiang 312000, China
`koujisheng@163.com`
[4] State Key Laboratory of Disaster Reduction in Civil Engineering, Tongji University, Shanghai 200092, China
`yuhaitao@tongji.edu.cn`
[5] The School of New Energy, China University of Petroleum (East China), Qingdao, Shaodong 266580, China
`tao.zhang@upc.edu.cn`
[6] Department of Geotechnical Engineering, College of Civil Engineering, Tongji University, Shanghai 200092, China
`xiaoyingzhuang@tongji.edu.cn`

Abstract. In this paper, we consider the numerical simulation of gas storage in geological formation in the context of hydrogen underground storage and carbon dioxide geological sequestration. We construct two energy-stable numerical schemes: one based on the energy factorization approach, which rigorously preserves the energy dissipation principle and combines discontinuous Galerkin approximations with mixed finite elements for spatial discretization; the other based on a stabilization approach, which conserves the original energy functional, has an adaptive stabilization parameter and time-stepping strategy, and ensures the boundedness of molar density. Through numerical experiments with methane gas, our schemes are validated in terms of capturing coupled hydro-mechanical processes, handling strong nonlinearities, and maintaining conservation properties.

Keywords: Efficiency of numerical schemes · Gas storage in geological Formation · Temporal discretization · Long-time accurate simulation · Spatial discretization · Robustness

© The Author(s), under exclusive license to Springer Nature Switzerland AG 2025
M. H. Lees et al. (Eds.): ICCS 2025, LNCS 15905, pp. 213–227, 2025.
https://doi.org/10.1007/978-3-031-97632-2_15

1   Introduction

Geological sequestration of carbon dioxide and underground storage of hydrogen are two critical applications of geological storage technologies, with profound implications for global energy utilization and environmental protection [5,10]. Numerical simulation plays a pivotal role in geological storage research, but its accuracy and stability are influenced by various factors. For instance, higher-order numerical algorithms are required to improve computational precision while ensuring local mass conservation and avoiding numerical dispersion and non-physical oscillations. Additionally, large time-step calculations must balance computational efficiency with stability and reliability. Furthermore, when simulating multiphase flow in geological environments, the phase behavior of fluids is particularly critical, especially the impact of temperature, pressure, and salt concentration on interfacial tension, which further complicates the simulation [15,17].

This study focuses on the numerical simulation of single-phase gas flow with compressible gas and rock in the context of hydrogen underground storage and carbon dioxide geological sequestration, exploring the construction of governing equations, numerical challenges, and corresponding strategies. The numerical implementation of our physical model must strictly adhere to thermodynamic consistency, as methods violating this principle can lead to unphysical or unstable simulation results [5,11]. Various approaches have been developed to handle Helmholtz free energy in numerical schemes. The convex splitting method [9] has been widely used for both single- and multi-component systems, providing nonlinear energy-stable schemes. However, it requires solving nonlinear equations, often at high computational cost. Alternative strategies include stabilization methods [19], exponential time-differencing [7], and the Invariant Energy Quadratization (IEQ) [21] and Scalar Auxiliary Variable (SAV) approaches [18]. While the IEQ and SAV methods yield linear, easily implementable energy-stable schemes, their modified energy functionals deviate from the original. The Energy Factorization (EF) method [12,13] provides an alternative approach that maintains the original energy dissipation structure while offering computational efficiency. The stabilization method proposed by [14] integrates features from existing approaches, preserving the original energy functional and enabling linear energy-stable formulations. Maintaining the physical bounds of molar density $0 < c < \frac{1}{\beta}$ is critical in simulating compressible gas flow through porous media. Deviations from these bounds yield non-physical solutions, necessitating numerical schemes that intrinsically enforce these constraints. Recent advances in bounding techniques include the constraint enforcement methods such as Lagrange multiplier approaches [2] and variational formulations [8], a posteriori corrections such as post-processing [22] and cut-off strategies [20], and the energy-based methods such as nonlinear convex splitting [1,6].

The paper is organized as follows. In Sect. 2, we introduce the formulation of gas flow model with rock compressibility. In Sect. 3, we introduce two energy-stable numerical schemes and discuss the properties they satisfy. In Sect. 4, numerical results are presented to verify the features of the proposed scheme.

2   Mathematical Model of Gas Flow in Poroelastic Media

For the convenience of presentation, we focus on single-phase gas flow with compressible gas and rock. We consider a fully coupled thermodynamic mathematical model for single-phase gas flow in porous media, incorporating the compressibility of both gas and rock.

$$\nabla \cdot \sigma(\boldsymbol{u}_s, p) = 0, \qquad \text{in } \Omega_t =: \Omega \times (0, t), \qquad (1a)$$

$$\frac{\partial(\phi c)}{\partial t} + \nabla \cdot (\boldsymbol{u}_f c) = 0, \qquad \text{in } \Omega_t, \qquad (1b)$$

$$\boldsymbol{u}_f = -\lambda(\phi) c \nabla \mu, \qquad \text{in } \Omega_t, \qquad (1c)$$

$$p = c\mu(c) - f(c), \qquad \text{in } \Omega_t, \qquad (1d)$$

$$\frac{\partial \phi}{\partial t} = \frac{1}{N} \frac{\partial p}{\partial t} + \alpha \nabla \cdot \mathbf{v}_s, \qquad \text{in } \Omega_t, \qquad (1e)$$

where $\lambda(\phi) = \frac{\kappa(\phi)}{\nu}$ is the mobility, ν is the viscosity of gas, $\kappa(\phi) = \kappa_0 (\frac{\phi}{\phi_r})^3 (\frac{1-\phi_r}{\phi})^2$ is the permeability, κ_0 is the initial intrinsic permeability and ϕ_r is the porosity at the reference pressure. $\sigma(\boldsymbol{u}_s, p) = 2\eta\varepsilon(\boldsymbol{u}_s) + \gamma \operatorname{div}(\boldsymbol{u}_s)\mathbf{I} - \alpha p\mathbf{I}$ is the stress tensor, $\varepsilon(\boldsymbol{u}_s) = \frac{1}{2}(\nabla\boldsymbol{u}_s + \nabla\boldsymbol{u}_s^T)$ is the strain tensor, $\mathbf{I}$ is the unit tensor, $\boldsymbol{u}_s$ is the displacement of solid, $\mathbf{v}_s = \frac{\partial \boldsymbol{u}_s}{\partial t}$ is the velocity of solid, η and γ are the Lamé parameters, α is the Biot's coefficient, N is the Biot's modulus, c is the molar density of gas. The Helmholtz free energy density $f(c)$ and chemical potential μ determined by the Peng-Robinson equation of state have the following expression

$$f(c) = f_{ide}(c) + f_{rep}(c) + f_{att}(c), \qquad (2a)$$

$$f_{\text{ide}}(c) = cRT \ln(c), \quad f_{\text{rep}}(c) = -cRT \ln(1 - \beta c), \qquad (2b)$$

$$f_{att}(c) = \frac{b(T)c}{2\sqrt{2}\beta} \ln \left(\frac{1 + (1 - \sqrt{2})\beta c}{1 + (1 + \sqrt{2})\beta c} \right), \qquad (2c)$$

$$\mu(c) = f'(c). \qquad (2d)$$

The pressure is given by the volumetric EoS: $p = \frac{cRT}{1-\beta c} - \frac{bc^2}{1+2\beta c-\beta^2 c^2}$. In order to close the system, the following boundary conditions are imposed

$$\sigma(\boldsymbol{u}_s, p) \cdot \boldsymbol{n} = 0, \qquad \text{on } \partial\Omega, \qquad (3)$$

$$\boldsymbol{u}_f \cdot \boldsymbol{n} = 0, \qquad \text{on } \partial\Omega. \qquad (4)$$

The total free energy of (1) and (3) in Ω can be defined as

$$E(t) = \int_\Omega \left(\phi f(c) + \frac{1}{2}\sigma_e(\boldsymbol{u}_s) : \varepsilon(\boldsymbol{u}_s) + \frac{1}{2N}p^2 \right) d\boldsymbol{x}. \qquad (5)$$

3   Two Energy-Stable Numerical Schemes

For thermodynamically consistent models of gas flow in porous media, it is essential to construct numerical schemes that strictly adhere to the energy dissipation law, thereby improving computational stability and efficiency. This section introduces two energy-stable numerical schemes based on an improved energy factorization method and a stabilization strategy, respectively, to guarantee energy stability in numerical computations and optimize computational efficiency.

3.1   Energy-Stable Scheme Based on Energy Factorization Approach

The first numerical approach utilizes an energy factorization method for treating the Helmholtz free energy density while implementing a semi-implicit time-stepping formulation. This method rigorously preserves the energy dissipation principle by precisely computing the pressure field through chemical potential and Helmholtz free energy. The spatial discretization framework combines discontinuous Galerkin approximations [16] with mixed finite elements, incorporating upwind flux treatment to maintain both mass conservation properties and numerical stability.

The time semi-discretized chemical potential μ^{n+1} is derived through a thermodynamically consistent energy factorization methodology [13], framework for constructing energy-stable semi-discrete chemical potentials by leveraging the convexity and concavity properties of different components of the free energy function. we get the linearized semi-discretized chemical potential

$$\mu^{n+1} = \mu_{ir}\left(c^{n+1}, c^n\right) + \mu_{att}\left(c^{n+1}, c^n\right)$$

where

$$\mu_{ir}\left(c^{n+1}, c^n\right) = RT\left(\ln\left(c^n\right) - \ln\left(1 - \beta c^n\right)\right) + RTc^{n+1}\left(\frac{1}{c^n} + \frac{\beta}{1 - \beta c^n}\right), \quad (6a)$$

$$\mu_{att}\left(c^{n+1}, c^n\right) = \frac{b}{2\sqrt{2}\beta}\left(\ln\left(1 + (1 - \sqrt{2})\beta c^n\right) + \frac{(1 - \sqrt{2})\beta c^{n+1}}{1 + (1 - \sqrt{2})\beta c^n}\right) \quad (6b)$$
$$- \frac{b}{2\sqrt{2}\beta}\left(\ln\left(1 + (1 + \sqrt{2})\beta c^n\right) + \frac{(1 + \sqrt{2})\beta c^n}{1 + (1 + \sqrt{2})\beta c^n}\right).$$

Building upon the semi-discrete formulation of the chemical potential derived above, we rigorously establish the following free energy dissipation inequality that governs the thermodynamic consistency of the numerical scheme:

$$f_{ide}\left(c^{n+1}\right) + f_{rep}\left(c^{n+1}\right) - f_{ide}\left(c^n\right) - f_{rep}\left(c^n\right) \quad (7)$$
$$\leq \mu_{ir}\left(c^{n+1}, c^n\right)\left(c^{n+1} - c^n\right),$$

$$f_{att}\left(c^{n+1}\right) - f_{att}\left(c^n\right) \leq \mu_{att}\left(c^{n+1}, c^n\right)\left(c^{n+1} - c^n\right). \quad (8)$$

By systematically integrating the discrete chemical potential formulation derived through energy factorization method [13] with a semi-implicit Euler temporal discretization framework, we rigorously construct the following energy-stable semi-discrete system:

$$- \nabla \cdot \sigma(\boldsymbol{u}_s^{n+1}, p^{n+1}) = 0, \tag{9a}$$

$$D_\tau(\phi^{n+1} c^{n+1}) + \nabla \cdot (\boldsymbol{u}_f^{n+1} c^n) = 0, \tag{9b}$$

$$\boldsymbol{u}_f^{n+1} = -\lambda(\phi^n) c^n \nabla \mu^{n+1}, \tag{9c}$$

$$p^{n+1} = c^n \mu^{n+1} - f(c^n), \tag{9d}$$

$$D_\tau \phi^{n+1} = \frac{1}{N} D_\tau p^{n+1} + \alpha \nabla \cdot D_\tau \boldsymbol{u}_s^{n+1}. \tag{9e}$$

We employ a temporally uniform discretization parameter defined as $\tau = t_{n+1} - t_n$, where B^n represents the discrete approximation of molar density at temporal node t_n The backward difference operator is systematically defined for all primary variables $B \in \{\boldsymbol{u}_s, \phi, p, c, E\}$ through the discrete temporal derivative:

$$D_\tau B^{n+1} := \frac{B^{n+1} - B^n}{\tau}.$$

The energy dissipation property of system (9a), rigorously proven in Theorem 3.2 of [4].

Theorem 1. ([4]) Let $\sigma(\boldsymbol{u}_s^{n+1}, p^{n+1}) \cdot \boldsymbol{n} = 0, \boldsymbol{u}_f^{n+1} \cdot \boldsymbol{n} = 0$ on the boundary $\partial\Omega$, where $\boldsymbol{n}$ denotes the normal unit outward vector to $\partial\Omega$. We assume that $0 < c^n < \frac{\varrho}{\beta}, n \geq 0$. Then the scheme (9) follows an energy dissipation law as

$$D_\tau E^{n+1} \leq 0,$$

where

$$E^{n+1} = \int_\Omega \left(\phi^{n+1} f(c^{n+1}) + \frac{1}{2} \sigma_e(\boldsymbol{u}_s^{n+1}) : \varepsilon(\boldsymbol{u}_s^{n+1}) + \frac{1}{2N} |p^{n+1}|^2 \right) \, d\boldsymbol{x}$$

is the semidiscrete total energy at the time t_{n+1}.

Let $\mathcal{K}_h$ be a family of nondegenerate, quasi-uniform partitions of Ω composed of triangles or quadrilaterals if $d = 2$, or tetrahedra, prisms, or hexahedra if $d = 3$. Define $\mathcal{E}_h$ as the set of all faces ($d = 3$) or edges ($d = 2$) of $\mathcal{K}_h$, and let h_T be the diameter of any element $K \in \mathcal{K}_h$. The set of interior edges or faces in $\mathcal{E}_h$ is denoted by $\mathcal{E}_h^I$. The standard finite element space of d-dimensional vector fields, whose components are piecewise linear polynomials, is given by:

$$\mathcal{V}_h := \left\{ \psi \in [L^2(\Omega)]^d : \psi|_K \in \mathbb{P}_1^d(K), \forall K \in \mathcal{K}_h \right\}.$$

We now define the average and jump operators for $\psi \in \mathcal{V}$. Given two neighboring elements $K_i, K_j \in \mathcal{K}_h$ and an interface $e = \partial K_i \cap \partial K_j \in \mathcal{E}_h^I$ with outward unit normal vector $\boldsymbol{n}_e$ exterior to K_i, we define:

$$\{\psi\} := \frac{1}{2}((\psi|_{K_i})|_e + (\psi|_{K_j})|_e), \quad [\psi] := \left(\psi|_{K_i}\right)\big|_e - \left(\psi|_{K_j}\right)\big|_e,$$

here, $\psi|_{K_i}$ represents the value of ψ in K_i.

The inner product on an edge or face e is given by $\langle \cdot, \cdot \rangle_e$, the associated norms are denoted by $\| \cdot \|_{L^2(e)}$. Next, we introduce the lowest-order Raviart-Thomas (RT_0) mixed finite element space, which will be used in the spatial discretization. On a simplicial mesh, the space RT_0 is defined as: $RT_0 = [\mathbb{P}_0]^d + \boldsymbol{x}\mathbb{P}_0$, where $\mathbb{P}_0$ represents the space of piecewise constant functions. We define the following finite element spaces:

$$\mathcal{U}_h = \left\{ \mathbf{v} \in H(\mathrm{div}, \Omega) : \mathbf{v}|_K \in RT_0(K), \forall K \in \mathcal{K}_h \right\},$$
$$\mathcal{Q}_h = \left\{ q \in L^2(\Omega) : q|_K \in \mathbb{P}_0(K), \forall K \in \mathcal{K}_h \right\},$$

where $H(\mathrm{div}, \Omega)$ is defined as: $H(\mathrm{div}, \Omega) = \left\{ \mathbf{v} \in [L^2(\Omega)]^d : \nabla \cdot \mathbf{v} \in L^2(\Omega) \right\}$.

Additionally, we define the space with homogeneous normal boundary conditions:

$$\mathcal{U}_h^0 = \left\{ \mathbf{v} \in \mathcal{U}_h : \mathbf{v} \cdot \boldsymbol{n} = 0 \text{ on } \partial\Omega \right\}.$$

Next, we develop a fully discrete numerical scheme using the mixed finite element method with upwind scheme that strictly maintains both energy stability and mass conservation. The scheme combines an upwind treatment of advection terms with a carefully balanced implicit-explicit temporal discretization, ensuring robust performance.

For any $\mathbf{v}_h \in \mathcal{V}_h$, $\mathbf{w}_h \in \mathcal{U}_h^0$, $q_h, z_h, \varphi_h \in \mathcal{Q}_h$, we determine $\boldsymbol{u}_{s,h}^{n+1} \in \mathcal{V}_h$, $\boldsymbol{u}_{f,h}^{n+1} \in \mathcal{U}_h$, $c_h^{n+1}, \phi_h^{n+1}, p_h^{n+1} \in \mathcal{Q}_h$ such that:

$$\mathcal{A}(\boldsymbol{u}_{s,h}^{n+1}, p_h^{n+1}, \mathbf{v}_h) = 0, \tag{10a}$$

$$(D_\tau(\phi_h^{n+1} c_h^{n+1}), q_h) + \sum_{e \in \mathcal{E}_h^I} \langle c_h^{n*} \boldsymbol{u}_{f,h}^{n+1} \cdot \boldsymbol{n}, [q_h] \rangle_e = 0, \tag{10b}$$

$$(\lambda^{-1}(\phi_h^n) \boldsymbol{u}_{f,h}^{n+1}, \mathbf{w}_h) = \sum_{e \in \mathcal{E}_h^I} \langle [\mu_h^{n+1}], c_h^{n*} \mathbf{w}_h \cdot \boldsymbol{n} \rangle_e, \tag{10c}$$

$$(p_h^{n+1}, z_h) = (c_h^n \mu_h^{n+1} - f(c_h^n), z_h), \tag{10d}$$

$$(D_\tau \phi_h^{n+1}, \varphi_h) = \frac{1}{N}(D_\tau p_h^{n+1}, \varphi_h) + \alpha(D_\tau(\nabla \cdot \boldsymbol{u}_{s,h}^{n+1}), \varphi_h) \tag{10e}$$
$$- \alpha \sum_{e \in \mathcal{E}_h^I} \langle \{\varphi_h \boldsymbol{n}_e\}, [D_\tau \boldsymbol{u}_{s,h}^{n+1}] \rangle_e.$$

The bilinear form $\mathcal{A}$ is defined as:

$$\mathcal{A}(\boldsymbol{u}_{s,h}, p_h, \mathbf{v}_h) := \sum_{K \in \mathcal{K}_h} (\sigma_e(\boldsymbol{u}_{s,h}), \varepsilon(\mathbf{v}_h))_K - \sum_{e \in \mathcal{E}_h^I} \langle \{\sigma_e(\boldsymbol{u}_{s,h}) \boldsymbol{n}_e\}, [\mathbf{v}_h] \rangle_e \tag{11}$$
$$- \alpha \sum_{K \in \mathcal{K}_h} (p_h, \nabla \cdot \mathbf{v})_K + \alpha \sum_{e \in \mathcal{E}_h^I} \langle \{p_h \boldsymbol{n}_e\}, [\mathbf{v}_h] \rangle_e$$
$$- \sum_{e \in \mathcal{E}_h^I} \langle [\boldsymbol{u}_{s,h}], \{\sigma_e(\mathbf{v}_h) \boldsymbol{n}_e\} \rangle_e + \sum_{e \in \mathcal{E}_h^I} \frac{\varsigma_2}{h_e} \langle [\boldsymbol{u}_{s,h}], [\mathbf{v}_h] \rangle_e.$$

Here, ς_2 is a penalty parameter and $h_e = |e|^{\frac{1}{d-1}}$.

To handle rock compressibility, the scheme (10) forms a nonlinear system, which we solve iteratively. Given c_h^n, $\boldsymbol{u}_{s,h}^n$, $\boldsymbol{u}_{f,h}^n$, ϕ_h^n, and p_h^n, we initialize:

$$c_h^{n+1,0} = c_h^n, \quad \boldsymbol{u}_{s,h}^{n+1,0} = \boldsymbol{u}_{s,h}^n, \quad \boldsymbol{u}_{f,h}^{n+1,0} = \boldsymbol{u}_{f,h}^n,$$
$$\phi_h^{n+1,0} = \phi_h^n, \quad p_h^{n+1,0} = p_h^n.$$

For $l \geq 0$, we solve the linear system:

$$\mathcal{A}(\boldsymbol{u}_{s,h}^{n+1,l+1}, p_h^{n+1,l+1}, \mathbf{v}_h) = 0,$$

$$(D\tau(\phi_h^{n+1,l} c_h^{n+1,l+1}), q_h) + \sum_{e \in \mathcal{E}_h^I} \langle c_h^{n*} \boldsymbol{u}_{f,h}^{n+1,l+1} \cdot \boldsymbol{n}, [q_h] \rangle_e = 0,$$

$$(\lambda^{-1}(\phi_h^n) \boldsymbol{u}_{f,h}^{n+1,l+1}, \mathbf{w}_h) = \sum_{e \in \mathcal{E}_h^I} \langle [\mu_h^{n+1,l+1}], c_h^{n*} \mathbf{w}_h \cdot \boldsymbol{n} \rangle_e,$$

$$(p_h^{n+1,l+1}, z_h) = (c_h^n \mu_h^{n+1,l+1} - f(c_h^n), z_h),$$

$$(D_\tau \phi_h^{n+1,l+1}, \varphi_h) = \frac{1}{N}(D_\tau p_h^{n+1,l+1}, \varphi_h) + \alpha(D_\tau(\nabla \cdot \boldsymbol{u}_{s,h}^{n+1,l+1}), \varphi_h)$$

$$- \alpha \sum_{e \in \mathcal{E}_h^I} \langle \{\varphi_h \boldsymbol{n}_e\}, [D_\tau \boldsymbol{u}_{s,h}^{n+1,l+1}] \rangle_e.$$

Theorem 2. *([4]) Let $\sigma(\boldsymbol{u}_{s,h}^{n+1}, p_h^{n+1}) \cdot \boldsymbol{n} = 0, \boldsymbol{u}_{f,h}^{n+1} \cdot \boldsymbol{n} = 0$ on the boundary $\partial\Omega$, where $\boldsymbol{n}$ denotes the normal unit outward vector to $\partial\Omega$. We assume that $0 < c_h^n < \frac{\varrho}{\beta}, n \geq 0$. Then the scheme (10) follows an energy dissipation law as*

$$D_\tau E_h^{n+1} \leq - \sum_{K \in \mathcal{K}_h} \int_K \lambda^{-1}(\phi_h^n) |\boldsymbol{u}_{f,h}^{n+1}|^2 \, d\boldsymbol{x} \leq 0.$$

3.2   Stabilization-Based Energy-Stable Scheme Approach

The second numerical formulation implements a stabilization methodology that rigorously conserves the original energy functional while establishing provably stable linear discretizations. During each temporal iteration, the stabilization parameter undergoes adaptive modification through an explicit closed-form relation, ensuring strict adherence to fundamental energy dissipation principles. To optimize computational performance, the algorithm incorporates an adaptive temporal discretization framework that strategically allocates computational resources while maintaining solution accuracy and boundedness. The time step size explicitly derived from local solution characteristics at each linear iteration. Spatial discretization is achieved through a mixed finite element formulation with consistent upwind flux approximation, ensuring discrete stability while exactly preserving mass conservation properties.

The discrete chemical potential incorporates a dynamically adjusted stabilization term $\mu^{n+1} = \mu(c^n) + \theta_n RT \underbrace{\dfrac{c^{n+1} - c^n}{c^n(1 - \beta c^n)^2}}_{\zeta^{n+1}}$, where ζ^{n+1} represents the sta-

bilization term, θ_n is the stabilization parameter. By the Taylor expansion and assuming that ξ is a number between c^n and c^{n+1}, we have $f\left(c^{n+1}\right) - f\left(c^n\right) = \mu\left(c^n\right)\left(c^{n+1} - c^n\right) + \frac{f''(\xi)}{2}\left(c^{n+1} - c^n\right)^2$. In view of the above stabilized chemical potential, we further obtain

$$f\left(c^{n+1}\right) - f\left(c^n\right) = \mu^{n+1}\left(c^{n+1} - c^n\right) \tag{12}$$
$$+ \left(\frac{f''(\xi)}{2} - \frac{\theta_n RT}{c^n\left(1 - \beta c^n\right)^2}\right)\left(c^{n+1} - c^n\right)^2.$$

By choosing a suitable stabilization parameter θ_n, we can obtain the following inequality.

$$f\left(c^{n+1}\right) - f\left(c^n\right) \leq \mu^{n+1}\left(c^{n+1} - c^n\right), \tag{13}$$

then, we can get

$$\phi^{n+1} f\left(c^{n+1}\right) - \phi^n f\left(c^n\right) = f\left(c^n\right)\left(\phi^{n+1} - \phi^n\right) \tag{14}$$
$$+ \phi^{n+1}\left(f\left(c^{n+1}\right) - f\left(c^n\right)\right)$$
$$\leq f\left(c^n\right)\left(\phi^{n+1} - \phi^n\right) + \phi^{n+1}\mu^{n+1}\left(c^{n+1} - c^n\right),$$

the specific choice of the stabilization term θ_n can be found in paper [3].

Building upon the semi-discrete stabilization framework for the chemical potential, we now present the complete space-time discrete formulation. This scheme systematically integrates the stabilization strategy with a mixed finite element discretization in space and an implicit-explicit temporal scheme, ensuring both thermodynamic consistency and numerical robustness: For any $\mathbf{v}_h \in \mathcal{V}_h, \mathbf{w}_h \in \mathcal{U}_h^0, q_h, z_h, \varphi_h \in \mathcal{Q}_h$, we find $\boldsymbol{u}_{s,h}^{n+1} \in \mathcal{V}_h, \boldsymbol{u}_{f,h}^{n+1} \in \mathcal{U}_h, c_h^{n+1}, \phi_h^{n+1}, p_h^{n+1} \in \mathcal{Q}_h$ such that

$$\mathcal{A}(\boldsymbol{u}_{s,h}^{n+1}, p_h^{n+1}, \mathbf{v}_h) = 0, \tag{15a}$$

$$(D_\tau(\phi_h^{n+1} c_h^{n+1}), q_h) + \sum_{e \in \mathcal{E}_h^I} \langle c_h^{n*} \boldsymbol{u}_{f,h}^{n+1} \cdot \boldsymbol{n}, [q_h] \rangle_e \tag{15b}$$

$$+ \sum_{e \in \mathcal{E}_h^I} \frac{\varsigma_1}{h_e} \langle [\mu_h^{n+1}], [q_h] \rangle = 0,$$

$$(\lambda^{-1}(\phi_h^n)\boldsymbol{u}_{f,h}^{n+1}, \mathbf{w}_h) = \sum_{e \in \mathcal{E}_h^I} \langle [\mu_h^{n+1}], c_h^{n*} \mathbf{w}_h \cdot \boldsymbol{n} \rangle_e, \tag{15c}$$

$$(p_h^{n+1}, z_h) = (c_h^n \mu_h^{n+1} - f(c_h^n), z_h), \tag{15d}$$

$$(D_\tau \phi_h^{n+1}, \varphi_h) = \frac{1}{N}(D_\tau p_h^{n+1}, \varphi_h) + \alpha(D_\tau(\nabla \cdot \boldsymbol{u}_{s,h}^{n+1}), \varphi_h) \tag{15e}$$

$$- \alpha \sum_{e \in \mathcal{E}_h^I} \langle \{\varphi_h \boldsymbol{n}_e\}, [D_\tau \boldsymbol{u}_{s,h}^{n+1}] \rangle_e,$$

Since the equation (15b) is nonlinear, we use the linear iteration method to solve the equations (15).

$$A(\mathbf{u}_{s,h}^{n+1,l+1}, p_h^{n+1,l+1}, \mathbf{v}_h) = 0, \tag{16}$$

$$\left(\frac{\phi_h^{n+1,l} c_h^{n+1,l+1} - \phi_h^n c_h^n}{\tau}, q_h\right) + \sum_{e \in \mathcal{E}_h^I} \langle c_h^{n*} \mathbf{u}_{f,h}^{n+1,l} \cdot \mathbf{n}, [q_h] \rangle_e \tag{17}$$

$$+ \sum_{e \in \mathcal{E}_h^I} \frac{\varsigma_1}{h_e} \langle [\mu_h^{n+1,l+1}], [q_h] \rangle_e = 0,$$

$$(\lambda^{-1}(\phi_h^n) \mathbf{u}_{f,h}^{n+1,l+1}, \mathbf{w}_h) = \sum_{e \in \mathcal{E}_h^I} \langle [\mu_h^{n+1,l+1}], c_h^{n*} \mathbf{w}_h \cdot \mathbf{n} \rangle_e, \tag{18}$$

$$(p_h^{n+1,l+1}, z_h) = (c_h^n \mu_h^{n+1,l+1} - f(c_h^n), z_h), \tag{19}$$

$$(D_\tau \phi_h^{n+1,l+1}, \varphi_h) = \frac{1}{N}(D_\tau p_h^{n+1,l+1}, \varphi_h) + \alpha(D_\tau (\nabla \cdot \mathbf{u}_{s,h}^{n+1,l+1}), \varphi_h) \tag{20}$$

$$- \alpha \sum_{e \in \mathcal{E}_h^I} \langle \{\varphi_h \mathbf{n}_e\}, [D_\tau \mathbf{u}_{s,h}^{n+1,l+1}] \rangle_e.$$

As rigorously proven in reference [3], the fully discrete scheme adopted in this paper achieves dynamic adaptive time step adjustment while ensuring the boundedness of the molar density.

Theorem 3. *([3]) Assume that $0 < \varrho_0 \leq \beta c_h^n \leq \varrho < 1$ and the boundary condition (3) holds. For $n \geq 0$ and given constants $0 < \delta_1 < 1$ and $0 < \delta_2 < 1$, if the time step size τ_n^l satisfies*

$$\tau_n^l = \min_{K \in \mathcal{K}_h} \left(\frac{\left(\phi_h^{n+1,l} c_h^n (1 - \beta c_h^n)^2 \delta_1 - (\phi_h^{n+1,l} - \phi_h^n) c_h^n\right) |K|}{\sum\limits_{e \in \partial K_{\mathbf{u}_{f,h}}^+} c_h^n \mathbf{u}_{f,h}^{n+1,l} \cdot \mathbf{n}|e| + \sum\limits_{e \in \partial K_\mu^+} \frac{\varsigma_1}{h_e}[\mu(c_h^n)]|e| + \epsilon}, \right. \tag{21}$$

$$\left. \frac{\left(\phi_h^{n+1,l} c_h^n (1 - \beta c_h^n)^2 \delta_2 + (\phi_h^{n+1,l} - \phi_h^n) c_h^n\right) |K|}{-\left(\sum\limits_{e \in \partial K_{\mathbf{u}_{f,h}}^-} c_h^{n*} \mathbf{u}_{f,h}^{n+1,l} \cdot \mathbf{n}|e| + \sum\limits_{e \in \partial K_\mu^-} \frac{\varsigma_1}{h_e}[\mu(c_h^n)]|e|\right) + \epsilon}, \tau_{max} \right),$$

where $\epsilon > 0$ is a very small constant to avoid zero denominator, τ_{max} is the allowed maximum time step size to guarantee the accuracy of numerical solutions and $\partial K_{\mathbf{u}_{f,h}}^+ = \{e \in \partial K : \mathbf{u}_{f,h}^{n+1} \cdot \mathbf{n}|_e > 0, \forall K \in \mathcal{K}_h\}$, $\partial K_{\mathbf{u}_{f,h}}^- = \{e \in \partial K : \mathbf{u}_{f,h}^{n+1} \cdot \mathbf{n}|_e < 0, \forall K \in \mathcal{K}_h\}$, $\partial K_\mu^- = \{e \in \partial K : [\mu(c_h^n)] < 0, \forall K \in \mathcal{K}_h\}$, $\partial K_\mu^+ = \{e \in \partial K : [\mu(c_h^n)] > 0, \forall K \in \mathcal{K}_h\}$. Then $c_h^{n+1,l+1}$ satisfies

$$0 < (1 - \delta_1 (1 - \beta c_h^n)^2) c_h^n \leq c_h^{n+1,l+1} \leq (1 + \delta_2 (1 - \beta c_h^n)^2) c_h^n < \frac{1}{\beta}.$$

It is also proved in paper [3] that the fully discrete numerical format based on the stabilization method satisfies the properties of energy dissipation.

Theorem 4. ([3]) *Assume that the boundary condition* (3) *holds,* $0 < \epsilon \le c_h^n \le \frac{\varrho}{\beta} < \frac{1}{\beta}$. *The stabilization parameter* θ_n *is taken as follows:*

$$\theta_n = \max_{K \in \mathcal{K}_h} \left\{ 1, \frac{(1 - \beta c_h^n)^2}{\chi_1^n (1 - \chi_1^n \beta c_h^n)^2}, \frac{(1 - \beta c_h^n)^2}{\chi_2^n (1 - \chi_2^n \beta c_h^n)^2} \right\}, \tag{22}$$

where $\chi_1^n = 1 - \delta_1 (1 - \beta c_h^n)^2$, $\chi_2^n = 1 + \delta_2 (1 - \beta c_h^n)^2$. *The total free energy generated by the scheme* (10) *is dissipated as*

$$D_\tau E_h^{n+1} \le 0,$$

where

$$E_h^{n+1} = \sum_{K \in \mathcal{K}_h} \int_K \left(\phi_h^{n+1} f(c_h^{n+1}) + \frac{1}{2} \sigma_e(\boldsymbol{u}_{s,h}^{n+1}) : \varepsilon(\boldsymbol{u}_{s,h}^{n+1}) \right) dx \tag{23}$$

$$+ \sum_{e \in \mathcal{E}_h^I} \frac{\varsigma_1}{2h_e} \langle [\boldsymbol{u}_{s,h}^{n+1}], [\boldsymbol{u}_{s,h}^{n+1}] \rangle_e - \sum_{e \in \mathcal{E}_h^I} \langle \{ \sigma_e(\boldsymbol{u}_{s,h}^{n+1}) \boldsymbol{n}_e \}, [\boldsymbol{u}_{s,h}^{n+1}] \rangle_e.$$

4   Numerical Examples

We design numerical experiments to validate the proposed computational framework for modeling compressible gas flow in poroelastic media. The test cases are specifically constructed to evaluate: the model's capability in capturing coupled hydro-mechanical processes, the algorithm's performance in handling strong nonlinearities and the numerical scheme's conservation properties. The simulations consider methane gas with physical properties listed in Table 1, maintaining a constant temperature of 330 K throughout all test cases.

Table 1. Physical properties of methane.

P_c(bar)	T_c(K)	Acentric factor	M_w(g/mole)	Viscosity (Pa · s)
45.99	190.56	0.011	16.04	10^{-5}

4.1   Example 1

In this example, we investigate the gas flow dynamics within a closed domain to rigorously validate both the physical model's accuracy and the first numerical scheme's performance in simulating compressible gas flow. The simulation serves to verify three fundamental physical principles: strict mass conservation, thermodynamic consistency in energy dissipation, and boundedness of molar density. The domain is initialized with a high-concentration zone (300 mol/m³ molar density) at the center, surrounded by a low-concentration region (10 mol/m³).

A distinctive cross-shaped high-permeability zone ($\kappa = 100$ md) is embedded within the central area, while the remaining domain maintains low permeability ($\kappa = 1$ md). This configuration, as illustrated in Fig. 1.

Figure 2 provides critical validation of the proposed numerical scheme's ability to preserve fundamental physical properties, demonstrating strict adherence to energy dissipation principles, exact mass conservation, and molar density boundedness. Figure 3 presents the temporal evolution of molar density distribution, demonstrating the dynamic mass transport characteristics. The corresponding thermodynamic driving forces are visualized in Fig. 4 through chemical potential contours, which reveal the non-equilibrium processes governing molecular diffusion.

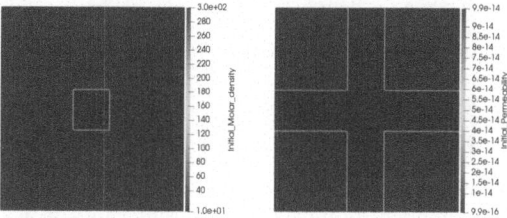

Fig. 1. Example 1: Distributions of initial molar density and permeability. Left: initial molar density. Right: initial permeability.

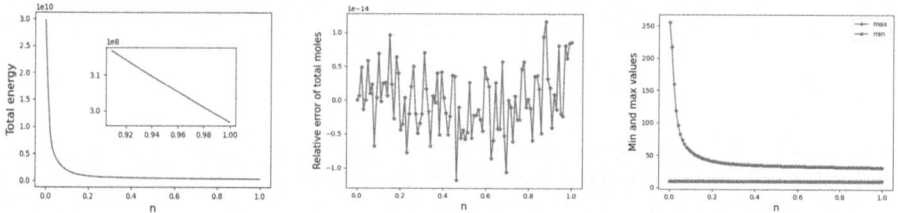

Fig. 2. Example 1: Left: System energy at different time steps. Middle: Mass conservation at different time steps. Right: Minimum and maximum values of molar density.

4.2 Example 2

To validate the performance of our second numerical scheme, we conduct the following test. We use the Perlin noise method to generate random permeability in order to simulate realistic geological scenarios, The initial molar density is obtained using $c^0 = c_0 + \text{rand}(\boldsymbol{x}) \cdot (c_1 - c_0)$, where $\text{rand}(\boldsymbol{x})$ is a function for generating random numbers within the range [0,1], $c_0 = 100$ mol/m^3, $c_1 = 300$

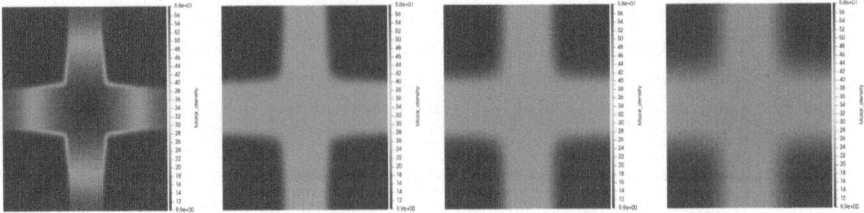

Fig. 3. Distributions of molar density at different times in Example 1. From left to right: The first: t = 0.1 h. The second: t = 0.4 h. The third: t = 0.8 h. The fourth: t = 2 h.

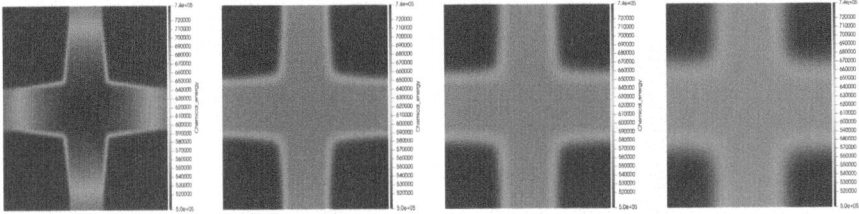

Fig. 4. Distributions of chemical potential at different times in Example 1. From left to right: The first: t = 0.1 h. The second: t = 0.4 h. The third: t = 0.8 h. The fourth: t = 2 h.

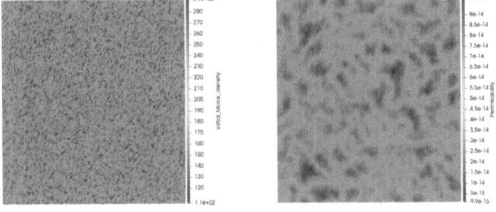

Fig. 5. Example 2: Distributions of initial molar density and permeability. Left: initial molar density. Right: initial permeability.

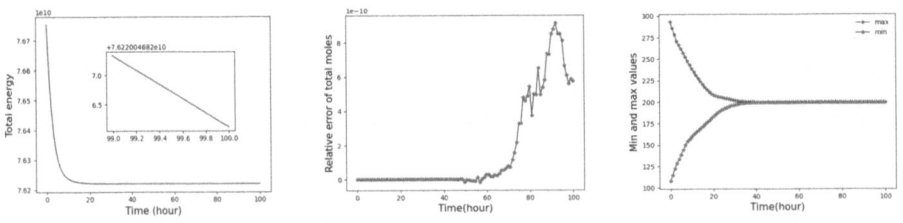

Fig. 6. Example 2: Left: Distributions of energy at different time steps. Middle: Mass conservation at different time steps. Right: Minimum and maximum values of molar density.

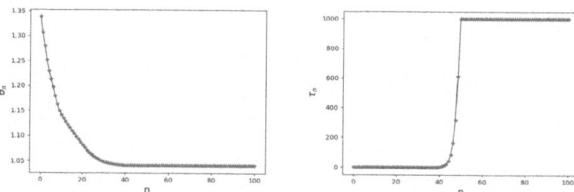

Fig. 7. Example 2: Left: Adaptive values of the stabilization parameter at different time steps. Right: Adaptive values of the time step size.

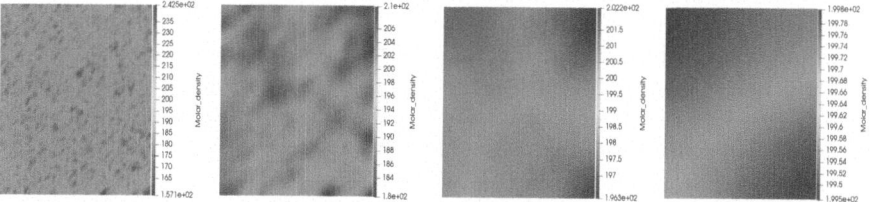

Fig. 8. Distributions of molar density at different times in Example 2. From left to right: The first: n = 10. The second: n = 30. The third: n = 30. The fourth: n = 40.

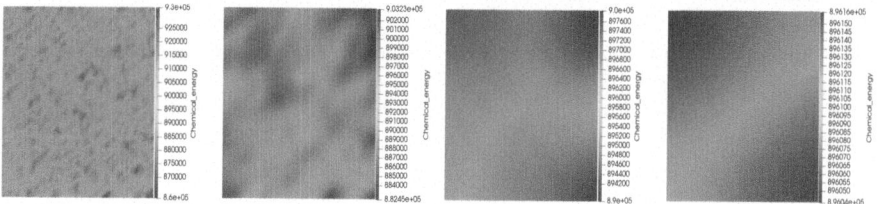

Fig. 9. Distributions of chemical potential at different times in Example 2. From left to right: The first: n = 10. The second: n = 30. The third: n = 30. The fourth: n = 40.

mol/m^3. In this test, we choose $N = 10^{15}$ Pa, $\gamma = 10^{15}$ Pa , $\eta = 10^{11}$ Pa, $\delta = 0.5$. Figure 5 shows the initial molar density and permeability distribution. Figure 6 demonstrates the proposed scheme's ability to maintain energy dissipation, mass conservation, and molar density bounds. The left picture shows the energy evolution, the right two picture verify strict molar density preservation within physical bounds and mass conservation. Figure 7 illustrates the adaptive parameters: the left picture displays the dynamic stabilization parameter adjustment, while the right picture shows the time step size continuously increases as the system approaches equilibrium until reaching the preset maximum time step size $\tau_{max} = 1000$, demonstrating computational efficiency. Figure 8, Fig. 9 illustrate the temporal evolution of molar density, chemical potential energy, respectively.

5 Conclusions

This paper focuses on the numerical simulation of single-phase gas flow with compressible gas and rock in the context of underground storage of hydrogen and carbon dioxide geological sequestration. It first formulates a fully coupled thermodynamic mathematical model considering the compressibility of gas and rock. Then, two energy-stable numerical schemes are proposed. The first scheme is based on an energy factorization approach, which combines a semi-implicit time-stepping formulation with a spatial discretization using discontinuous Galerkin approximations and mixed finite elements, ensuring energy dissipation and mass conservation. The second scheme, based on a stabilization approach, conserves the original energy functional, has an adaptive temporal discretization framework, and can dynamically adjust the stabilization parameter. Both schemes are proven to satisfy the energy dissipation law. Through numerical examples, including simulations in a closed domain with specific initial conditions and in a domain with randomly generated permeability, the proposed computational framework is validated. The examples demonstrate the model's ability to capture coupled hydro-mechanical processes, the algorithm's performance in handling strong nonlinearities, and the numerical scheme's conservation properties such as mass conservation and bounded molar density. This study provides a consistent and stable modeling and simulation framework with theoretical support for the engineering practice of carbon dioxide sequestration and underground storage of hydrogen.

References

1. Chen, W., Wang, C., Wang, X., Wise, S.M.: Positivity-preserving, energy stable numerical schemes for the Cahn-Hilliard equation with logarithmic potential. J. Comput. Phys. **3**, 100031 (2019)
2. Cheng, Q., Shen, J.: A new Lagrange multiplier approach for constructing structure preserving schemes, I. Positivity preserving. Comput. Methods Appl. Mech. Eng. **391** (6), 114585 (2022)
3. Chen, H., Chen, Y., Kou, J., Sun, S.: An Energy-Stable Adaptive Time-Stepping Method for Modeling Compressible Gas Flow in Poroelastic Media. Preprint (2025)
4. Chen, H., Chen, Y., Kou, J.: Energy stable finite element approximations of gas flow in poroelastic media. Comput. Methods Appl. Mech. Engrg. **428**, 117082 (2024)
5. Chen, Z., Huan, G., Ma, Y.: Computational Methods for Multiphase Flows in Porous Media. SIAM, Philadelphia (2006)
6. Dong, L., Wang, C., Wise, S.M., Zhang, Z.: A positivity-preserving, energy stable scheme for a ternary Cahn-Hilliard system with the singular interfacial parameters. J. Comput. Phys. **442**, 110451 (2021)
7. Du, Q., Ju, L., Li, X., Qiao, Z.: Maximum bound principles for a class of semilinear parabolic equations and exponential time-differencing schemes. SIAM Rev. **63**, 317–359 (2021)
8. Joshi, V., Jaiman, R.K.: A positivity preserving and conservative variational scheme for phase-field modeling of two-phase flows. J. Comput. Phys. **360**, 137–166 (2018)

9. Eyre, D.J.: Unconditionally gradient stable time marching the Cahn-Hilliard equation, in Computational and Mathematical Models of Microstructural Evolution (San Francisco, CA, 1998). Mater. Res. Soc. Sympos. Proc. 529, MRS, Warrendale, PA, 39–46 (1998)

10. Firoozabadi, A.: Thermodynamics of Hydrocarbon Reservoirs. McGraw-Hill, New York (1999)

11. Kou, J., Sun, S.: Thermodynamically consistent modeling and simulation of multicomponent two-phase flow with partial miscibility. Comput. Methods Appl. Mech. Engrg. **331**, 623–649 (2018)

12. Kou, J., Sun, S., Wang, X.: A novel energy factorization approach for the diffuse-interface model with Peng-Robinson equation of state. SIAM J. Sci. Comput. **42**, B30–B56 (2020)

13. Kou, J., Wang, X.H., Du, S.G.: Energy stable and mass conservative numerical method for gas flow in porous media with rock compressibility. SIAM J. Sci. Comput. **44**, B938–B963 (2022)

14. Kou, J., Wang, X.H., Chen, H.X., Sun, S.: An efficient bound-preserving and energy stable algorithm for compressible gas flow in porous media. J. Comput. Phys. **473**, 111751 (2023)

15. Niessner, J., Helmig, R.: Multi-scale modelling of two-phase–two-component processes in heterogeneous porous media. Numer. Linear Algebra Appl. **13**(9), 699–715 (2006)

16. Phillips, P.J., Wheeler, M.: A coupling of mixed and discontinuous Galerkin finite-element methods for poroelasticity. Comput. Geosci. **12**, 417–435 (2008)

17. Qi, R., LaForce, T.C., Blunt, M.J.: A three-phase four-component streamline-based simulator to study carbon dioxide storage. Comput. Geosci. **13**(4), 493–509 (2009)

18. Shen, J., Xu, J., Yang, J.: The scalar auxiliary variable (SAV) approach for gradient flows. J. Comput. Phys. **353**, 407–416 (2018)

19. Xu, C., Tang, T.: Stability analysis of large time-stepping methods for epitaxial growth models. SIAM J. Numer. Anal. **44**, 1759–1779 (2006)

20. Yang, J., Yuan, Z., Zhou, Z.: Arbitrarily high-order maximum bound preserving schemes with cut-off postprocessing for Allen-Cahn equations. J. Sci. Comput. **90**, 76 (2022)

21. Yang, X., Ju, L.: Efficient linear schemes with unconditionally energy stability for the phase field elastic bending energy model. Comput. Methods Appl. Mech. Eng. **315**, 691–712 (2017)

22. Zhang, X., Shu, C.-W.: On positivity-preserving high order discontinuous Galerkin schemes for compressible Euler equations on rectangular meshes. J. Comput. Phys. **229**(23), 8918–8934 (2010)

A Hybrid Approach for Medical Deepfake Detection Using Depth-Wise Convolutions in Vision Transformer and Frequency Domain Analysis

R. Dhanyalakshmi[1], Alexander Zakharov[2], Natalia Romanchuk[2], J. Anitha[1], and Jude Hemanth[1(✉)]

[1] Department of ECE, Karunya Institute of Technology and Science, Coimbatore, India
judehementh@karunya.edu
[2] Neurosciences Research Institute, Samara State Medical University, Samara, Russia

Abstract. These days, identifying medical deepfakes is crucial for preventing fraudulent activity, to avoid inaccurate diagnoses, as well as to uphold patient confidence. The massive increase in the production of realistic synthetic medical images presents significant challenges for clinical decision-making, highlighting the need for effective detection techniques. This proposed method offers a hybrid deepfake detection model which incorporates a lightweight Depth-Wise Convolution module in a Vision Transformer (DWConv-ViT) and a Fast Fourier Transform (FFT) module to improve feature extraction in the deepfake detection process. In contrast to conventional models which primarily use either frequency-based analysis or spatial analysis, our method integrates both feature types to increase resilience against malicious attacks. The proposed model was trained and tested using two datasets consisting of real knee X-ray images and GAN-generated osteoarthritis X-ray images. By utilizing both spatial and frequency-based details, our approach improves generalization and robustness against sophisticated deepfake approaches. Therefore, this work helps to ensure the reliability and validity of medical diagnoses.

Keywords: Medical Deepfake Detection · Deepfake Detection · Hybrid CNN-Transformer Model · Vision Transformer · Medical AI · Fast Fourier Transform · Depthwise Convolution

1 Introduction

Deepfake is a cutting-edge technology that makes it possible to create realistic medical images, including CT scans, X-rays, MRIs, and more, in addition to creating artificial human photos, videos and so on [1]. These artificially generated images have become crucial for data augmentation [2], which protects patient privacy while allowing AI models to train on a variety of datasets. Furthermore, by producing realistic simulations of complicated procedures, medical deepfakes strengthen training in surgical procedures [3]. Deepfake-based avatars are used in telemedicine to improve personalized and

M. H. Lees et al. (Eds.): ICCS 2025, LNCS 15905, pp. 228–240, 2025.
https://doi.org/10.1007/978-3-031-97632-2_16

interactive doctor-patient communication, which ultimately makes remote consultations more accessible [4]. Additionally, the primary issue in medical research is the difficulty of collecting or obtaining data on rare diseases. Whereas, deepfake has become a life-saver in this case [5], allowing for the creation of more trustworthy diagnostic models. Medical deepfakes have transformed the medical and healthcare sector by strengthening training methodology, diversifying data, ensuring privacy protection, and enhancing data privacy [6].

Despite its importance in the medical sector, medical deepfake carries serious concerns that could endanger patient safety and treatment [7]. Since deepfakes induce abnormalities are indistinguishable from real ones, it could mislead doctors and healthcare professionals, leading to inappropriate therapies. Therefore, Concerns regarding various possible misdiagnoses are raised by the creation of extremely realistic synthetic medical images. The threat was illustrated in a real-world scenario by a team of researchers in 2019 [8]. They used 3D conditional GAN to successfully modify CT scans of patients by adding or deleting lung cancer indications. This demonstrated that deepfake technology can bring false positives or negatives in medical diagnosis, which can lead to a life-threatening situation.

Further, these deepfakes generated medical images and scans can be exploited maliciously to create fake medical records to manipulate the diagnoses or to commit insurance fraud, which can result in monetary losses and violations of ethics. In light of these risks, it is imperative to create sophisticated techniques for identification of medical deepfakes to prevent artificial images from jeopardizing patient confidence or clinical accuracy. Research in this area must be accelerated in order to preserve the integrity of medical diagnostics and decision-making, as medical deepfakes carry similarly serious consequences to those of media deepfakes in terms of misinformation and fraud against identities. Key Focus Areas for Medical Deepfake detection are

- Maximizing Diagnostic Accuracy
- Detecting Medical Irregularities
- Safeguarding Patient Privacy.
- Improving Performance with Limited Data

The problem of identifying medical deepfakes is addressed through our study by introducing a hybrid deep learning architecture that incorporates ViT-Small [9], Depth-Wise Convolution (DWConv) [10], and Fast Fourier Transform (FFT) [11]. We demonstrated this study by training and testing the model with manipulated knee osteoarthritis X-rays images. Although all the components of a detection system play an important role, starting from preprocessing to detection, the feature extraction part majorly influences the decision-making process. So, we have exclusively designed a feature extraction module that performs well on a limited dataset by capturing global, local and spectral features.

Here, DWConv along with ViT facilitates in capturing both local texture features and global anatomical structure with minimal computational load. Additionally, the FFT branch is incorporated to investigate the spectral anomalies that are typically introduced by generative models, enhancing robustness against adversarial attacks. Unlike conventional methods that solely depend on either spectral analysis or frequency analysis, our approach utilizes both frequency and spectral information to classify real and deepfake images. Crucially, the majority of medical deepfake detection models that have been developed to date require a substantial quantity of training data in order to be trained. Our approach, on the other hand, can do better with a smaller dataset.

2 Related Work

2.1 General Deepfake Generation

The creation of realistic deepfake data is significantly impacted by the advancements in Generative Adversarial Networks (GANs) [12], autoencoders [13], NeRFs [14], and diffusion models. These models aid in producing deepfake text data, voices, and human images (facial and entire body) [15, 16]. To execute face swapping and full body or face reenactment, these models are often trained using a large dataset that captures fine details of speech, motion, and facial expression [17–20]. Deepfakes were once simple to identify with a human eye, but in the past few years, as computer hardware and software have advanced, it has become increasingly challenging for both people and machines to distinguish the difference between the real and the fake. Now it has become possible to generate deepfake with a few or one image of the target individual [21]. Furthermore, real-time deepfake synthesis, driven by efficient neural networks, has broadened its use in live streaming and interactive media. While these advancements have enhanced creativity in filmmaking, gaming, and virtual communication, they have also heightened concerns about digital deception, cybercrimes involving deepfakes, and declining trust in visual content, underscoring the importance of stringent regulations and advanced detection methods.

2.2 Medical Deepfake Generation

Tools for Generating Medical Deepfakes
To generate medical deepfake, researchers utilize the most advanced deep learning models like GAN, Variational Autoencoders (VAE), diffusion model, transformers and so on. Style GAN and CycleGAN are widely used GAN models in medical image manipulation or generation. GAN-based frameworks like CycleGAN and StyleGAN can effortlessly generate or manipulate medical data like MRI, CT, X-ray, mammography, and much more since they are capable of learning the features of medical images and scans in an unsupervised manner [1]. Similar to GAN models, VAE plays a major role in medical image reconstruction and anomaly detection, which makes them useful for producing controlled modifications to artificial medical datasets. The most advanced deep learning developments like transformers and diffusion models have recently become an excellent substitute, providing high-quality medical image synthesis with reduced artifacts and enhanced feature retention [22].

Categories of Medical Deepfakes
Medical deepfakes fall under three major categories. First is the generation of an entirely new medical image or scan called synthetic medical image generation. Second is image-to-image translation, where models like Pix2Pix are utilized to modify the existing

medical data. When there is very little data available, as in rare diseases cases, this is most frequently utilized [23]. The final technique is inpainting based modification, in which these manipulations emphasize specific modifications like adding or removing lesions, tumors, or scars. These manipulations are often made possible by attention-based technology, which ensures seamless interaction with the surrounding anatomical structures [24]. These realistic synthetic data create issues of disinformation in medical diagnosis, underscoring the necessity for careful validation and ethical measures to prevent abuse in clinical settings, even though they have promise for medical research and education.

2.3 Deepfake Detection

Fernandes et al. investigated the application of Neural Ordinary Differential Equations (Neural-ODEs) for deepfake detection by estimating heart rates, revealing a notable distinction between authentic and manipulated videos [25]. Although their approach utilizes physiological signals, its effectiveness may be impacted by inconsistencies in video quality and subject movement, potentially reducing heart rate estimation accuracy. In [26] the research introduced a deep learning-based convolutional neural network designed to automatically detect diabetic retinopathy and macular edema in retinal fundus images, demonstrating strong sensitivity and specificity. However, its effectiveness depends on extensively annotated datasets and high-resolution images, which may restrict its usability across varied clinical environments. Solaiyappan et al. [1] examined medical deepfake detection using eight machine learning models, demonstrating high accuracy in detecting manipulated CT scans. However, their dependence on pre-trained models and feature extraction may restrict adaptability to emerging manipulation techniques.

In [27] the author examines the effectiveness of different YOLO models in identifying medical deepfakes within Knee Osteoarthritis X-rays and lung CT scans. The results indicate promising performance, though variations exist across datasets. While the study underscores the potential of YOLO models, inconsistencies in detection accuracy highlight the need for further refinement. In [28] the study investigates deep learning models, such as CNNs and patch-based networks, for identifying deepfake medical images, focusing on skin cancer images generated via stable diffusion. The findings demonstrate the models' effectiveness in differentiating real and synthetic images, with histogram analysis uncovering significant color distribution shifts. However, challenges remain in establishing a consistent classification threshold, and the models exhibit limitations in generalizing across datasets.

3 Methodology

The proposed DWConv-ViT + FFT architecture ensures enhanced medical deepfake detection by capturing both local and global features through integrating depth-wise convolution module in vision transformer. The DWConv is integrated with vision transformer in a plug and play concept, without modifying any of the internal components of the transformer, including MHSA and FFN as shown in Fig. 3 with minimal computational load. Additionally, as GAN synthesized images suffer from spectral artifacts, a lightweight FFT module is incorporated in the detection network. The fusion of spatial and spectral characteristics guarantees excellent accuracy while preserving computational economy, making it lightweight, flexible, and ideal for clinical applications.

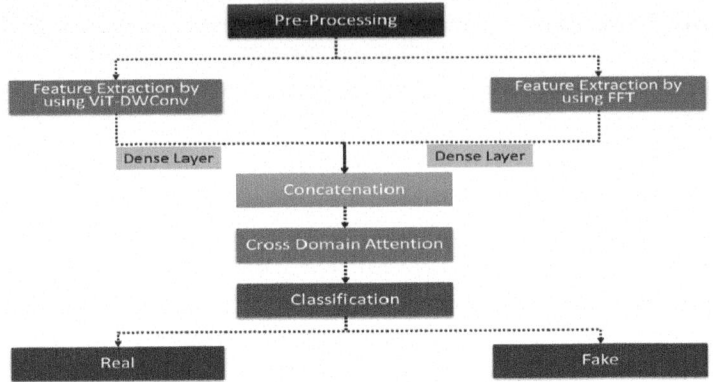

Fig. 1. Proposed Hybrid DWConv-ViT + FFT flowchart

3.1 Preprocessing

It is important to maintain the anatomical consistency of the medical image to ensure high-quality detection. By standardizing knee X-ray images via the preprocessing workflow, it is made possible. To avoid the distortion and to maintain the original aspect of the X-ray image, it was compressed to 256 × 256 pixels with padding. To account for real-time variation between different X-ray scanners, values of pixel intensity are standardized within a range of [0,1]. Even while performing augmentation, excessive rotation and flipping are avoided, as this could produce inaccurate and misleading medical data for the model's training. To bridge the gap between natural images and grayscale X-rays, the input X-rays are converted into pseudo-RGB images and undergo histogram matching to normalize their intensity distribution. In Sect. 4.1, preprocessing procedures were described in depth. Figure 1 illustrates the workflow of the proposed Hybrid DWConv-ViT + FFT model.

3.2 Feature Extraction in DWConv-ViT Variant

The DWConv-ViT model enhances feature extraction by utilizing a pretrained Vision Transformer (DINOv2 ViT-S/14)[1], which is specifically adapted to handle the unique characteristics of medical images. To retain the characteristic of the vision transformer, the first seven transformer layers are frozen, allowing the remaining layers to be fine-tuned for capturing domain-specific details in medical images, including bone textures and subtle anatomical structures.

---

[1] https://dl.fbaipublicfiles.com/dinov2/dinov2_vits14/dinov2_vits14_pretrain.pth.

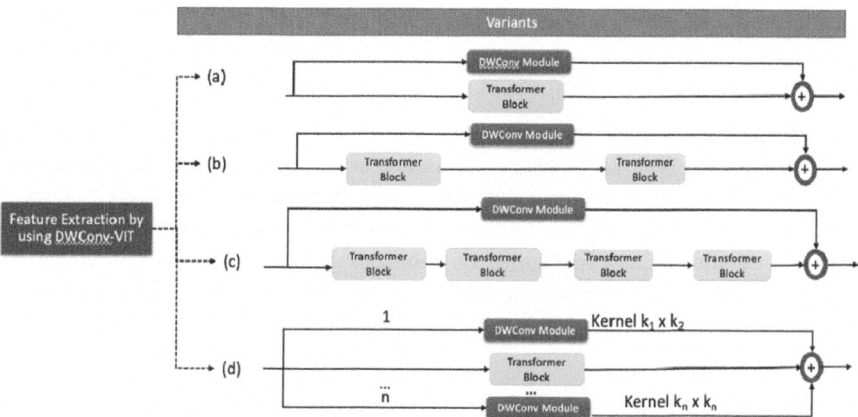

Fig. 2. Feature extraction using different variants of DWConv-VIT

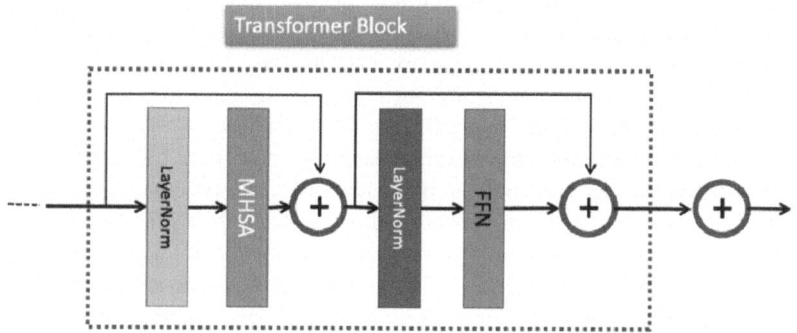

Fig. 3. Architecture of the Vision transformer used DWConv-VIT model

DWConv-ViT improves vision transformers by incorporating depth-wise convolutions (DWConv), which enhance local spatial feature extraction while preserving the global contextual modeling of transformers. In the base variant, each transformer block is paired with a DWConv module, where 1D patch tokens are temporarily reshaped into 2D feature maps. These maps undergo 3×3 depth-wise convolution, batch normalization, and GELU activation before reintegrating with the transformer's global self-attention output. This mechanism ensures the model captures fine-grained textures like bone structures and synthetic noise, which standard self-attention may overlook. In addition to the base variant, three optimized variants shown in Fig. 2 refine this approach:

1. Base Variant: Each transformer block is individually paired with a DWConv module.
2. 2-Block Bypass: A single DWConv module serves two transformer blocks, striking a balance between parameter efficiency and sensitivity to intricate medical features.
3. 4-Block Bypass: One DWConv module spans four blocks, maximizing computational efficiency while maintaining spatial awareness, making it suitable for edge deployment.

4. Parallel Multi-Kernel: Multiple DWConv branches with different kernel sizes (e.g., 3×3 and 5×5) operate simultaneously, improving detection of both small-scale GAN artifacts and larger anatomical distortions.

All variants maintain the transformer's multi-head self-attention (MHSA) and feed-forward network (FFN) layers, ensuring seamless compatibility with pretrained weights and minimal computational overhead. In bypass variants, the shared DWConv module acts as a persistent local memory, mitigating the tendency of deep transformers to lose fine-grained details. This hybrid approach enhances global anatomical coherence while effectively detecting local synthetic artifacts, making it particularly effective for medical deepfake detection.

3.3 Feature Extraction Using Fast Fourier Transform (FFT)

The FFT-based extraction module processes medical images in the frequency domain, uncovering synthetic artifacts that might be imperceptible in spatial analysis. Applying a 2D FFT to an input X-ray produces a magnitude spectrum that highlights high-frequency patterns linked to generative models (e.g., GANs, diffusion models), such as repetitive edges, grid-like distortions, and irregular texture harmonics. This transformation shifts the image into its frequency representation, recenters the zero-frequency component, and employs logarithmic scaling to amplify subtle anomalies.

The FFT module strengthens medical deepfake detection by identifying high-frequency artifacts that spatial analysis may overlook. The process starts with applying a 2D Fast Fourier Transform (FFT) to the grayscale X-ray image, followed by generating a log-scaled magnitude spectrum to highlight subtle inconsistencies in the frequency domain. A lightweight convolutional neural network (CNN) then extracts spectral features, incorporating batch normalization, GELU activation, and adaptive pooling for effective dimensionality reduction. These spectral features are then fused with spatial representations from the Vision Transformer with Depth-Wise Convolution (DWConv-ViT) through a dense transformation layer, ensuring a comprehensive and robust analysis.

3.4 Feature Fusion and Classification

The fusion and classification stages combine spatial (DWConv-ViT) and spectral (FFT) features through cross-domain attention, dynamically balancing their influence based on input properties. Spatial features that capture anatomical consistency are refined by spectral features that highlight synthetic artifacts, enhancing the model's focus on medically significant patterns. The fused representation is processed through a compact dense network with Mish activations and strong regularization to minimize noise while retaining subtle synthetic markers. Focal loss emphasizes difficult cases, while weight normalization ensures stable training on limited data. The final sigmoid layer produces calibrated probabilities, boosting AUC-ROC and reducing false positives for dependable medical application.

3.5 Loss Function

The proposed loss function combines multiple elements to improve deepfake detection by tackling class imbalance, challenging samples, and overly confident predictions. At its core, it utilizes Binary Cross-Entropy (BCE) loss from Eq. (1):

$$\mathcal{L}_{\text{BCE}} = -\left[y\,log(p) + (1-y)log(1-p)\right] \tag{1}$$

where y denotes the actual class (real or synthetic) and p represents the predicted probability. To emphasize hard-to-classify cases, Focal Loss in Eq. (2) extends BCE with a modulation factor $(1\text{-pt})^\gamma$, giving more importance to misclassified instances:

$$\mathcal{L}_{\text{Focal}} = -\alpha_t(1-p_t)^\gamma log\,(p_t) \tag{2}$$

where γ regulates this effect. To balance real-class distributions, KL-Grade Weighting assigns a weighted factor $\alpha\,{}^{KL}_{Real}$ in Eq. (3) to real samples based on their frequency:

$$\alpha\,{}^{KL}_{Real} = \alpha_{\text{Real}} \times (N_{\text{total}}/(N_{\text{KL-grade}})) \tag{3}$$

where $N_{\text{KL-grade}}$ denotes the sample count for a specific severity level. Additionally, Label Smoothing in Eq. (4) prevents the model from making excessively confident predictions by adjusting the target labels:

$$y_{\text{smooth}} = y \times (1-\epsilon) + \frac{\epsilon}{2} \tag{4}$$

where ϵ defines the smoothing intensity. The final loss function integrates these components into a unified formulation in Eq. (5):

$$\mathcal{L} = \frac{1}{N}\sum_{i=1}^{n}\alpha\,{}^{(i)}_t\,(1-p\,{}^{(i)}_t)\gamma.\mathcal{L}_{\text{BCE}}(y_{smooth}^{(i)}, p^{(i)}) \tag{5}$$

allowing the model to effectively learn from both real and synthetic data while minimizing bias and enhancing generalization.

4 Experimental Result

4.1 Dataset

The process of constructing a clinically meaningful and artifact-rich osteoarthritis (OA) X-ray dataset starts by collecting 9,786 authentic knee X-rays from Chen's dataset [29]. All KL grades are treated uniformly and labeled as "real" to focus solely on image authenticity. A matching set of 9,786 synthetic X-rays is then randomly drawn from a pool of 320,000 GAN-generated images by Prezja et als [3], resized to 256×256 pixels, normalized, and labeled as "fake". To introduce controlled visual distortions, 30% (2,936) of these synthetic images are modified with artifacts such as grid patterns (simulating GAN upsampling flaws), Fourier-based high-frequency noise (to mimic spectral irregularities), and Gaussian blur patches over joints (replicating copy-paste errors). Augmentation strategies differ by image type. Real images undergo transformations like elastic warping, KL progression simulation, and realistic noise; synthetic ones are further modified with channel dropout, overlayed DICOM tags, and the artifact injections mentioned earlier.

To enhance variability in synthetic data, StyleGAN3 is trained on 50,000 OAI X-rays[2] to generate 4,893 new synthetic samples. These can be blended with either the original or artifact-injected Prezja images to reduce reliance on a single GAN source. Hybrid images are then synthesized by embedding 64 × 64 synthetic patches (e.g., knee joints) into real X-rays using seamless blending through imgaug, with the results still marked as fake. A stratified train/validation/test split is performed: real samples are grouped by KL grade, and synthetic samples by type (original, artifact-injected, or StyleGAN3-based). The dataset is divided into 6,850 real and 6,850 synthetic images for training, 1,468 of each for validation, and 1,468 of each for testing. An extra 500 synthetic images from an unseen diffusion-based model are appended to the test set to test generalization. This pipeline ensures a dataset rich in artifact variety, realistic hybrid cases, balanced class distributions, and resilience to novel synthetic image types.

4.2 Analysis

A comparison of the model variants with other models in Table 1 highlights the 2-Block Bypass as the top performer, achieving the highest accuracy (91.5%), AUC-ROC (92.0%), and synthetic recall (92.5%), along with a low false positive rate (1.8%) and fast inference time (24 ms). The Base Model delivers stable results but falls short in both recall and accuracy. In contrast, the 4-Block Bypass records the lowest metrics for accuracy (84.0%) and recall (83%), though it benefits from the quickest inference speed. The Parallel Multi-Kernel variant presents a solid compromise, offering competitive AUC and recall performance, albeit with a slightly slower processing time. The 2-Block Bypass is the most effective in balancing performance, detection reliability, and speed.

As shown in Table 2 and the confusion matrices in Fig. 4, the 2-Block Bypass model outperforms the others, delivering the highest synthetic recall of 92.5% and the lowest false positive rate of 1.8%, highlighting its strong suitability for clinical applications. In contrast, the 4-Block Bypass emphasizes speed but compromises on accuracy, evidenced by a much higher number of false negatives (1,664) and a lower recall rate of 83%. While the Parallel Multi-Kernel model improves artifact detection with an 89% recall, it introduces greater computational demands. Overall, the 2-Block Bypass offers the most effective trade-off between performance and efficiency, making it the preferred option for detecting medical deepfakes.

Table 3 illustrates that the combination of Depthwise Convolution and Fast Fourier Transform markedly enhances the effectiveness of deepfake detection in medical imaging. While DWConv targets localized texture irregularities, FFT focuses on high-frequency signal anomalies. Together, they boost synthetic recall by 13.5% (rising from 79% to 92.5%) and improve AUC-ROC by 7.8% (from 84.2% to 92.0%) over the base

---

[2] https://nda.nih.gov/oai.

Table 1. Comparison of Variants on X-Ray Dataset with different models

Variant	Acc	AUC-ROC	Synthetic Recall	FP Rate (Real)	Inference Speed
Base Model	88%	89%	86%	2.5%	28 ms
2-Block Bypass	**91.5%**	**92.0%**	**92.5%**	**1.8%**	24 ms
4-Block Bypass	84.0%	87.0%	83%	3.2%	**20 ms**
Parallel Multi-Kernel	89.5%	91.5%	89%	2.3%	29 ms
VGG19	82%	84%	78%	6.5%	35 ms
InceptionV2	85%	86%	81%	5.0%	30 ms

Table 2. Confusion Matrices for Different Model Variants (Balanced Test Set, N = 19,572)

Model Variant	TN (Real Detected)	FP (Real as Fake)	FN (Fake as Real)	TP (Fake Detected)	FP Rate (%)	Synthetic Recall (%)
Base Model	9,541	245	1,370	8,416	2.5%	86%
2-Block Bypass	9,610	176	734	9,052	1.8%	92.5%
4-Block Bypass	9,473	313	1,664	8,122	3.2%	83%
Parallel Multi-Kernel	9,561	225	1,076	8,710	2.3%	89%

Table 3. Ablation Configurations

Model Variant	AUC-ROC	Synthetic Recall	FP Rate(Real)	Grade4 Recall
Base ViT	84.2%	79%	5.1%	72%
ViT + DWConv	88.5%	85%	3.5%	80%
ViT + FFT	86.1%	82%	4.2%	75%
DWConv-ViT + FFT (2 block bypass)	92.0%	92.5%	1.8%	89%

ViT model. Additionally, incorporating KL-grade weighting significantly improves performance, raising Grade 4 recall by 17% (from 72% to 89%), thus supporting consistent detection across various osteoarthritis grades. With an excellent trade-off between precision (92.0% AUC-ROC) and a low false positive rate (1.8%), the DWConv-ViT + FFT model stands out as the most reliable option for clinical deployment.

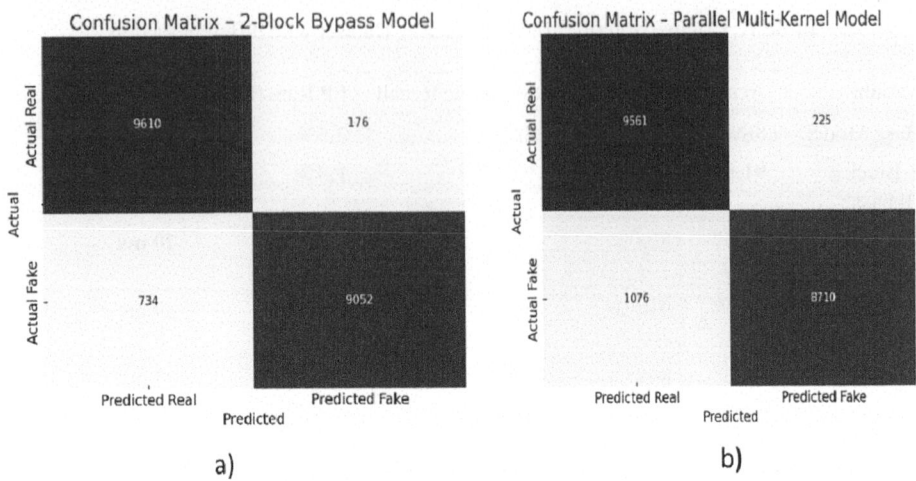

Fig. 4. Confusion Matrix: a) 2-Block bypass model and b) Parallel Multi-Kernel

5   Conclusion and Future Directions

The proposed DWConv-ViT + FFT-based deepfake detection model demonstrates outstanding performance in identifying medical deepfakes, owing to its ability to extract both spectral and spatial features. It effectively uncovers subtle inconsistencies that conventional methods often miss, thereby strengthening defenses against malicious manipulations. This research presents a robust hybrid deep learning architecture—DWConv-ViT combined with FFT, specifically tailored to detect synthetic osteoarthritis X-ray images, addressing the rising threat of medical image forgery. The model leverages depthwise convolutions and vision transformers to capture intricate texture patterns along with larger anatomical structures. Incorporating the Fast Fourier Transform (FFT) module further refines its sensitivity to frequency-domain artifacts commonly introduced by GAN-based generation techniques. Among various configurations evaluated, the 2-Block Bypass variant emerged as the most effective. This approach holds promise for broader application in detecting deepfakes across diverse medical imaging formats, including CT and MRI, where adversarial anomalies may differ. Additionally, integrating this model with explainable AI tools in real-time clinical environments could boost interpretability and foster greater trust in AI-assisted diagnostics.

References

1. Solaiyappan, S., Wen, Y.: Machine learning based medical image deepfake detection: a comparative study. Mach. Learn. Appl. **8**(April), p. 100298 (2022). https://doi.org/10.1016/j.mlwa.2022.100298
2. Waqas, N., Safie, S.I., Kadir, K.A., Khan, S., Kaka Khel, M.H.: DEEPFAKE image synthesis for data augmentation. IEEE Access **10**, 80847–80857 (2022). https://doi.org/10.1109/ACCESS.2022.3193668

3. Prezja, F., Paloneva, J., Pölönen, I., Niinimäki, E., Äyrämö, S.: DeepFake knee osteoarthritis X-rays from generative adversarial neural networks deceive medical experts and offer augmentation potential to automatic classification. Sci. Rep. **12**(1), 1–16 (2022). https://doi.org/10.1038/s41598-022-23081-4

4. Yang, H.C., Rahmanti, A.R., Huang, C.W., Jack Li, Y.C.: How can research on artificial empathy be enhanced by applying deepfakes? J. Med. Internet Res. **24**(3), 1–8 (2022). https://doi.org/10.2196/29506

5. Falahkheirkhah, K., et al.: Deepfake histologic images for enhancing digital pathology. Lab. Invest. **103**(1), 100006 (2023). https://doi.org/10.1016/j.labinv.2022.100006

6. Coyner, A.S., et al.: Synthetic medical images for robust, privacy-preserving training of artificial intelligence: application to retinopathy of prematurity diagnosis. Ophthalmol. Sci. **2**(2), 100126 (2022). https://doi.org/10.1016/j.xops.2022.100126

7. Stokel-Walker, C.: Deepfakes and doctors: how people are being fooled by social media scams. BMJ (2024). https://doi.org/10.1136/bmj.q1319

8. Mirsky, Y., Mahler, T., Shelef, I., Elovici, Y.: CT-GAN: malicious tampering of 3D medical imagery using deep learning. In: Proceedings of the 28th USENIX Conference on Security Symposium, in SEC'19. USA: USENIX Association, pp. 461–478, 2019

9. Azad, R., et al.: Advances in medical image analysis with vision transformers: a comprehensive review. Med. Image Anal. **91**, 103000 (2024). https://doi.org/10.1016/j.media.2023.103000

10. Guo, Y., Li, Y., Wang, L., Rosing, T.: Depthwise convolution is all you need for learning multiple visual domains. In: 33rd AAAI Conference on Artificial Intelligence AAAI 2019, 31st Innov. Appl. Artificial Intelligence Conference IAAI 2019 9th AAAI Symp. Educ. Advance Artificial Intelligence EAAI 2019, pp. 8368–8375, 2019, https://doi.org/10.1609/aaai.v33i01.33018368

11. Gao, J., Xia, Z., Marcialis, G.L., Dang, C., Dai, J., Feng, X.: DeepFake detection based on high-frequency enhancement network for highly compressed content. Expert Syst. Appl. **249**(March) (2024). https://doi.org/10.1016/j.eswa.2024.123732

12. Zhang, L., Yang, H., Qiu, T., Li, L.: AP-GAN: improving attribute preservation in video face swapping. IEEE Trans. Circuits Syst. Video Technol. **32**(4), 2226–2237 (2022). https://doi.org/10.1109/TCSVT.2021.3089724

13. Li, Z., et al.: Identity-aware variational autoencoder for face swapping. IEEE Trans. Circuits Syst. Video Technol. **PP**(July 2023), 1 (2024). https://doi.org/10.1109/TCSVT.2024.3349909

14. Liu, X., Xu, Y., Wu, Q., Zhou, H., Wu, W., Zhou, B.: Semantic-aware implicit neural audio-driven video portrait generation. In: Avidan, S., Brostow, G., Cissé, M., Farinella, G.M., Hassner, T. (eds.) Computer Vision – ECCV 2022. ECCV 2022. LNCS, vol. 13697, pp. 106–125. Springer, Cham (2022). https://doi.org/10.1007/978-3-031-19836-6_7

15. Sha, T., Zhang, W., Shen, T., Li, Z., Mei, T.: Face, pose and cloth synthesis. J. ACM **37**(4) (2018)

16. Li, T., Zhang, W., Song, R., Li, Z., Liu, J.: PoT-GAN: pose transform GAN for person image synthesis. IEEE Trans. Image Process. **30**, 7677–7688 (2021). https://doi.org/10.1109/TIP.2021.3104183

17. Nirkin, Y., Keller, Y., Hassner, T.: FSGANv2: improved subject agnostic face swapping and reenactment. IEEE Trans. Pattern Anal. Mach. Intell. **45**(1), 560–575 (2023). https://doi.org/10.1109/TPAMI.2022.3155571

18. Chen, X., Ni, B., Liu, Y., Liu, N., Zeng, Z., Wang, H.: SimSwap++: towards faster and high-quality identity swapping. IEEE Trans. Pattern Anal. Mach. Intell. **46**(1), 576–592 (2024). https://doi.org/10.1109/TPAMI.2023.3307156

19. Waseem, S., Abu Bakar, S.A.R.S., Ahmed, B.A., Omar, Z., Eisa, T.A.E., Dalam, M.E.E.: DeepFake on face and expression swap: a review. IEEE Access **11**, 117865–117906 (2023). https://doi.org/10.1109/ACCESS.2023.3324403

20. Wen, X., Wang, M., Richardt, C., Chen, Z.-Y., Hu, S.-M.: Photorealistic audio-driven video portraits. IEEE Trans. Vis. Comput. Graph. **26**(12), 3457–3466 (2020). https://doi.org/10.1109/TVCG.2020.3023573

21. Thies, J., Elgharib, M., Tewari, A., Theobalt, C., Nießner, M.: Neural Voice Puppetry: Audio-driven Facial Reenactment, pp. 1–16

22. Dash, A., Ye, J., Wang, G.: A review of generative adversarial networks (GANs) and its applications in a wide variety of disciplines: from medical to remote sensing. IEEE Access **12**(December 2023), 18330–18357 (2024). https://doi.org/10.1109/ACCESS.2023.3346273

23. Chen, J.S., et al.: Deepfakes in ophthalmology: applications and realism of synthetic retinal images from generative adversarial networks. Ophthalmol. Sci. **1**(4), 100079 (2021). https://doi.org/10.1016/j.xops.2021.100079

24. Yeh, C.-H., Yang, H.-F., Chen, M.-J., Kang, L.-W.: Image inpainting based on GAN-driven structure- and texture-aware learning with application to object removal. Appl. Soft Comput. **161**, 111748 (2024). https://doi.org/10.1016/j.asoc.2024.111748

25. Fernandes, S., et al.: Predicting heart rate variations of deepfake videos using neural ODE. In: 2019 IEEE/CVF International Conference on Computer Vision Workshop (ICCVW), pp. 1721–1729, 2019. https://doi.org/10.1109/ICCVW.2019.00213

26. Gulshan, V., et al.: Development and validation of a deep learning algorithm for detection of diabetic retinopathy in retinal fundus photographs. JAMA **316**(22), 2402–2410 (2016). https://doi.org/10.1001/jama.2016.17216

27. Karaköse, M., Yetiş, H., Çeçen, M.: A new approach for effective medical deepfake detection in medical images. IEEE Access **12**(March), 52205–52214 (2024). https://doi.org/10.1109/ACCESS.2024.3386644

28. Arshed, M.A., Mumtaz, S., Gherghina, Ş.C., Urooj, N., Ahmed, S., Dewi, C.: A deep learning model for detecting fake medical images to mitigate financial insurance fraud. Computation **12**(9), 173 (2024). https://doi.org/10.3390/computation12090173

29. Chen, P.: Knee Osteoarthritis Severity Grading Dataset, Mendeley Data, 2018. https://doi.org/10.17632/56rmx5bjcr.1

Static Load Balancing for Molecular-Continuum Flow Simulations with Heterogeneous Particle Systems and on Heterogeneous Hardware

Amartya Das Sharma[1]($\boxtimes$) (ID), Louis Viot[1], Piet Jarmatz[1] (ID), Hauke Preuß[1] (ID), and Philipp Neumann[2,3] (ID)

[1] High Performance Computing, Helmut Schmidt University, Hamburg, Germany
das-sharma@hsu-hh.de
[2] High Performance Computing and Data Science, Universität Hamburg, Hamburg, Germany
[3] Deutsches Elektronen-Synchrotron (DESY), Hamburg, Germany
philipp.neumann@desy.de

Abstract. Load balancing in particle simulations is a well-researched field, but its effect on molecular-continuum coupled simulations is comparatively less explored.

In this work, we implement static load balancing into the macro-micro-coupling tool (MaMiCo), a software for molecular-continuum coupling, and demonstrate its effectiveness in two classes of experiments by coupling with the particle simulation software ls1 mardyn. The first class comprises a liquid-vapour multiphase scenario, modelling evaporation of a liquid into vacuum and requiring load balancing due to heterogeneous particle distributions in space. The second class considers execution of molecular-continuum simulations on heterogeneous hardware, running at very different efficiencies. After a series of experiments with balanced and unbalanced setups, we find that, with our balanced configurations, we achieve a reduction in runtime by 44% and 55% respectively.

Keywords: Coupled Simulations · Multiphase Simulations · Load Balancing · Molecular-Continuum · Heterogeneous Architecture

1 Introduction

Coupled multiscale simulations are a viable way of reducing compute times while retaining simulation details. Since they are often still computationally expensive, leveraging the full capabilities of hardware and software is greatly desirable.

One example for coupled multiscale simulations are molecular-continuum systems: molecular dynamics (MD) simulation is employed in areas where greater granularity is desired, and a classical computational fluid dynamics (CFD) solver is used everywhere else. One way that MD simulations gain performance in heterogeneous situations is through effectively balancing computational load in a

M. H. Lees et al. (Eds.): ICCS 2025, LNCS 15905, pp. 241–255, 2025.
https://doi.org/10.1007/978-3-031-97632-2_17

distributed environment. In this work, we extend the molecular-continuum coupling tool MaMiCo with static load balancing. We demonstrate the effectiveness using well-established community codes, such as the open-source MD library ls1 mardyn, which leverages the node-level library AutoPas in its kernel, or the CFD package OpenFOAM [27], using the popular coupling tool preCICE [4,5] as a bridge to MaMiCo. Heterogeneity is considered in two ways: a multiphase scenario modelling evaporation, introducing heterogeneity due to inhomogeneous particle distributions, and a heterogeneous-architecture scenario with the simulation running on a mix of x86 and aarch64 hardware.

In Sect. 2, the dominant software stack of ls1 mardyn + AutoPas + MaMiCo is introduced, and related work is discussed. Coupling and load balancing are explained in Sect. 3. We then describe and analyse the evaporation scenario in Sect. 4 and a homogeneous Couette flow scenario on heterogeneous distributed hardware in Sect. 5. We close with a short discussion and an outlook to future work in Sect. 6.

2   Background

2.1   Load Balancing

Particle simulations such as MD are often parallelised by using spatial domain decomposition. The computational domain is divided into subdomains, and each parallel process computes particle updates independently, exchanging relevant information with neighbouring subdomains at the domain boundaries. As shown in Fig. 1a, a regular decomposition can be suboptimal in many cases, such as those with varied particle density throughout the domain. *Load balancing* allows to distribute the computational load more equally (as shown in Fig. 1b) and can dramatically increase the performance of the simulation. Balancing can be static (ratios defined at the beginning of the simulation), which is the focus in this work, or dynamic (ratios calculated and subdomains adjusted based on minimising some predefined "load", such as runtime, or particle imbalance). Static imbalances occur from constant workload distribution, while dynamic imbalances signify time-dependent changes in the scenario [3].

Load balancing in MD has been extensively explored since the 90s [11]. A more recent review of load balancing techniques is given in [8]. Many community codes such as GROMACS [1] or LAMMPS [16] offer static and dynamic load balancing.

Load balancing in coupled simulations is relatively less explored, with a few representative examples provided in the following. Ko et al. [10] introduce a coordinated job submission API for computing clusters, so that coupled solvers may execute synchronously and independently. They introduce load balancing by changing the resources available to each coupled solver, assigning them more or fewer nodes until their execution times match. Niemöller et al. [14] demonstrate a coupled CFD-CAA (Computational Aero-acoustics) simulation. They implement load balancing with a partitioning approach based on space-filling curves (SFC [24]). Pour et al. [17] discuss their aero-acoustic simulation coupling three separate elements: an innermost domain where fully compressible Navier-Stokes equations are solved, surrounded by a subdomain where inviscid Euler equations are solved, and a far-field where linearized Euler equations are solved. They use the SFC-based algorithm SPartA [7] for load balancing. Besseron et al. [2] couple a discrete elements method (DEM) solver XDEM with the CFD solver OpenFOAM, using the preCICE library for coupling. Load balancing occurs by assigning MPI ranks manually between the two solvers, allowing the solvers to operate in a black-box fashion and divide the computational load internally over whatever resources it receives.

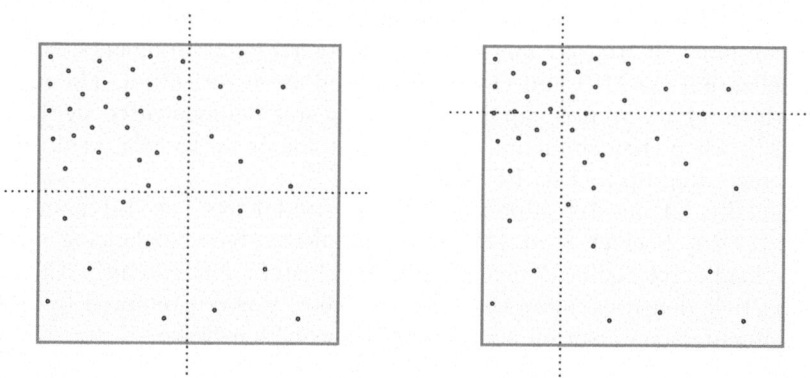

(a) Regular grid, with unequal particle distribution

(b) Irregular rectilinear grid, with a
more balanced particle distribution

Fig. 1. An illustration of 2D load balancing in MD. Dotted lines indicate process boundaries, dividing the domain into four subdomains. Solid lines show global boundaries, and dots indicate particles.

2.2 AutoPas

AutoPas[1] [6,19] is a node-level particle simulation library. It supports a variety of algorithmic MD configuration options and *auto-tuning* by sampling the performance of configurations and dynamically switching between them without intervention. These configurations include particle containers (linked lists, Verlet lists, etc.), shared-memory parallel particle traversal schemes or data layouts. In this work, we determine a configuration and keep it invariant, so as to not have to consider additional variables when comparing results from our static load balancing.

In our work, we use rigid single-site molecules and short-range Lennard-Jones interactions [12], and allow AutoPas to only use the linked cell container, resulting in a linear run time complexity $O(N)$ with N being the number of molecules.

---

[1] https://github.com/AutoPas/AutoPas.

We allow sliced cell traversal, structure-of-arrays layout of data, and keep the optimization `newton3` enabled, which halves the number of force calculations.

2.3   ls1 mardyn

ls1 mardyn[2] is an open-source MD solver for simulating small rigid molecules at large time and length scales [15]. It focuses on thermodynamics and nanofluidics, and its features include easy extensibility using a flexible plugin system (which MaMiCo employs for coupling [9]) . Newer versions of ls1 mardyn allow the usage of AutoPas at node level, and then ls1 mardyn handles the inter-node orchestration, communication and balancing, leading to even better computational performance [6,19]. ls1 mardyn is built for HPC environments, supporting parallelization via MPI and OpenMP, as well as vectorisation. The massive scale at which ls1 mardyn is capable of running was demonstrated by Tchipev et al. in [23], when they simulated 20 trillion atoms at up to 88% weak scaling efficiency, recording up to 1.33 PFLOPS.

Additionally, ls1 mardyn supports diffusive and kd-tree based dynamic load balancing [19,20]. Seckler et al. [21] used the kd-tree based balancing method to balance load across heterogeneous hardware clusters; for systems with homogeneous particle densities, more than one rebalance was not required to find an optimal domain decomposition across heterogeneous hardware.

2.4   MaMiCo

MaMiCo[3] is an open-source software designed to couple microscopic (e.g. MD) solvers with macroscopic (e.g. CFD) solvers in a domain decomposition sense, enabling fully three-dimensional, transient molecular-continuum flow simulation on strongly coupled time scales (i.e., one CFD time step corresponds to O(100) MD time steps [9,13]). MaMiCo achieves this by defining an overlap domain within a larger CFD region (where the MD simulation is embedded) and by exchanging data between the solvers within this overlap. The exchange is facilitated by Cartesian *coupling cells*, which cover the entire coupling domain. Using these cells, MaMiCo extracts relevant data (averaged mass and momentum values from the micro solvers, and the mass flux and boundary conditions from the macro solvers), and then converts and exchanges this data between the solvers. These cells, however, need to (at minimum) cover the whole micro solver region. Hence, MaMiCo needs to be aware of the distribution of the micro region across processes in a distributed environment, to be able to map the coupling cells to linked cells of the micro solver. Since the run time of a coupled simulation is typically dominated by the micro simulation, and since more micro-level data than macro-level data needs to be accessed and modified for the coupling, MaMiCo always runs on the processes of the micro solver and manages communication to

---

[2] https://github.com/ls1mardyn/ls1-mardyn.
[3] https://github.com/HSU-HPC/MaMiCo.

the macro simulation processes from there, for efficiency and performance reasons. This means that if the process distribution on micro side changes, then the allocation for the MaMiCo coupling cells should also be altered correspondingly.

For (Brownian) noise reduction in the MD region, MaMiCo provides noise filters and supports parallel independent MD simulations all simulating the same region, from which the required properties are averaged out and sent to the macroscopic solver [13]. This approach, known as *ensemble averaging* or *multi-instance sampling*, helps reduce the inherent randomness of particle simulations. These independent simulations are also subdivided and run in parallel. A sample coupling setup that MaMiCo supports, relevant to the experiments in this paper, is shown in Fig. 2.

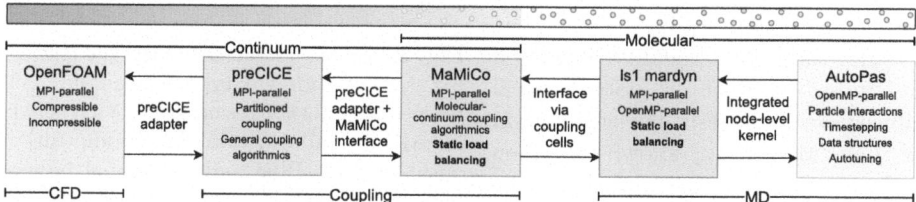

Fig. 2. A coupling setup with MaMiCo, showing our typical software stack. The Open-FOAM + preCICE + MaMiCo coupling and the MaMiCo + ls1 mardyn + AutoPas coupling have been separately validated [9,25]. New implementations in the current work are in bold.

3   Methods and Implementation

3.1   MaMiCo: Indexing System and Rectilinear Decompositions

MaMiCo's coupling strategy relies on the coupling cells as the main data structure, for interfacing with the underlying simulation libraries and for iterating over subdomains. These coupling cells form a regular grid over the 3D domain spanning the global MD volume. In practice, it is often required to iterate over a subset of coupling cells, spanning a certain volume or satisfying certain criteria. For example, one may need to iterate over only the region where mass transfer from CFD to MD occurs. Both scalar and vector indices for cells are desirable: unsigned scalar indices are better for direct element access in array-based underlying data structures, and they enable efficient and optimised iteration. Signed vector indices allow to find neighbour cells more quickly, they can be converted more easily and are needed for isolating layers and shells.

To fulfil these criteria, a flexible and robust *indexing type system* was developed for MaMiCo. It operates on sixteen cell index types, which are instantiated at compile time using a template meta-programming based approach for faster execution times. Each type refers to a subset of coupling cells with a rectangular cuboid shape as illustrated in Fig. 3. The types arise from four boolean traits of each index type that are relevant for interpretation of indices and their conversion: locality, halo, scalarity, and direction. Locality decides if the index is local or global. Local indices are relative to the subdomain of their MPI rank, shown in green and pink in Fig. 3. Global indices are shown in blue and red. The halo trait decides whether an additional ghost layer, which is used for boundary handling and data communication, should be included (blue) or excluded (red). Scalarity decides whether the index is scalar (denoted by numbers in the grid) or vector (denoted by components at the right side next to the cell grid). The data direction index trait decides whether the cells contain data to be transferred from the micro to macro solver (green and orange indices, blue area in Fig. 3), or vice-versa (red area in Fig. 3). Notable features of the indexing system include compile time type safety for cell indices, and automatic conversion between the index types.

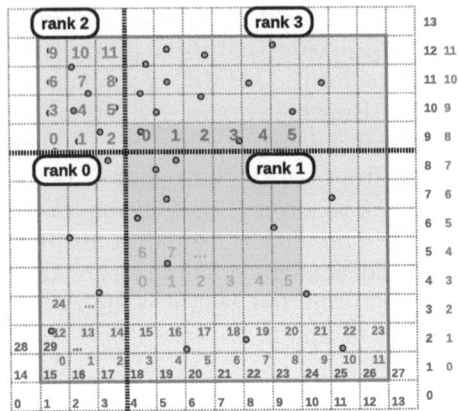

Fig. 3. Load balancing with the indexing system: different types of coupling cell indices on a 2D Cartesian grid with a generalized non-uniform domain decomposition with 4 ranks. Blue and red global indices include and exclude a ghost layer, pink are the subdomain of rank 2, green and orange are the MD-to-macro data transfer domain of MD rank 3 and 1. (Colour figure online)

Furthermore, we extended the indexing system to account for non-regular rectilinear decompositions of the MD domain, to enable load balancing. Now, we can define a static breakdown of the 3D domain into subdomains, and this breakdown is propagated to the MD solver (currently supported by ls1 mardyn) and is also used by the indexing system to define the coupling cells and index conversions in each dimension. This extension itself was easily facilitated by the flexibility of this new indexing system, which itself (in contrast) required a semi-complete overhaul of all iteration logic in the entire codebase, changing 10,000+ lines of code.

3.2   ls1 mardyn: Static Load Balancing

Although ls1 mardyn already implements diffusive and kd-tree based dynamic load-balancing [19,20], we require static load balancing as a first step towards compatibility with MaMiCo. This balancing needs to be able to split the 3D domain of particles into subdomains in a way that is predictable and reproducible, so that the balancing can be propagated to every MD simulation instance within an ensemble.

This required feature was implemented by adding a new domain decomposition class to ls1 mardyn that generalises the existing regular domain decomposition, and now the user can specify a specific breakdown of the 3D domain in the x, y, and z axes. This meets the aforementioned compatibility criteria with MaMiCo's multi-instance sampling, and the rectilinear subdomains of arbitrary dimensions map seamlessly onto the coupling cells of MaMiCo as before, allowing sampling and mass + velocity transfer. We have validated that the state of the coupled particle system modelled by ls1 mardyn in a load balanced setup matches the behaviour of a non-balanced regular subdomain decomposition. This was done for several configurations of one-way coupled setups with a variety of flow scenarios, including a multi-phase droplet scenario, and the Couette flow startup [9].

4   Experiment 1: Coupled Multiphase Simulation

4.1   Introduction

Molecular-continuum coupling offers significant utility in multiphase scenarios, since different phases (such as vapour or liquid phase) can be simulated with specialised solvers that may then be coupled. This way, we can run particle simulations in specific areas of interest in a scenario, but then also conserve the effects of the overall flow field by simulating the surrounding volumes via a less resource-intensive solver. This resultant complexity reduction can also be considered as a form of load balancing, which only becomes possible in coupled simulations. However, the surface where the phase change occurs is often the most interesting part of these simulations, and hence more than one phase are often present in the higher-granularity solver's domain. Here it becomes necessary to have support for load balancing, since density imbalances severely affect the performance of particle simulations.

4.2   Scenario Description

We adapt the evaporation scenario described in Homes et al. [22]. A cuboid with vacuum on one end and a Lennard-Jones fluid on the other end are set up, and the fluid is allowed to evaporate into the vacuum. This leads to a coexistence of vapour and liquid phases at a fixed temperature [26].

In our coupled setup, we define a similar cuboid, as a domain of size $100 \times 360 \times 100$ and divide it into two subregions: an initial $100 \times 40 \times 100$ region filled with liquid acting as the reservoir, simulated using the CFD solver OpenFOAM, and the remainder (containing the liquid-vapour interface) simulated with ls1 mardyn+AutoPas. Since an OpenFOAM-preCICE coupling [5], a preCICE-MaMiCo coupling [25], and a MaMiCo-ls1 mardyn coupling [9] already exist, we are able to couple OpenFOAM and ls1 mardyn+AutoPas using MaMiCo and preCICE as a combined bridge.

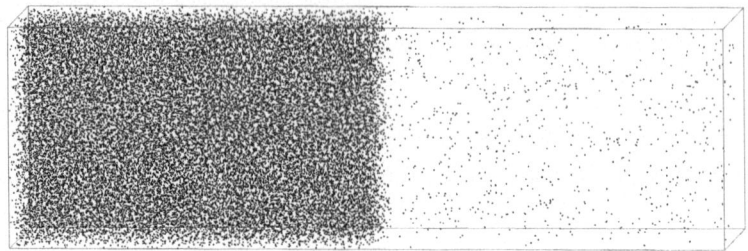

Fig. 4. The MD region of the multiphase evaporation scenario. The denser liquid phase is on the left, and the vapour phase on the right.

The particle region is evenly split into liquid ($\rho_{liq} \approx 0.7302$) and vapour ($\rho_{vap} \approx 0.0198$). The values are chosen to ensure vapour-liquid coexistence at temperature $T = 0.8$ [26]).

4.3   Experiment Setup

The MD region, cf. Fig. 4, is of dimensions $100 \times 320 \times 100$. It is entirely covered by coupling cells of size 2.5^3. We run the experiments on one node of the cluster HSUper[4], containing two sockets with an Intel Xeon Platinum 8360Y processor per socket with up to 36 cores. We use 32 MPI ranks, in a $2 \times 8 \times 2$ distribution. The simulation is run for 100 coupling cycles, with 50 MD timesteps per cycle. We use only one MD instance as part of our multi-instance sampling.

We run two experiments, one with the domain along the y axis divided equally between the liquid and vapour phases (termed the *4+4* setup) shown in Fig. 5a, and one with all ranks save one simulating the liquid phase (termed the *7+1* setup) shown in Fig. 5b. From a simple ratio of the maximum number of coupling cells in a rank in each phase, we predict a speedup of approximately 60% in the 7+1 case compared to the 4+4 case.

4.4   Results

For the 4+4 case, we achieve a time of 61.11 s per coupling cycle. For the 7+1 case, we record 34.18 s per coupling cycle, implying a 44% reduction in runtime from this static load balancing. However this value is far from our initial estimate of 60%, signifying that there are overheads related to the coupling which are not yet accounted for, such as insertion/deletion of particles. Checking the simulation files, we see that each coupling step adds an average overhead of 10 s, however this is independent of the rank breakdown of the simulation, and thus is a constant overhead which skews the ratio of runtimes. However, the speedup obtained is still significant, and emphasises the importance of load balancing in multiphase scenarios.

---

[4] https://portal.hpc.hsu-hh.de/documentation/hsuper.

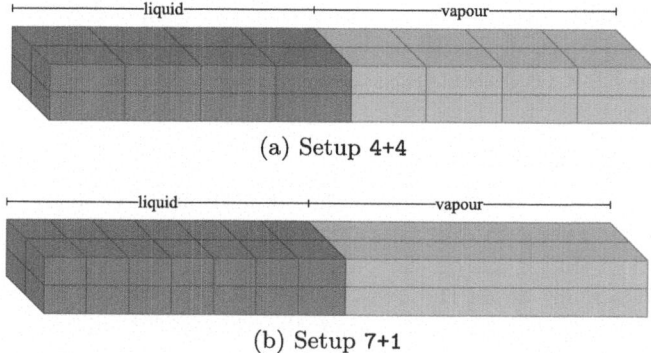

(a) Setup 4+4

(b) Setup 7+1

Fig. 5. Diagrams illustrating the rank breakdown of the setups. The configuration is named according to the number of ranks per phase in the y axis. Green and orange denote the liquid and vapour phase respectively. (Colour figure online)

5   Experiment 2: Handling Heterogeneous Architectures

5.1   Introduction

To demonstrate handling truly distributed molecular-continuum simulations with MaMiCo, Viot et al. in [25] successfully coupled a CFD simulation running on a local workstation with an MD simulation running on a cluster, communicating using TCP/IP. With the added support for static load balancing, we can explore another way of running a coupled simulation on a mix of hardware. This may be done for various reasons, such as hardware availability, GPU support and offloading, or energy requirements. Load balancing becomes even more important in this framework, to offset the performance difference between the different hardware. This allows users to take advantage of hardware-specific optimisations (vectorisation support, compiler optimisations, mixed CPU-GPU computing etc.) In this experiment, we demonstrate this capability by executing a coupled simulation on two different compute nodes at a time, namely an AMD node and an ARM node.

5.2   Scenario Description

We adopt a Couette flow scenario for this experiment. This scenario describes a liquid at homogeneous density trapped between two infinite surfaces with differing relative velocities. In our case, one of the two surfaces is motionless and the other slides in the $+x$ direction at a constant velocity. We set up a Couette flow, with the gap between the surfaces being 150 units. We simulate a 150^3 volume with a CFD simulation, inside which we have a 120^3 MD region. We fix the MaMiCo coupling cell size to 5^3, and we run two MD instances. Ten coupling cycles are run, each for 100 MD timesteps. We use a simple in-house Lattice-Boltzmann (LB) solver on the CFD side (running on one rank) and ls1 mardyn+AutoPas on the MD side. The fluid density is ≈ 0.813.

We use one AMD EPYC 7763 as our AMD node, with 64 hardware cores, and one Fujitsu A64FX as our ARM node, with 48 cores. The nodes are connected via InfiniBand EDR (100Gbit/s).

5.3   Rank Migration Test

Setup. We start with a setup where the simulation fully saturates one AMD node (64 ranks). Then, we migrate ranks one by one onto the ARM node, and observe the effect that it has on the simulation speed. We divide the domain into $4 \times 4 \times 4$ subdomains along the axes, giving us 6^3 coupling cells per rank. We also conduct the experiment in the other direction i.e. by migrating ranks from a fully saturated ARM node to the AMD node. In this case, since we only have 48 ranks, the domain breakdown is $4 \times 4 \times 3$ along the xyz axes. In both cases, the experiments are run thrice, and the average value is reported.

From an earlier report utilising ls1 mardyn+AutoPas on the same hardware [18], we expect the simulation to run significantly slower on ARM.

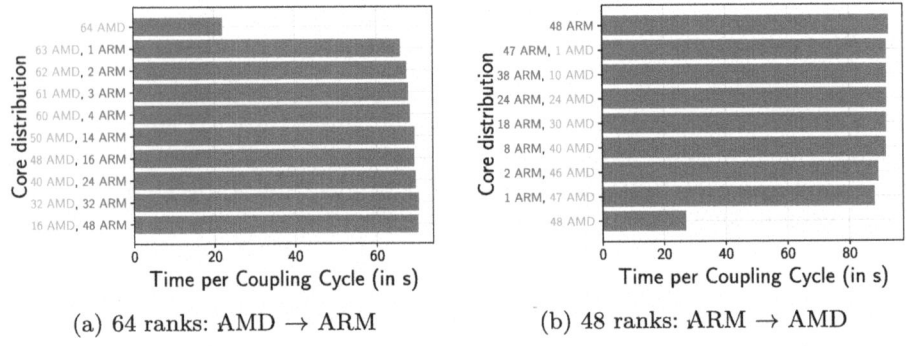

(a) 64 ranks: AMD → ARM   (b) 48 ranks: ARM → AMD

Fig. 6. Chosen results (simulation speed) from migration experiments.

Results. Selected results of the AMD→ARM migration experiment are shown in Fig. 6a, and results of the ARM→AMD migration experiment are shown in Fig. 6b. The performance is severely affected as soon as a single subdomain is migrated to ARM, and it remains relatively stable afterwards, only worsening marginally with the migration of more ranks. As more ranks move to ARM, more subdomains are dependent on their ARM neighbours to finish their computations each MD timestep, hence the marginal increase in walltime can be explained by compounding delays due to synchronisation.

Another conclusion drawn from Fig. 6b is that communication overhead between the ARM and AMD node is negligible, since there is an immediate improvement from migrating a single rank from ARM to AMD, and no additional (network-related) walltime increases can be observed.

Since the scenario is of homogeneous density, we can determine the expected runtime per coupling cell from these results. As each rank has $6^3 = 216$ coupling cells in the case with 64 ranks, we find that the time per coupling cell per cycle for AMD $t_{\text{AMD,cell}} \approx 0.102$ s, and similarly for ARM we have $t_{\text{ARM,cell}} \approx 0.305$ s. We notice that

$$t_{\text{ARM,cell}} \approx 3 \times t_{\text{AMD,cell}} \tag{1}$$

Denoting the predicted time per coupling cycle as t_{pred}, we have

$$\begin{aligned}
t_{\text{pred}} &= \max\left(t_{\text{AMD,pred}}, t_{\text{ARM,pred}}\right) \\
&= \max\left(t_{\text{AMD,cell}} \times n_{\text{AMD}}, t_{\text{ARM,cell}} \times n_{\text{ARM}}\right) \\
&= \max\left(0.102 n_{\text{AMD}}, 0.305 n_{\text{ARM}}\right)
\end{aligned} \tag{2}$$

where $n_{\text{AMD}}, n_{\text{ARM}}$ are the maximum number of coupling cells in a rank on AMD and ARM respectively. We use the value of t_{pred} to perform optimal load balancing by adjusting the values of n and verify our configurations experimentally. The constants $t_{\text{AMD,cell}}, t_{\text{ARM,cell}}$ account for two MD instances with 100 MD timesteps per instance, and hence are not generalisable beyond this context unless those factors are accounted for.

As we only use the 64 rank case in future experiments, we omit the calculations to derive time values for the 48 rank case here.

5.4   Balancing Tests

With the reference values in mind, we set up two load balancing experiments, using 64 ranks: one where only one rank (and thus only one subdomain) of the MD simulation resides on ARM and the remaining 63 ranks run on AMD (termed *63+1* setup), and one where the domain (and therefore the 64 ranks) are evenly split among the nodes (termed *32+32* setup). The balancing is done keeping eq. (1) in mind, and only along the x axis for simplicity.

Setup. We name our configurations in this section using the number of coupling cells in the x direction. Since we have 4 ranks across each axis, all configurations are four comma-separated integers. In figures and diagrams, cells running on AMD are denoted in thistle, and cells running on ARM are denoted in royal purple. Thus, the default, evenly distributed configuration for the 32+32 setup is 6,6,6,6 (6 coupling cells along x-direction for fixed y and z per rank).

For the 63+1 setup, we choose four configurations. The default configuration is shown in Fig. 7a. Firstly, from eq. (1), we allocate $n_{AMD} = 3 \times n_{ARM}$ with config. 6,6,9,3. Next, we even out the load on the AMD side with config. 7,7,7,3, illustrated in Fig. 7b. Finally, we attempt to create a bottleneck on the AMD side instead, with config. 7,7,8,2 and config. 8,8,7,1.

Then, for the 32+32 setup, we choose four configurations again, with the default 6,6,6,6 shown in Fig. 7b. Once again, we allocate $n_{AMD} = 3 \times n_{ARM}$ with config. 9,9,3,3, and try to create a bottleneck on the AMD side with config. 10,10,2,2, shown in Fig. 7d, and also with a config. 11,11,1,1. We

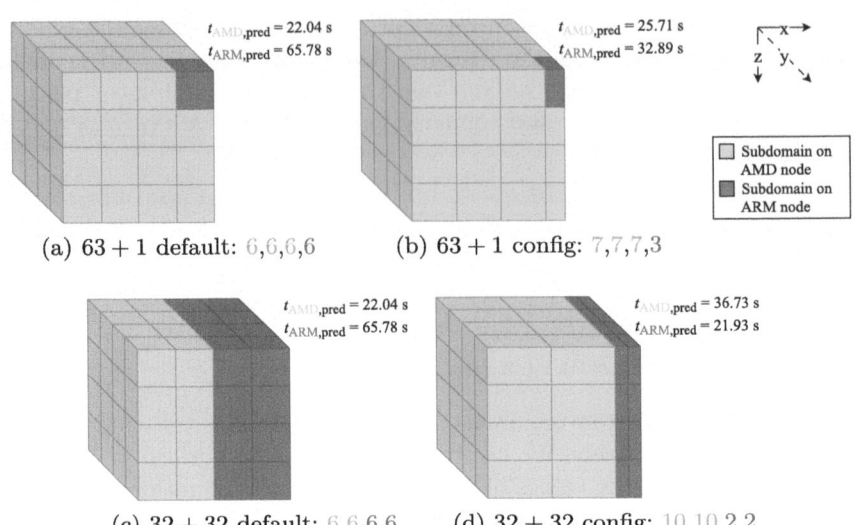

(a) 63 + 1 default: 6,6,6,6   (b) 63 + 1 config: 7,7,7,3

(c) 32 + 32 default: 6,6,6,6   (d) 32 + 32 config: 10,10,2,2

Fig. 7. Diagrams illustrating the rank breakdown of the setups. The configuration is named counting the allocation of cells along the x axis.

include an additional config. 8,8,4,4 to illustrate a gradual improvement over the default setup.

Results. In Fig. 8a we show the results of the experiments involving the 63+1 setup, and in Fig. 8b we show the same for the 32+32 setup. We see that proper balancing halves the time required per coupling cycle, as the bottleneck created by the ARM node is somewhat alleviated. For the 63+1 setup, the time per coupling cycle is reduced from 65.78 s to 29.18 s ($\approx$ 55.6% reduction of runtime) in the best case, and similarly, in the 32+32 setup we see an improvement from 70.12 s to 34.94 s ($\approx$ 50.2% reduction of runtime).

We see that proper balancing halves the time required per coupling cycle, as the bottleneck created by the ARM node is somewhat alleviated. For the 63+1 setup, the time per coupling cycle is reduced from 65.78 s to 29.18 s ($\approx$ 55.6% reduction of runtime) in the best case, and similarly, in the 32+32 setup we see an improvement from 70.12 s to 34.94 s ($\approx$ 50.2% reduction of runtime).

Metrics reported by ls1 mardyn per simulation show that this improvement can be attributed in a large part to the reduction of the runtime per coupling cycle in the force calculation step of the simulation ($\approx$55.07% reduction from 11.214 s to 5.038 s in the 63+1 setup, $\approx$46.71% reduction from 11.258 s to 6.002 s in the 32+32 setup). A deeper performance analysis may yield further insight.

The experimental results lie within a 5.7% margin of our predictions, with the notable outlier being the 7,7,8,2 configuration in the 64+1 setup, with a discrepancy of −11.76%. We can conclude that it is still not ideal to run the

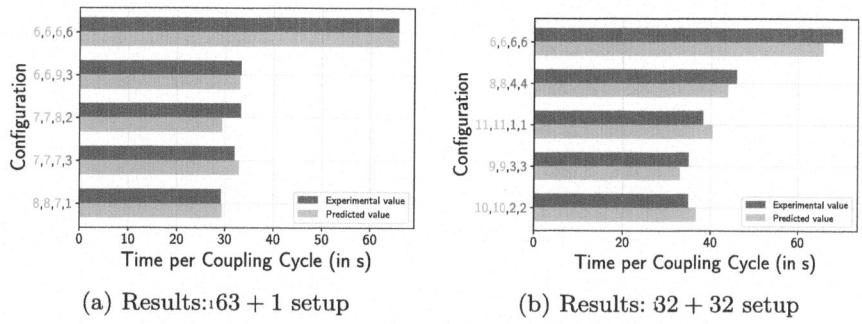

(a) Results: 63 + 1 setup   (b) Results: 32 + 32 setup

Fig. 8. Simulation speed results from load balancing.

simulation on mixed hardware without further hardware-specific optimisation, but proper load balancing successfully reduces the overall runtime.

6   Conclusion

From our experiments, we have verified the importance of load balancing in coupled scenarios. We have implemented a static load balancer for ls1 mardyn, and have extended MaMiCo to be able to interface with the resultant rectangular rectilinear grids. We have tested our implementation with both a heterogeneous phase and heterogeneous hardware setup, both of which are highly relevant to coupled simulations. The results are favourable, leading to a 44% to 55% reduction of runtime across the fastest configurations. These results provide ample motivation for further work in this area.

The first goal for future work should be the automatic selection of an optimal configuration of static load balancing at the beginning of the simulation. This could be done with expert knowledge, or from estimates (from previous work, or from automatic mini-benchmarks). Then MaMiCo should be further extended to support dynamic load balancing, driven by the MD simulation; it should be able to adapt to changing subdomain sizes without further interference from the user. Finally, the balancing support should be extended towards other particle dynamics solvers, such as LAMMPS or GROMACS.

Acknowledgments. A. Das Sharma and L. Viot acknowledge financial support by the projects MaST and hpc.bw. Computational resources (HPC cluster HSUper, ARM Minicluster, AMD Minicluster) have been provided by the project hpc.bw. MaST and hpc.bw have been funded by dtec.bw - Digitalization and Technology Research Center of the Bundeswehr; dtec.bw is funded by the European Union - NextGenerationEU. The authors further acknowledge the provision of computational resources for more numerical studies at HLRS (project GCS-MDDC). They thank Simon Homes and Prof. Jadran Vrabec from TU Berlin for their fruitful correspondence and invaluable support regarding the implementation of the evaporation scenario.

Disclosure of Interests. The authors have no competing interests to declare that are relevant to the content of this article.

References

1. Berendsen, H., van der Spoel, D., van Drunen, R.: GROMACS: a message-passing parallel molecular dynamics implementation. Comput. Phys. Commun. **91**(1), 43–56 (1995)
2. Besseron, X., Adhav, P., Peters, B.: Parallel multi-physics coupled simulation of a midrex blast furnace. In: Proceedings of the International Conference on High Performance Computing in Asia-Pacific Region Workshops, pp. 87–98. HPCAsia '24 Workshops, Association for Computing Machinery, New York, NY, USA (2024)
3. Böhme, D., Wolf, F., Geimer, M.: Characterizing load and communication imbalance in large-scale parallel applications. In: 2012 IEEE 26th International Parallel and Distributed Processing Symposium Workshops & PhD Forum, pp. 2538–2541. IEEE (2012)
4. Chourdakis, G., et al.: preCICE v2: A sustainable and user-friendly coupling library [version 2; peer review: 2 approved]. Open Research Europe **2**(51) (2022)
5. Chourdakis, G., Schneider, D., Uekermann, B.: OpenFOAM-preCICE: Coupling OpenFOAM with external solvers for multi-physics simulations. OpenFOAM® J. **3**, 1–25 (2023)
6. Gratl, F.A., Seckler, S., Tchipev, N., Bungartz, H.J., Neumann, P.: AutoPas: auto-tuning for particle simulations. In: 2019 IEEE International Parallel and Distributed Processing Symposium Workshops (IPDPSW), pp. 748–757 (2019)
7. Harlacher, D.F., Klimach, H., Roller, S., Siebert, C., Wolf, F.: Dynamic load balancing for unstructured meshes on space-filling curves. In: 2012 IEEE 26th International Parallel and Distributed Processing Symposium Workshops & PhD Forum, pp. 1661–1669. IEEE (2012)
8. Hirschmann, S., Pflüger, D., Glass, C.W.: Towards understanding optimal load-balancing of heterogeneous short-range molecular dynamics. In: 2016 IEEE 23rd International Conference on High Performance Computing Workshops (HiPCW), pp. 130–141 (2016)
9. Jarmatz, P., et al.: MaMiCo 2.0: an enhanced open-source framework for high-performance molecular-continuum flow simulation. SoftwareX **20**, 101251 (2022)
10. Ko, S.H., Kim, N., Kim, J., Thota, A., Jha, S.: Efficient runtime environment for coupled multi-physics simulations: dynamic resource allocation and load-balancing. In: 2010 10th IEEE/ACM International Conference on Cluster, Cloud and Grid Computing, pp. 349–358 (2010)
11. Kohring, G.: Dynamic load balancing for parallelized particle simulations on MIMD computers. Parallel Comput. **21**(4), 683–693 (1995)
12. Lennard-Jones, J.E.: Cohesion. Proc. Phys. Soc. **43**(5), 461 (1931)
13. Neumann, P., Flohr, H., Arora, R., Jarmatz, P., Tchipev, N., Bungartz, H.J.: MaMiCo: software design for parallel molecular-continuum flow simulations. Comput. Phys. Commun. **200**, 324–335 (2016)
14. Niemöller, A., Schlottke-Lakemper, M., Meinke, M., Schröder, W.: Dynamic load balancing for direct-coupled multiphysics simulations. Comput. Fluids **199**, 104437 (2020)
15. Niethammer, C., et al.: ls1 mardyn: the massively parallel molecular dynamics code for large systems. J. Chem. Theory Comput. **10**(10), 4455–4464 (2014)

16. Plimpton, S.: Fast parallel algorithms for short-range molecular dynamics. J. Comput. Phys. **117**(1), 1–19 (1995)
17. Pour, N.E., Krupp, V., Klimach, H., Roller, S.: Load balancing for immersed boundaries in coupled simulations. In: Resch, M.M., Kovalenko, Y., Bez, W., Focht, E., Kobayashi, H. (eds.) Sustained Simulation Performance 2018 and 2019, pp. 185–201. Springer International Publishing, Cham (2020)
18. Preuß, H., et al.: hpc.bw benchmark report 2022–2024 (2024)
19. Seckler, S., Gratl, F., Heinen, M., Vrabec, J., Bungartz, H.J., Neumann, P.: AutoPas in ls1 mardyn: massively parallel particle simulations with node-level auto-tuning. J. Comput. Sci. **50**, 101296 (2021)
20. Seckler, S., et al.: Load balancing and auto-tuning for heterogeneous particle systems using ls1 mardyn. In: Nagel, W.E., Kröner, D.H., Resch, M.M. (eds.) High Performance Computing in Science and Engineering '19, pp. 523–536. Springer International Publishing, Cham (2021)
21. Seckler, S., Tchipev, N., Bungartz, H.J., Neumann, P.: Load balancing for molecular dynamics simulations on heterogeneous architectures. In: 2016 IEEE 23rd International Conference on High Performance Computing (HiPC), pp. 101–110 (2016)
22. Simon Homes, Matthias Heinen, J.V., Fischer, J.: Evaporation driven by conductive heat transport. Mol. Phys. **119**(15–16), e1836410 (2021)
23. Tchipev, N., et al.: TweTriS: twenty trillion-atom simulation. Int. J. High Perform. Comput. Appl. **33**(5), 838–854 (2019)
24. Teresco, J.D., Devine, K.D., Flaherty, J.E.: Partitioning and dynamic load balancing for the numerical solution of partial differential equations. In: Numerical Solution of Partial Differential Equations on Parallel Computers, pp. 55–88. Springer (2006)
25. Viot, L., Piel, Y., Neumann, P.: From desktop to supercomputer: computational fluid dynamics augmented by molecular dynamics using MaMiCo and preCICE. In: Bienz, A., Weiland, M., Baboulin, M., Kruse, C. (eds.) High Performance Computing, pp. 567–576. Springer Nature Switzerland, Cham (2023)
26. Vrabec, J., Kedia, G.K., Fuchs, G., Hasse, H.: Comprehensive study of the vapour-liquid coexistence of the truncated and shifted Lennard-Jones fluid including planar and spherical interface properties. Mol. Phys. **104**(09), 1509–1527 (2006)
27. Weller, H.G., Tabor, G., Jasak, H., Fureby, C.: A tensorial approach to computational continuum mechanics using object-oriented techniques. Comput. Phys. **12**(6), 620–631 (1998)

Joint Spatial-Temporal Representation for Host Intrusion Detection System

Hao Li[1,2], Zehui Wang[1,2], Shang Shang[1,2], Zhengwei Jiang[1,2], Qiuyun Wang[1,2], Fangli Ren[1,2(✉)], and Baoxu Liu[1,2]

[1] Institute of Information Engineering, Chinese Academy of Sciences,
Beijing 100093, China
{lihao,renfangli}@iie.ac.cn
[2] School of Cyber Security, University of Chinese Academy of Sciences,
Beijing 100049, China

Abstract. Host-based Intrusion Detection Systems (HIDS) collect host system logs and generate alerts when the host is attacked. However, existing research fails to adequately capture the spatiotemporal relationships within host behaviors, limiting the accuracy of their representation and modeling. To address this, we propose a spatiotemporal graph representation learning method. This method extracts key data from system logs to construct provenance graphs. A spatiotemporal joint encoder decomposes features along spatial and temporal dimensions independently, then aggregates them to capture spatiotemporal dependencies, explicitly modeling these relationships in host behavior. Experiments on the Streamspot and DARPA-Theia datasets show that the proposed method effectively captures interaction patterns and outperforms baseline models in recall rate, false positive rate, and other evaluation metrics.

Keywords: Network Attack Detection · Graph Deep learning · Spatial-Temporal Representation

1 Introduction

Network attack detection has been a key focus in computer security research. As hosts are primary targets, these attacks often leave traceable footprints within the host systems. Host-based Intrusion Detection Systems (HIDS) can effectively capture these traces and identify attack behaviors, playing a critical role in network security defense [1].

To better represent the contextual information of host events, existing research models log data as a Directed Acyclic Graph (DAG), known as the provenance graph, as shown in Fig. 1. In this graph, nodes represent system entities, and edges represent system events. The Provenance Graph organizes the spatial (semantic) and temporal (interaction) information of each system entity, offering a crucial framework for capturing complex behavior patterns [2–8].

However, existing methods typically represent semantic and temporal information implicitly, which limits effective modeling of system behavior. In log

M. H. Lees et al. (Eds.): ICCS 2025, LNCS 15905, pp. 256–269, 2025.
https://doi.org/10.1007/978-3-031-97632-2_18

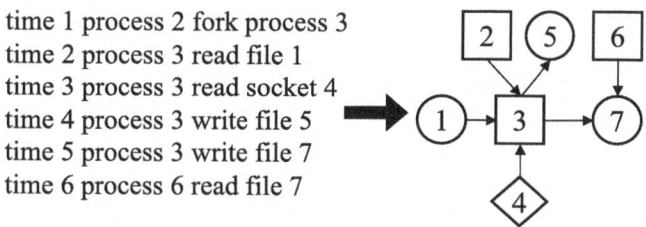

time 1 process 2 fork process 3
time 2 process 3 read file 1
time 3 process 3 read socket 4
time 4 process 3 write file 5
time 5 process 3 write file 7
time 6 process 6 read file 7

Fig. 1. System Log to Provenance Graph

data, substantial semantic information is often sparse, providing limited details on activity. Additionally, temporal interactions exhibit multiple structural forms, complicating the learning of underlying spatiotemporal dependencies.

In this study, we construct a provenance graph by extracting key information from logs and explicitly model the semantic and interaction data of entities using Graph Attention Networks (GAT) [9] and Temporal Graph Networks (TGN) [10]. A spatiotemporal joint encoder decomposes features along spatial and temporal dimensions, followed by joint aggregation to capture spatiotemporal dependencies. Attack behaviors are assessed by combining reconstruction errors from the decoder with anomalous node associations. Experimental results demonstrate that the proposed method significantly outperforms baseline models across several public datasets.

The main contributions of this study can be summarized as follows:

- Provenance Graph Construction: We extracted key entity information from system logs and constructed a provenance graph to describe system behavior.
- Joint Spatial-Temporal Representation: We independently decomposed features along spatial and temporal dimensions, then aggregated them jointly to capture spatiotemporal dependencies, explicitly modeling these relationships in host behavior.
- Attack Behavior Association: During the training phase, we use an encoder to learn normal behaviors. In the testing phase, the system assigns higher Reconstruction Error (RE) scores to deviations from the known baseline behavior and associates attack behaviors through queue linking. This approach improves the accuracy and precision of the alert system.

The paper is structured as follows: Sect. 2 reviews related work on host anomaly detection; Sect. 3 introduces the proposed model and its technical details; Sect. 4 presents data processing and experimental results and discusses the findings; and the final section concludes the paper and outlines future work.

2   Related Work

In recent years, Advanced Persistent Threats (APT) have become a significant challenge in cybersecurity. Intrusion Detection Systems (IDS) play a crucial role in detecting attack behaviors within host systems. IDS are primarily classified into two categories: anomaly-based detection [2–6,11–16] and heuristic rule-based detection [17–23]. Effectively detecting APT attacks is challenging, as these attacks often exploit zero-day vulnerabilities. Defense systems relying on known threat intelligence struggle to identify new or unknown attack strategies. Current research mainly focuses on anomaly-based IDS, which analyze system behavior data—such as network traffic, process activity, and file operations—to establish baselines of normal behavior. This method is effective in identifying deviations from normal host activities.

Recently, embedding techniques have been widely applied in Intrusion Detection Systems. These methods typically use machine learning models, including neural networks and n-gram models, to convert logs into vector representations. Unicorn [4] employs graph sketching technology to summarize system behavior over long periods, addressing slow attacks that occur over extended time spans. Deeplog [27] models system logs as natural language sequences using Long Short-Term Memory (LSTM) networks. Attack2Vec [28] uses a time-based word embedding model to simulate attack steps. Flash [29] combines Graph Neural Networks (GNN) and Word2Vec embeddings in a lightweight classifier to detect potential malicious activities.

However, existing research often fails to explicitly model the spatiotemporal relationships within host behaviors, making it difficult to learn and accurately represent system activities. As a result, current methods need further enhancement to explicitly capture spatiotemporal dependencies. This paper proposes a spatiotemporal joint encoder method that independently extracts features in both the spatial and temporal dimensions, followed by joint aggregation to capture spatiotemporal dependencies. The decoder's reconstruction error and its correlation with anomalous nodes are then used to accurately detect attack behaviors.

3 Proposed Method

Figure 2 illustrates the construction process of Joint Spatial-Temporal Host Intrusion Detection System (JST-HIDS), designed to analyze host behavior across spatial and temporal dimensions. It covers behavioral logs, such as network, file, and process activities. Through spatiotemporal fusion representation, JST-HIDS learns host behavior patterns effectively, enhancing anomaly detection and generating precise alerts by correlating attack behaviors.

The process begins by constructing a comprehensive traceability graph from the log data. Next, the spatiotemporal encoder independently extracts features from both spatial and temporal dimensions. Through joint aggregation, it captures spatiotemporal dependencies. The decoder identifies anomalous events by predicting and reconstruction error from the true values. Finally, event associations are analyzed to detect attack behaviors, thereby enhancing the accuracy of attack analysis.

3.1   Construction of the Provenance Graph

This study constructs the provenance graph by extracting log data from the host's kernel auditing systems. These systems track all operational activities within the host, including processes, files, network addresses, and other entities.

In the provenance graph, nodes represent system entities (e.g., processes, files, network addresses), while edges indicate control flow (e.g., process 1 calling process 2) and data flow (e.g., process 1 writing to file 1) between entities. We focus on system events related to attack steps, as listed in Table 1. The provenance graph organizes and presents the causal relationships and contextual information of each system entity.

Table 1. System Events and Type

events	type
process and file	read,write,create,chmod,rename
process and process	fork,clone,execve,pipe
process and ip	sendto,recvfrom,recvmsg,sendmsg

In-depth analysis of the provenance graph reveals correlations between attack behaviors and their temporal sequence, offering a comprehensive view for constructing an attack chain. This analysis aids in accurately identifying attack patterns and provides essential support for attack tracing and the development of defense strategies.

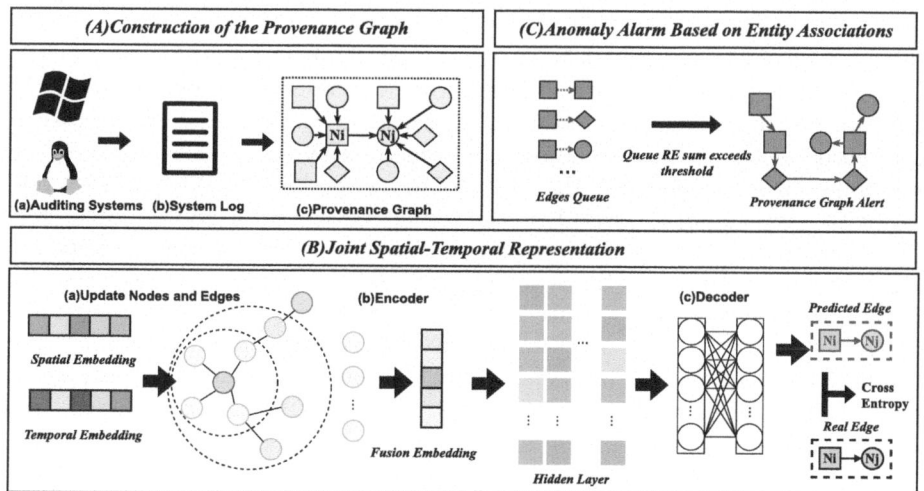

Fig. 2. Construction Process of JST-HIDS

3.2 Spatiotemporal Joint Representation

The provenance graph contains both temporal and spatial information, as shown in Fig. 2. In the anomaly detection process, a benign provenance graph is used to simultaneously train the encoder and decoder. The goal of the training is to minimize the error between the actual edge type (when a new edge appears in the graph) and the type predicted by the decoder from the embedded vector of the edge. This error is referred to as the Reconstruction Error (RE).

During the testing phase, if the graph structure encoded by the edge's embedding closely matches the structure observed from benign system activities in a similar temporal context, the decoder assigns a small reconstruction error. Otherwise, the decoder assigns a larger reconstruction error, with the magnitude of the error reflecting the degree of deviation from the normal baseline. Host behaviors in the test samples with a reconstruction error exceeding a predefined threshold are identified by the detection module as anomalous behaviors.

Spatial Semantic Embedding of Entities in the Provenance Graph.
Entities in the provenance graph are categorized into three types: processes, files, and network addresses. First, the features of these entities (such as command lines, file paths, and IP addresses) are converted into natural language sentences. For example, the file path "/usr/local/bin/app" is transformed into the sentence "usr local bin app". Next, the Natural Language Processing (NLP) technique [30] is applied to eliminate meaningless non-natural language components, such as the hash string found in the path. Finally, FastText [31] is used to project these sentences into numerical vectors, generating the semantic embedding f_i for each node i.

Entity Temporal Interaction Information Embedding. To capture the evolving characteristics of entity interactions in a dynamic provenance graph, node states record the interaction history of each node. At time t, the historical information of the provenance graph is represented by the state vector $s_i(t)$ for each node i, which compresses the entity's interaction history. When a new node is added to the graph, its state is initialized as a zero vector and updated as it interacts with other nodes. For instance, at time t, when a new edge e_{ij} appears, indicating an interaction between nodes i and j, the message m transmitted by edge e_{ij}, such as a process writing to a file, is calculated. This message updates the state vectors $s_i(t)$ and $s_j(t)$ for nodes i and j, respectively. The messages are computed from the perspectives of both the source node and the destination node:

$$m_i(t_1) = msg\left(s_i(t_0), s_j(t_0), \Delta t, e_{ij}(t_1)\right) \tag{1}$$

$$m_j(t_1) = msg\left(s_j(t_0), s_i(t_0), \Delta t, e_{ij}(t_1)\right) \tag{2}$$

To thoroughly capture the characteristics of dynamic provenance graphs as they change over time, node states are used to record each node's interaction history. The Gated Recurrent Unit (GRU) model [32] is employed to continuously update the node states through iterative processing:

$$s_i(t_1) = GRU(m_i(t_1), s_i(t_0)) \tag{3}$$

The encoder uses a Temporal Graph Network (TGN) [10] architecture to encode source graph features into edge embeddings. At time t, the model, based on Graph Neural Networks (GNNs), generates edge embeddings f_{e_t} for new edges:

$$f_i(t) = Encoder(i, t) \tag{4}$$

$$f_i(t) = \sum_{j \in \Gamma_i^K([0,t])} h(s_i(t), s_j(t), e_{ij}) \tag{5}$$

$$f_{e_t} = f_{v_{src}} + f_{v_{des}} \tag{6}$$

This formula represents the embedding of an edge f_{e_t} at time t in a dynamic graph. The edge embedding is computed as the sum of the embeddings of the source node $f_{v_{src}}$ and the destination node $f_{v_{des}}$.

Learning Trajectories with Spatiotemporal Joint Embedding. Estimating probability distributions related to time and semantic spaces is a complex task, but it plays a crucial role in capturing spatiotemporal patterns of host behavior. However, traditional methods extract the semantic space information distribution $p(s|t)$ under time conditions. However, the representations learned from $p(s|t)$ are often implicit and limited, leading to a reduction in expressive power. To address this issue, we integrate semantic space and interaction temporal features, learning the spatiotemporal correlations. To effectively capture the correlations in both spatial and temporal domains, we adopt a generalized graph neural network (GNN) model with a three-layer structure [33], which aids in learning the joint spatiotemporal distribution. Specifically, the information update process of vertex v is as follows:

$$m_{vn}^i = ReLu\left(f_n^i + f_{e_{vn}}^i + f_{\omega_{vn}}^i\right) + \epsilon \tag{7}$$

The message m_{vn}^i is updated by combining the features of node n, the edge e_{vn}, and the normalized edge weight $f_{\omega_{vn}}$, followed by a ReLU activation and adding noise ϵ.

$$m_{e_{vn}}^i = MLP\left(Concate\left(f_v^i, f_n^i\right)\right) \tag{8}$$

The edge message $m_{e_{vn}}^i$ is updated by concatenating the features of nodes v and n, and passing the result through Multi-Layer Perceptron(MLP) [33].

$$m_v^i = \sum_{n \in V} \frac{exp(\alpha m_{vn})}{\sum_{j \in N(V)} exp(\alpha m_{vj})} \tag{9}$$

The equation computes the aggregated message m_v^i for node v at the i-th iteration, where m_{vn} represents the information exchanged between node v and its neighboring nodes n. This is done by performing a weighted sum of the information from all neighboring nodes n in the graph, with the weights α determined by an attention mechanism based on the information.

$$f_{e_{vn}}^{i+1} = MLP(f_{e_{vn}}^i + c \cdot \left\lVert f_{e_{vn}}^i \right\rVert_2 \cdot \frac{m_{e_{vn}}^i}{\left\lVert m_{e_{vn}}^i \right\rVert_2}) \tag{10}$$

The equation describes the feature update process for edge e_{vn} at the $i+1$-th iteration. $f_{e_{vn}}^i$ represents the feature of edge e_{vn} at the i-th iteration, while $m_{e_{vn}}^i$ denotes the information exchanged between node v and its neighboring node n. Both $\left\lVert f_{e_{vn}}^i \right\rVert_2$ and $\left\lVert m_{e_{vn}}^i \right\rVert_2$ are normalized using their respective L2 norms. The constant c is used to regulate the normalization factor. All these components are processed through MLP, resulting in the updated feature $f_{e_{vn}}^{i+1}$ for edge e_{vn}.

$$f_v^{i+1} = MLP(f_v^i + c \cdot \left\lVert f_v^i \right\rVert_2 \cdot \frac{m_v^i}{\left\lVert m_v^i \right\rVert_2}) \tag{11}$$

This equation updates the feature f_v^{i+1} of node v, following a similar approach as the previous equation. Here, f_v^i represents the feature of node v at the i-th iteration, and m_v^i denotes the aggregated message for node v. Both $\left\lVert f_v^i \right\rVert_2$ and $\left\lVert m_v^i \right\rVert_2$ are normalized using their respective L2 norms. After normalizing these features and messages, they are processed through MLP, resulting in the updated feature f_v^{i+1} for node v.

$$R_p(e_{vn}) = MLP(concat(f_v, f_{e_{vn}}, f_n)) \tag{12}$$

This equation is used to make the final prediction for edge e_{vn}. First, the feature vectors of the two nodes v and n along with the edge feature vector $f_{e_{vn}}$ are concatenated to form a unified vector containing all relevant information. This concatenated vector is then processed through MLP to predict the edge, with the resulting output representing the probability distribution over the possible edge types.

$$RE = CrossEntropy(R_p(e_{vn}), R_r(e_{vn})) \tag{13}$$

The output of the decoder is a vector $R_p(e_{vn})$, representing the model's predicted probabilities for edge e_t being one of the edge types. During training, the model's optimization objective is to minimize the RE between $R_r(e_{vn})$ and the observed edge type $R_p(e_{vn})$ from the benign provenance graph. During testing, for edges whose structural and temporal context information is similar to those learned from the benign provenance graph, the model assigns a lower RE score. Conversely, if there are edges in the graph that significantly deviate from the known normal system baseline behavior, the model assigns a higher RE score.

3.3   Anomaly Alarm Based on Entity Associations

After obtaining the reconstruction errors for each event, we construct a queue based on the associative relationships between system entities to accurately pinpoint anomalous events. The queue stores the related events, and anomalies are detected by calculating the sum of the RE values of the nodes within the queue. An associative queue Q is used to store events based on the association between system entities. If there is a strong association between events E_i and E_j, they will be stored in the same queue:

$$Q = (E_i, E_j) \mid \text{Edge}(E_i, E_j) = 1 \tag{14}$$

where $\text{Edge}(E_i, E_j) = 1$ indicates that there is a direct edge or calling relationship between events E_i and E_j, meaning that one event calls or triggers the other.

$$RE_{\text{queue}} = \sum_{(E_i, E_j) \in Q} RE(E_i) \tag{15}$$

If $RE_{\text{queue}} \geq T_{\text{alarm}}$, then trigger alarm. $\tag{16}$

When the total RE exceeds a predefined threshold, the system triggers an alarm. The threshold is set with careful consideration of the system's fault tolerance and normal operating range, ensuring that alarms are triggered only when a node's error significantly deviates from normal values and exhibits strong associations with other anomalous nodes. This method effectively prevents false alarms caused by minor errors in local or unrelated nodes, thereby improving the accuracy and precision of the alarm system.

4   Evaluation and Discussion

In this chapter, we compare the proposed method with the existing state-of-the-art methods [4,29], demonstrating the superiority and effectiveness of the approach presented in this paper.

4.1   Experimental Setup

All experiments were conducted on a server running Ubuntu 18.04 with 64GB of RAM and a 2.20GHz 20-core Intel Xeon CPU. The node feature embedding dimension is 16, the node state dimension is 100, the neighborhood size is 20, and the edge embedding dimension is 200. We recorded the outliers (reconstruction errors) for each event in different systems and labeled the anomalous queues based on the event reconstruction errors (the process is shown in the Fig. 2). The threshold is determined by the upper limit when testing for partial benign behavior. Finally, we computed the accuracy, recall, and F1 score, with the formulas as follows:

$$Accuracy = \frac{TP + TN}{TP + TN + FP + FN} \tag{17}$$

$$Recall = \frac{TP}{TP + FN} \tag{18}$$

$$Precision = \frac{TP}{TP + FP} \tag{19}$$

$$F1 = 2 \times \frac{Precision \times Recall}{Precision + Recall} \tag{20}$$

4.2 Datasets

We used two publicly available datasets, StreamSpot and Darpa-e3-Theia, as shown in Table 2. The following provides a detailed description of these datasets:

The StreamSpot dataset contains system logs from six different scenarios, with five normal scenarios, including YouTube, email detection, download tasks, CNN, and VGame. The attack scenario involves downloading a program from a malicious URL and exploiting a flash vulnerability to gain system administrator privileges.

The DARPA TC dataset is from the Transparent Computing (TC) project by the Defense Advanced Research Projects Agency (DARPA). In this project, red teams and blue teams conduct offense-defense exercises. During the process, fine-grained system behavior data is collected for attack detection and forensic tracing, which is then used to generate reports.

Table 2. Overview of Datasets

Dataset	Nodes	Edges	Attack Edges	Proportion
STREAMSPOT	999,999	89.8 millions	2,842,345	3.165%
E3-THEIA	690,105	32.4 millions	3,119	0.010%

4.3 Visualization

We visualize the extracted embedded features (Fig. 3) and reconstruction errors (Fig. 4) on the StreamSpot dataset to intuitively demonstrate the effectiveness of our proposed method.

The Fig. 3 show the feature representations of attack events after PCA dimensionality reduction, marked by red areas. Our proposed spatiotemporal joint representation method effectively captures the inherent spatiotemporal dependencies in the data. The encoder distinguishes attack events from benign ones with high accuracy, while the decoder produces higher reconstruction errors for anomalous interaction patterns. This approach enhances the system's ability to differentiate between attack and normal behaviors. As illustrated, previously unseen anomalous patterns complicate the original detector's interpretation of attack events, leading to higher reconstruction errors.

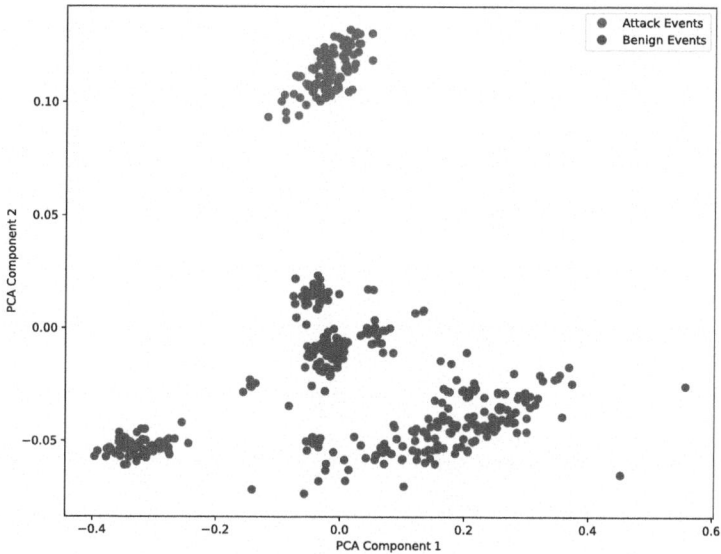

Fig. 3. Host Event Feature Representation After PCA Dimensionality Reduction in StreamSpot Dataset

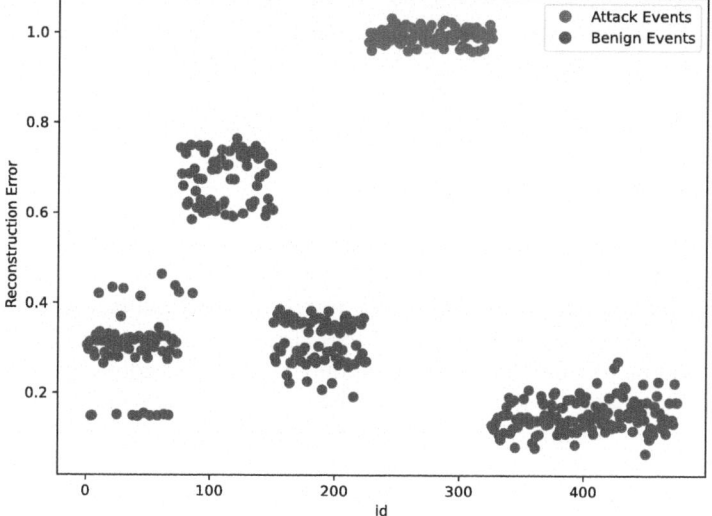

Fig. 4. Reconstruction Error of Host Events in StreamSpot Dataset

4.4   Metrics

We selected two HIDS [4, 29] as baseline methods for comparison and conducted ablation experiments, the JST-S system uses only spatial features, the JST-T system uses only temporal features, and the JST system uses the spatiotemporal fusion features proposed in this paper. The experimental results are shown in the Table 3 below.

Table 3. Performance Comparison of HIDS

Streamspot Dataset			
System	Accuracy (%)	Recall (%)	F1 Score
Flash	98.0	95.0	97.4
Unicorn	96.2	93.1	95.4
JST-S	100.0	100.0	100.0
JST-T	100.0	100.0	100.0
JST	100.0	100.0	100.0
E3-THEIA Dataset			
System	Accuracy (%)	Recall (%)	F1 Score
Flash	35.9	99.8	23.3
Unicorn	36.4	100.0	24.4
JST-S	95.6	100.0	64.3
JST-T	84.1	88.9	30.8
JST	96.9	100.0	72.0

The proposed method effectively identifies potential entity interaction patterns and significantly outperforms all baseline models across various evaluation metrics, including recall and false positive rates. Specifically, while both baseline models can detect nearly all attack behaviors in the E3-THEIA dataset, they exhibit high false positive rates by misclassifying benign behaviors as attacks. Observations reveal that the baseline models adopt a broader concept of attack nodes, which reflects their reliance on only semantic information and shallow temporal features, resulting in imprecise modeling and numerous false positives. In ablation experiments, although the performance declines when only temporal or semantic features are used, the proposed method consistently outperforms baseline models. This indicates that the relationships among attack nodes helps reduce false positives while maintaining a high recall rate.

We tested different thresholds on the E3-THEIA dataset and observed variations in True Positive Rate (TPR) and True Negative Rate (TNR), which represent the accuracy of detecting attack and benign behaviors, respectively. As shown in the Fig. 5, increasing the threshold reduces the false positive rate but results in more missed detections. The optimal threshold for balancing these metrics should be determined based on specific environmental conditions, emphasizing the importance of precise modeling of system behavior.

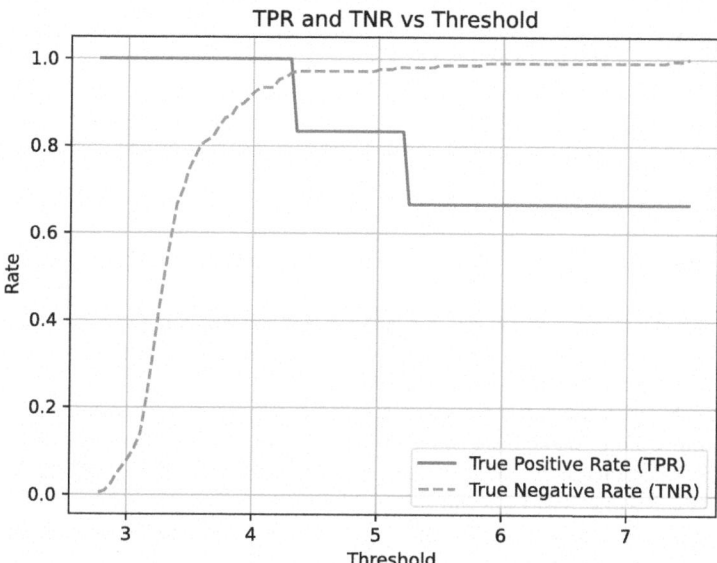

Fig. 5. Relationship between TPR and TNR across Different Thresholds in Theia Dataset

Although the proposed method performs best in terms of false positive rate, false positives remain an unresolved issue. This is particularly true in practical applications, where the large volume of data may prevent timely processing of the generated false positives [34]. The main cause of this phenomenon is the potential misclassification of benign activity patterns as anomalies, which is a common challenge in anomaly detection systems. Leveraging the world knowledge of large language models [35] is a promising avenue for reducing false positives. We plan to explore the integration of large model techniques in our future work to address this issue.

5   Conclusion

To more accurately extract the embedding representations of host behaviors, this paper proposes a joint spatiotemporal graph representation learning method that explicitly models the spatiotemporal dependencies within host behaviors. The method uses a spatiotemporal joint encoder to independently extract features in both spatial and temporal dimensions, while a joint aggregation mechanism effectively captures spatiotemporal dependencies. Based on the relationships between anomalous behaviors and host activities, an alarm queue is constructed, which generates a comprehensive alarm. Experimental results on the Streamspot and DARPA-Theia datasets demonstrate that the proposed method effectively identifies potential entity interaction patterns and significantly outperforms all baseline models across multiple evaluation metrics, such as recall rate and false positive rate. Future work will integrate large language models for alarm analysis to further enhance the automated handling of false alarms, thereby improving the method's applicability in real-world scenarios.

Acknowledgments. We would like to express our sincere gratitude for the financial support provided by Youth Innovation Promotion Association, CAS (No.2023170).

References

1. Li, Z., Chen, Q. A., Yang, R., et al.: Threat detection and investigation with system-level provenance graphs: a survey. Comput. Secur.**106**(C), 102282 (2021). https://doi.org/10.1016/j.cose.2021.102282
2. Manzoor, E., Milajerdi, S. M., Akoglu, L.: Fast memory-efficient anomaly detection in streaming heterogeneous graphs. In: KDD '16: Proceedings of the 22nd ACM SIGKDD International Conference on Knowledge Discovery and Data Mining, pp. 1035–1044. Association for Computing Machinery, New York (2016)
3. Xie, Y., et al.: Pagoda: a hybrid approach to enable efficient real-time provenance-based intrusion detection in big data environments. IEEE Trans. Dependable Secure Comput. **17**(6), 1283–1296 (2020)
4. Han, X., et al.: UNICORN: runtime provenance-based detector for advanced persistent threats. In: 27th Annual Network and Distributed System Security Symposium, NDSS 2020, San Diego, California, USA, February 23–26, 2020. The Internet Society (2020)
5. Han, X., et al.: SIGL: securing software installations through deep graph learning. In: Security Symposium (USENIX Sec'21). USENIX (2021)
6. Wang, Q., et al.: You are what you do: Hunting stealthy malware via data provenance analysis. In: Proceedings 2020 Network and Distributed System Security Symposium (2020)
7. Li, S., et al.: NODLINK: an online system for fine-grained APT attack detection and investigation. In: Proceedings of the Network and Distributed System Security Symposium (2024)
8. Cheng, Z., et al.: KAIROS: practical intrusion detection and investigation using whole-system provenance. In: 2024 IEEE Symposium on Security and Privacy (2024)
9. Veličkovi'c, P., Cucurull, G., Casanova, A., Romero, A., Li'o, P., Bengio, Y.: Graph attention networks. In: International Conference on Learning Representations (2018)
10. Rossi, E., Chamberlain, B., Frasca, F., Eynard, D., Monti, F., Bronstein, M.: Temporal graph networks for deep learning on dynamic graphs. In: ICML 2020 Workshop on Graph Representation Learning (2020)
11. Han, X., et al.: FRAPPuccino: fault-detection through runtime analysis of provenance. In: HotCloud '17: Proceedings of the 9th USENIX Conference on Hot Topics in Cloud Computing, p. 18. USENIX Association, USA (2017)
12. Xie, Y., et al.: Unifying intrusion detection and forensic analysis via provenance awareness. Fut. Gener. Comput. Syst. **61**(C), 26–36 (2016)
13. Xie, Y., et al.: P-Gaussian: provenance-based Gaussian distribution for detecting intrusion behavior variants using high-efficiency and real-time memory databases. IEEE Trans. Dependable Secure Comput. **18**(6), 2658–2674 (2021)

14. Sun, X., et al.: Using Bayesian networks for probabilistic identification of zero-day attack paths. IEEE Trans. Inf. Forensics Secur. **13**(10), 2506–2521 (2018)
15. Li, Z., et al.: A hierarchical approach for advanced persistent threat detection with attention-based graph neural networks. Sec. Commun, Netw. (2021)
16. Ayoade, G., et al.: Evolving advanced persistent threat detection using provenance graph and metric learning. In: 2020 IEEE Conference on Communications and Network Security (CNS), pp. 1–9. IEEE (2020)
17. Crowdstrike: "Why Dwell Time Continues to Plague Organizations." (2019). https://www.crowdstrike.com/blog/why-dwell-time-continues-to-plague-organizations/
18. Gartner Peer Insights: "Endpoint Detection and Response Solutions Market." (2019). https://www.gartner.com/reviews/market/endpoint-detection-and-response-solutions
19. Hiroki, T., Yoshiaki, S., Koji, K., and Takayoshi, A.: Automated security intelligence (ASI) with auto detection of unknown cyber-attacks. NEC Tech. J. **11** (2016)
20. Fireeye: "Incident Investigation." (2019). https://www.fireeye.com/solutions
21. swimlane: "Automated Incident Response: Respond to Every Alert." (2019). https://swimlane.com/blog/automated-incident-response-respond-every-alert/
22. Malwarebytes Inc.: "Malwarebytes." (2022). https://www.malwarebytes.com/
23. Splunk Inc.: "splunk." (2018). https://www.splunk.com
24. Aljawarneh, S., Aldwairi, M., Yassein, M.B.: Anomaly-based intrusion detection system through feature selection analysis and building hybrid efficient model. J. Comput. Sci. **25** (2018)
25. Alsaheel, A., et al.: ATLAS: a sequence-based learning approach for attack investigation. In: USENIX Security Symposium (2021)
26. Maseer, Z.K., Yusof, R., Bahaman, N., Mostafa, S.A., Foozy, C.F.M.: Benchmarking of machine learning for anomaly based intrusion detection systems in the cicids2017 dataset. IEEE Access **9** (2021)
27. Du, M., Li, F., Zheng, G., Srikumar, V.: DeepLog: anomaly detection and diagnosis from system logs through deep learning. In: ACM Conference on Computer and Communications Security (CCS) (2017)
28. Shen, Y., Stringhini, G.: Attack2vec: leveraging temporal word embeddings to understand the evolution of cyberattacks. In: USENIX Security Symposium (2019)
29. Rehman, M.U., Ahmadi, H., Hassan, W.U.: Flash: a comprehensive approach to intrusion detection via provenance graph representation learning. In: 2024 IEEE Symposium on Security and Privacy (SP), San Francisco, CA, USA, 2024, pp. 3552–3570 (2024). https://doi.org/10.1109/SP54263.2024.00139.
30. Hucka, M.: Nostril: a nonsense string evaluator written in Python. J. Open Source Softw. **3**(25), 596 (2018)
31. Bojanowski, P., Grave, E., Joulin, A., Mikolov, T.: Enriching word vectors with subword information. Trans. Assoc. Comput. Linguist. **5**, 135–146 (2017)
32. Cho, K., et al.: Learning phrase representations using RNN encoder-decoder for statistical machine translation. Comput. Sci. (2014)
33. Li, G., Xiong, C., Thabet, A., Ghanem, B.: DeeperGCN: all you need to train deeper GCNs. In: ICLR 2022 Conference Withdrawn Submission (2022)
34. Zipperle, M., Gottwalt, F., Chang, E., Dillon, T.: Provenance-based intrusion detection systems: a survey. ACM Comput. Surv. **55**, 7 (2022). https://doi.org/10.1145/3539605. Article 135
35. Yang, Y., Tang, J., Xia, L., Zou, X., Liang, Y., Huang, C.: GraphAgent: Agentic Graph Language Assistant. arXiv preprint: arXiv:2412.17029 (2024)

Will It Blend? Mixing Numerical and Machine-Learned Physics Quantities for Accurate on-the-Fly Surrogate Modeling

Michael Tynes[1,2]([✉])(iD), Kyle Chard[1,2](iD), Ian Foster[1,2](iD), and Logan Ward[2](iD)

[1] Department of Computer Science, University of Chicago, Chicago, IL 60637, USA
{mtynes,chard}@uchicago.edu
[2] Data Science and Learning Division, Argonne National Laboratory, Lemont, IL, USA
lward@anl.gov

Abstract. Learning and deploying inexpensive replacements for computationally expensive subroutines "on-the-fly" (OTF) during a dynamic simulation offers potential advantages and unique drawbacks. In OTF learning, a machine learned surrogate function is trained to replace a target subroutine in a simulation as the dynamics evolve. The advantages of OTF learning include reducing simulation error and model training costs, but the weaknesses include the possibility of introducing artifacts when adaptively updating the physics of the simulation over time. Here, we enhance an existing control system, PROXIMA, which ensures that surrogates are used appropriately in time-independent state-sampling-based simulations, to time-dependent dynamical simulations by introducing a "blending" procedure that hides discontinuities when transitioning between original subroutine and surrogate. Our new control system, PROXIMA+BLEND, produces a blend of the surrogate and target functions according to the relationship between error and an uncertainty signal observed as the dynamic simulation evolves. We show that while the original control system can shorten application runtime and accurately capture some macro-scale observables of a molecular dynamics simulation, the addition of blending is necessary to avoid unphysical dynamics at shorter time and length scales and to correct observables derived from these dynamics. PROXIMA+BLEND delivers a 1.5x speedup over use of the target subroutine while producing solutions within 5% error in dynamical quantities, while the original PROXIMA algorithm has up to 80% error. Our implementation of PROXIMA+BLEND can be deployed by simply replacing the existing subroutine with a wrapper that includes a machine learning approach for surrogate training along with specifying control and uncertainty signals for the simulation of interest.

Keywords: Machine learning · dynamic simulations · uncertainty quantification

© The Author(s), under exclusive license to Springer Nature Switzerland AG 2025
M. H. Lees et al. (Eds.): ICCS 2025, LNCS 15905, pp. 270–284, 2025.
https://doi.org/10.1007/978-3-031-97632-2_19

1   Introduction

Machine learning (ML) surrogates can speed up physical simulations by orders of magnitude, granting access to previously unavailable length and time scales, but optimal methods for constructing such surrogates and integrating them into simulations are still being developed. A classic route is to invest resources into gathering data and training ahead of time, keeping the surrogate fixed during execution [25]. Keeping the model fixed simplifies deployment at the risk of poor performance if the simulation samples states unseen in the training data. On-the-fly (OTF) learning strategies mitigate this risk by continually updating surrogates during the course of a simulation [9,27]. However, while promising, OTF learning comes with its own uncertainties and challenges.

Existing OTF methods enhance a simulation by updating a surrogate over the course of a simulation by gradually training it on data acquired during execution. While simple in concept, running a simulation with a constantly changing surrogate presents subtle challenges. Reliably training machine learning models for a particular physics code is, of course, a challenge. Balancing the computational cost differences between target and surrogate requires non-trivial system design, but such challenges have been overcome in previous cases [7,20]. Even with a model with sufficient accuracy and a deployment strategy in hand, knowing whether the outcome of the simulation is correct is not assured.

For one, metrics to predict the quality of ML predictions are often unreliable [1,21]. The inability to detect when a simulation has begun to explore states different than those encountered in training inhibits a scientist's ability to know when they should be skeptical of an ML-driven simulation. At best, a strategy based on quality metric involves introducing a control parameter (e.g., a threshold on an uncertainty metric) which must be carefully selected and continually re-evaluated. Below, we further explore a strategy to automate auditing and adjusting such control parameters introduced by Zamora et al. [30].

Another, yet-unaddressed question with OTF learning is whether replacing a target function with a mutable, learning object can lead to issues in the enveloping application. In the ideal case, the machine learned surrogate is indistinguishable from the original function so any algorithms which employ it should be unaffected. However, achieving a perfect surrogate is impossible. While pleasingly straightforward, we posit that naively switching a surrogate on or off will lead to unphysical results in dynamical simulations by causing rapid changes to variables that are expected by numerical integrators to vary smoothly. Established methods which blend multiple fidelities of models together, such as the classic QM/MM technique, employ methods to blend differing levels across spatial dimensions [28]. Below, we demonstrate blending in time is needed as well.

In this work, we present a control system strategy which simplifies integrating surrogate models into dynamic simulations without modifying the original algorithms. We build on the existing PROXIMA approach [30], which uses a linear controller to determine when a surrogate model should be used. We propose that smoothly adjusting convex combinations of ML surrogate predictions and physics calculations will reduce unphysical results which stem from turning a surrogate on or off immediately. We demonstrate that our new method, Prox-IMA+BLEND, fixes instability issues found in molecular dynamics simulations and achieves accurate simulation outputs without sacrificing the accelerations provided by surrogates.

2   Methods

Our approach developed here, PROXIMA+BLEND, extends an existing method for replacing functions with machine-learned surrogates, PROXIMA [30]. PROX-IMA is an OTF method that uses a control system to ensure that a surrogate function is used only when it meets user-prescribed accuracy bounds by adaptively tuning an uncertainty quantification (UQ) threshold.

2.1   Existing PROXIMA Implementation

Here we formally describe the implementation of the existing PROXIMA algorithm [30], an abstract description of the simulation, surrogate and target functions, uncertainty quantification, and a control strategy. The goal of the algorithm is to maximize use the surrogate while ensuring that the average of the errors between the surrogate and target functions adhere to user-specified bound.

We let x_t be the state of the system under simulation at time t; F be a computationally expensive target function that is used to compute $y_t = F(x_t)$, a physical quantity in this system that determines the simulation evolution; $\hat{y}_t = S(x_t)$ be an ML surrogate estimate of this quantity; $\epsilon_t = |y_t - \hat{y}_t|$ be the error between S and F at t; ϵ_{bound} be the user-specified error bound; and UQ be a measure of the uncertainty of S; UQ_t^{thr} be the current uncertainty threshold. The UQ metric is commonly chosen to the variance of y_t among an ensemble of surrogate models or the distance between x_t and all points included in the training set for S.

Multiple computations occur at each step of the simulation:

1. *Reliability check* The UQ metric is computed and compared against the current threshold. If $UQ < UQ_t^{thr}$, then declare the surrogate unreliable. Optionally, we also declare a certain fraction of surrogate evaluations unreliable regardless of the outcome of the UQ test.
2. *Retraining* If the training set has grown by more than some threshold amount since the last training and the surrogate is declared reliable, the surrogate (S) is retrained.
3. *Evaluation* If the surrogate is determined to be reliable, S is used to compute y_t; otherwise, F is used and the resultant (x_t, y_t) point is stored.
4. *Threshold Update* If $UQ \geq UQ_t^{thr}$, UQ^{thr} is updated according to the error between the surrogate and target $\epsilon_t = |y_t - \hat{y}_t| = |S(x_t) - F(x_t)|$

PROXIMA generally, adjusts UQ_t^{thr} to be smaller (more conservative) when the error (ϵ_t) is high and larger (less conservative) when the error is low. The original PROXIMA implementation used a linear controller to adjust UQ_t^{thr},

$$UQ_{t+1}^{thr} = UQ_t^{thr} - \alpha_t(\epsilon_t - \epsilon_{target}) \tag{1}$$

where α_t is a regression coefficient fit to the accumulated dataset of UQ and observed errors.

2.2 PROXIMA+BLEND Implementation

To extend PROXIMA to use in dynamical simulations, we introduce a new approach that smoothly transitions between the target and surrogate functions over multiple timesteps. The blend is controlled by a mixing parameter, $\lambda \in [0, 1]$, which defines the amount of surrogate S and target function F used in evaluating the output:

$$\hat{y}_t^{\text{blend}} = \lambda_t S(x_t) + (1 - \lambda_t)F(x_t). \tag{2}$$

When $\lambda_t = 1$, the simulation does not require evaluating the target function.

Whereas the original implementation of PROXIMA transitions between two values of λ ($\{0, 1\}$), our new approach PROXIMA+BLEND varies transitions smoothly over several timesteps. The number of timesteps it takes for a full transition from $\lambda = 0$ to $\lambda = 1$ is a user-supplied parameter n. As detailed in Algorithm 1, the degree of mixing is represented by a blending index (i) that varies between 0 and n by ± 1 depending on whether PROXIMA's control algorithm decides the surrogate could be used. (The original PROXIMA algorithm is where $n = 1$.) The degree of mixing λ is finally determined transforming i/n through a smoothing function, f, which has gradients tending to zero at arguments 0 and 1, ensuring that blending away from full use of ML (S) or the target function (F) occurs slowly. In this work, we use

$$f(z) = -cos(\pi x) + 1, \tag{3}$$

but any sigmoidal function taking on values from 0 to 1 could be used.

2.3 Computational Evaluation Overview

We implement PROXIMA and PROXIMA+BLEND for atomistic simulations, where the system under study is a set of atoms distributed in 3-dimensional space. Specifically, we consider molecular dynamics (MD), where the positions and momenta of each atom are evolved over time by integrating a $6N$-dimensional set of ordinary differential equations which are primarily Newton's Laws of Motion. The expensive step in propagating the MD simulation is often computing the forces acting on each atom as a function of the relative positions of all other atoms. The gold standard for computing the forces on atoms is *ab initio* quantum mechanical calculations, such as those based on Density Functional Theory (DFT). The run duration of molecular dynamics simulations is typically dicated by the force compuations, with more accurate forces taking more time to compute and having worse algorithmic complexity, making the force computation a popular target for ML surrogate modeling [2,3,18,24].

Algorithm 1. Algorithm for blending physics and machine-learned quantities

---

if $UQ < UQ_t^{thr}$ then
    $i \leftarrow i + 1$
else
    $i \leftarrow i - 1$
end if
$i \leftarrow \text{clip}(i, 0, n)$
$\lambda_t \leftarrow f(i/n)$
if i = n then
    $\hat{y} \leftarrow S(x_t)$
else
    $\hat{y} \leftarrow \lambda_t S(x_t) + (1 - \lambda_t) F(x_t)$
end if
return $\hat{y}$

We implemented PROXIMA and PROXIMA+BLEND as a `Calculator` class in the Atomic Simulation Environment (ASE) [16]. ASE provides a common Python interface to various tools for atomic simulation, including many tools to perform *ab initio* calculations. ASE also provides classes for the numerical integration in MD, which depend on Calculator classes to produce forces at each simulation step. Our ASE `Calculator` class has as attributes two other ASE `Calculator` objects–one for S and one for F—and implements the logic from PROXIMA+BLEND to take convex combinations of their results, based on the varying the threshold UQ^{thr}, and to update λ as presented in Algorithm 1.

Our implementation is available online on github[1].

Further Details on Simulations. For our MD simulations, we used the ASE interface to CP2K [13] to perform DFT force evaluations. These calculations used the local density approximation functional and Gaussian plane wave basis set (LDA/GPW) as our target physics F. (Full details of our DFT configuration are available on GitHub[2]).

For our surrogate, we used the ANI neural network architecture [24], a widely used neural network approach for predicting energies and forces for atomistic systems, as implemented in the TorchANI package [11]. ANI takes as input a set of atomic species and coordinates, and predicts the potential energy of the atomic system and the forces acting on each atom, with the forces obtained by automatic differentiation of the energy contribution of each atom with respect to the atomic positions. We optimize a weighted sum of Huber loss terms for the energy, forces, and stressses. Stress is also determined by automatic differentiation of the predicted energy, in this case with respect to strain applied to each coordinate in each spatial direction. In our experiments, we trained our networks

---

[1] https://github.com/globus-labs/cascade.
[2] https://github.com/globus-labs/cascade/blob/main/cascade/files/cp2k-lda-template.inp.

for 32 epochs using the Adam optimizer with initial learning rate 10^{-3}. We used an ensemble of two ANI networks each trained using a different bootstrapped sample of available training data, with the intent of using differences between ensemble members as an uncertainty metric. More details on hyperparameters and our fitting procedure can be found on GitHub[3].

We configured PROXIMA and PROXIMA+BLEND to control the maximum error between the target DFT forces (our instance of F) and those predicted by the surrogate (S). The input control signal is the maximum deviation in the forces predicted by any model from the mean over all models. The error (ϵ_t) and the regression coefficient (α_t) are determined using averages over the last eight timesteps for which the target model was used.

Our simulations examine crystalline silicon at various temperatures. Specifically, our system under study is a lattice of 64 Silicon atoms under periodic boundary conditions, with a single atom removed to create a vacancy, allowing us to study the diffusion of this vacancy throughout the system: an important determinant of the mechanical properties of solids. We performed simulations at constant volume and temperature (NVT) and constant pressure and temperature (NPT) for various temperatures and pressures.

3   Results

The goals of our experiments are to (1) explore whether switching between a numerical routine and a machine-learned surrogate distorts the dynamics of a molecular dynamics simulation, and (2) evaluate strategies for mitigating distortions.

3.1   On-the-Fly Learning in a MD Simulation

We start by evaluating whether OTF learning techniques result in correct behavior in a classic dynamics problem: constant temperature and pressure molecular dynamics (often called NPT, for fixed **number** of particles, **pressure**, and **temperature**). We choose a standard approach using a variant of the popular NoséâĂŞHoover thermostat, which controls temperature through a coupling of the system with a virtual heat bath, and of the Parniello-Raman barostat, which controls pressure through the equilibrium between the stress state of the atoms and a virtual "piston" that represents external pressure [19]. The piston term includes momentum, which leads to oscillations around equilibrium and implies a strong dependence of the evolution of a system on its history. This history dependence creates a challenge that could expose issues in replacing a function with a surrogate.

The goal of our computation is to estimate the density of silicon at 800K. We do so by starting with a 63-atom cell of Si atoms at 0K, adding random velocities

---

[3] https://github.com/globus-labs/cascade/blob/main/2_proxima/0_run-serial-proxima.py.

corresponding to an initial temperature of 800K, and evolving Newton's laws of motion and the NPT thermostat and barostat over 2048 timesteps (a total of 2048 fs). The density of the system is related to the average volume over the last 512 timesteps, after giving the thermostat and barostat time to equilibrate the pressure and temperature. This computation is simple enough to solve without machine learning acceleration, and using the target numerical routine at every step yields an equilibrium density of 2.166 g/cc.

Introducing OTF learning involves replacing the calculation of the forces used in Newton's laws of motion and the stress used in the NPT with a surrogate. We start with a simple strategy: first gather 1024 data points from F before training an ensemble of neural networks S, allow the PROXIMA and PROXIMA+BLEND algorithms to decide when to use F, S, or a blend of each to evolve the dynamics without further training. We set the error target ϵ_{bound} to 0.9 eV/Å(well above the convergence threshold for the target method), add stochasticity such that the target method at least 20% of the time.

Viewing the resulting calculations at a coarse level, OTF learning has only minor differences in the simulation outcomes, giving an equilibrium density of 2.163 g/cc, only a 0.11% error with respect to the reference calculations, while yielding a 1.57x speedup (3.41 hrs/MD run vs. 2.18 h). We note that the runtimes are dominated by force evaluations and that we observed average timestep durations of 4 s when using DFT forces and 0.2 s when using ML forces (so roughly a 20x speedup), though with considerable variation depending on atomic configuration and the recent runtime history of the DFT routine.

Viewing the calculations at a finer level, the errors between surrogate and target ϵ_t have a median below the user-supplied bound of ϵ_{bound}, that is the error is below the bound slightly more than half the time (Fig. 1). The mean of ϵ_t is, however, above the bound. The discrepancy between the mean and median is driven by high-error outliers. On one hand, such outliers are to be expected because the variation in forces on atoms in MD simulations are very large, while on the other hand, this does present an opportunity to develop a more advanced control strategy. We will not pursue this further in this work, since the majority of errors are within the threshold and other problems must be solved first to use PROXIMA in dynamical simulations.

While PROXIMA accurately captures the density of the system, there are problems in the small-scale dynamics. We expect the volume of the system to fluctuate over time with oscillation periods much larger than the simulation timestep (1 fs). Instead, there are oscillations of with periods on the order of 10 fs (Fig. 2). The outcome of the simulation is correct on one metric, but alarmingly wrong on another.

3.2 Introducing Blending Into on-the-Fly ML-Driven Dynamics

Switching abruptly from the target to surrogate function creates a rapid, non-smooth change in the system over time, which is not only unphysical but could lead to long-lasting effects in the dynamics of the system. Thus, we propose a gradual transition between one function and another. Detailed in Sect. 2.2, this

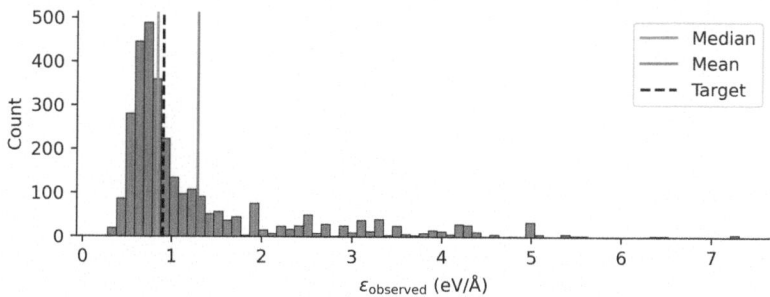

Fig. 1. A histogram of observed errors between the surrogate and target functions, their mean and median, and the target error from an example trajectory.

method slowly introduces the surrogate over time when uncertainty is below the threshold established by the PROXIMA algorithm, and likewise slowly reduces surrogate usage when the uncertainty is above the threshold.

The goal of our blending method is to mitigate the changes in the dynamics introduced by rapid switching, and this is what we see in our results. Figure 2 shows the volume oscillation over a segment of one of these trajectories. We can see that once PROXIMA begins switching between the surrogate and the target a high-frequency oscillation is introduced to the PROXIMA trajectory which is not present in the target-only trajectory and the amplitude of the lower frequency oscillation increases, both indicating a departure from the expected dynamics of the system caused by rapid switching. When using PROXIMA+BLEND, no such oscillations are present, indicating that our blending method has the intended effect.

The insight gained from inspecting an individual trajectory above is further supported by quantifying the high-frequency oscillations. Rapid oscillations will be associated with large time derivatives, so we calculate the maximum absolute gradient of the volume over time across all trajectories; in doing so we see in Fig. 3 that PROXIMA has much larger gradients than the trajectories generated by the target method, while PROXIMA+BLEND is nearly identical to the target method, indicating a major reduction in rapid oscillations.

While we expect blending to reduce artifacts introduced by switching between target and surrogate, we expected it to be slower than the un-blended PROX-IMA due to the possibility of more calls to the target method being performed. However, we observed that the speedup of PROXIMA+BLEND over DFT is only slightly less than that of PROXIMA without blending, as shown in Fig. 4, indicating that the additional evaluations of F when the UQ is under the threshold but $\lambda < 1$ are infrequent enough to make PROXIMA+BLEND a useful tool for accelerating calculations.

The promise of the blending method is that it does not require altering the original algorithms. There are, indeed, algorithms for molecular dynamics which are less sensitive to the abrupt changes of introducing surrogates. We found that molecular dynamics performed with Berendsen barostat does not exhibit the

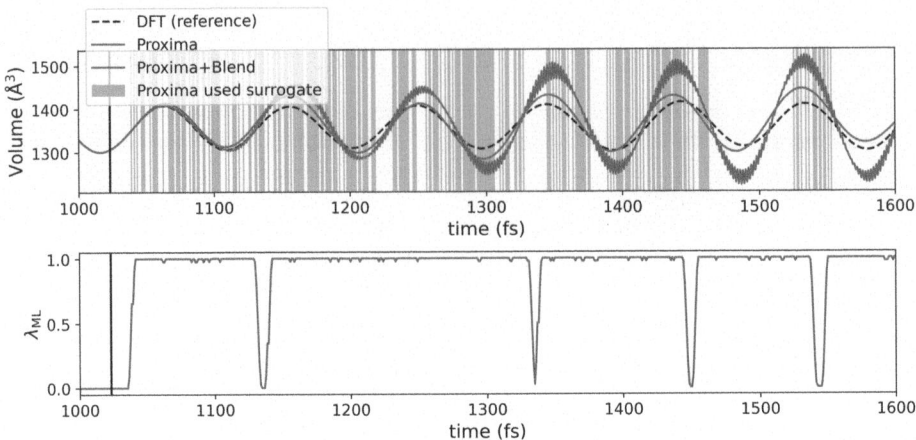

Fig. 2. The top panel shows the fluctuation of the volume of the simulation cell over time across conditions from a selected velocity configuration. The dashed line is derived from the trajectory that used DFT forces at every time step. The blue and red lines are derived from the PROXIMA and PROXIMA+BLEND trajectories, respectively. The vertical black line indicates when the first surrogate was trained and could be used by PROXIMA or PROXIMA+BLEND. The blue shaded regions indicate when PROXIMA (without blending) used ML to advance the trajectory. The bottom pane shows the value of λ with PROXIMA+BLEND. (Color figure online)

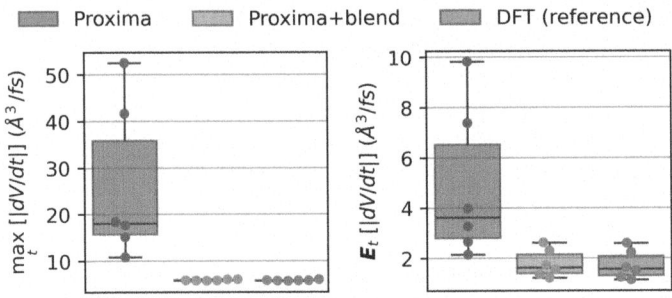

Fig. 3. A box plot of the maximum gradient in volume with respect to time ($\mathring{A}^3$/fs), derived from the trajectories generated by DFT, PROXIMA, and PROXIMA+BLEND.

fluctuations, perhaps because the volume changes depend only on the pressure at the present timestep [4]. Our method eliminates the need to carefully examine whether OTF learning is appropriate for a new simulation process. Validation and adjustment of the blending timescales are necessary, but there are fewer limits to which algorithms can tolerate changing between original and surrogate on-the-fly.

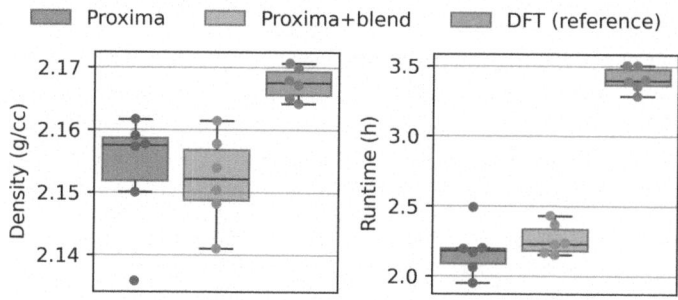

Fig. 4. A box plot of the runtimes of the simulations by condition.

3.3 Blending Corrects Fine-Trained Dynamics

Eliminating the unphysical oscillations of cell volume using PROXIMA+BLEND is a positive result; however, volume is only one of the 373 degrees of freedom in our molecular dynamics calculation. We further explore the reliability of OTF learning by studying the dynamics of the remaining degrees of freedom (atom positions, momenta) by measuring how often the system transitions between different states. Specifically, we perform a simulation where one atom is removed from an otherwise regular crystal lattice and then we measure the frequency that another atom "hops" into the newly-vacant site (known as a vacancy). Vacancy hopping is controlled by interactions and motion of many individual atoms and, thus, requires accurate dynamics to model correctly.

To examine the effects of switching and blending on vacancy diffusion, we performed a set of NVT (that is, fixed number, volume, and temperature) molecular dynamics simulations at the equilibrium volume determined from previous NPT simulations. NVT is often used when examining diffusion due to effects of the barostat on such bevavior. We used a higher temperature (1573K) to increase the hopping rate. The key metric of our calculation is the time between hops, which we detect by determining when the neighbors associated with each atom change. We detect neighbors by counting the number of atoms within a distance slightly larger than the equilibrium bond distance in Si.

Our results show that hop-rates derived from PROXIMA+BLEND agree with those generated by the target function F within 5%, while those from PROXIMA (without blending) disagree by more than 80%, further indicating that our addition of blending removes artifacts and makes simulations more accurate. As shown in Table 1, using PROXIMA to control the surrogate estimates a hop rate nearly twice as fast as the surrogate-free, DFT-only calculation.

A remaining challenge highlighted by the hop-rate study is that the erroneous hop rate is not simple to find. As with density calculation, PROXIMA's controller achieved the target level of error between surrogate and target function. The unphysical dynamics, however, were not apparent until comparing hop rate with a full-fidelity calculation. The ability to perform a full-fidelity calculation is unavailable in many cases and techniques for directly relating error in a specific subroutine to larger scale observables are limited. As such, we recommend users of OTF learning—with PROXIMA or otherwise—to carefully examine the effect of algorithm parameters on their target outcome. Even with such a limitation, adding PROXIMA+BLEND reduces the chance for unphysical behavior and improves agreement with full-fidelity dynamics compared to PROXIMA.

Table 1. Estimates of the Si vacancy hop rate estimated from trajectories generated by DFT, PROXIMA, and PROXIMA+BLEND, respectively

Method	Hop rate (SE) [1/ps=THz]
DFT-only	2.68 (0.0834)
PROXIMA	5.06 (0.114)
PROXIMA+BLEND	2.80 (0.0852)

4   Related Work

4.1   Machine-Learned Surrogates for Expensive Computations

Developing cheap approximates for expensive calculations is a widespread activity in modern computational sciences and is becoming even more effective with advancements in machine learning. Within our focus domain, electronic-scale physics, cheap approximates for computing atomic forces have nearly a century of practice [17] and are undergoing a renaissance with advanced machine learning techniques [3]. The machine learning models create surrogates with many more degrees of freedom than those created by humans, which are then able to recover more nuances of the function being approximated.

The scope of surrogates for simulations extends well past learning subroutines which compute forces on atoms. Tools which approximate specific subroutines are growing in impact for climate simulations, in particular [22,23]. Methods for learning a faster set of dynamic equations rather than just replacements for evaluating specific are also emerging [6,15].

The tradeoff for the intricate surrogates available from machine learning is the lack of a clear understanding of the limitations of the model. As such, approaches for estimating the uncertainty of the model are of key importance. Techniques range from evaluating differences between models trained with varied training sets or directly training a confidence signal that will be output with every prediction [26]. Regardless of method, the quality of the confidence interval must be assumed to be imperfect and must be calibrated to produce meaningful results [14].

4.2 Offline Training of Surrogates

The conventional practice for machine-learned surrogates is to separate training a surrogate from using it to solve a science problem. The workflow for creating a surrogate includes *enumerating* the space of potential inputs to a subroutine, *evaluating* many to produce a training set, and *training* the surrogate.

Enumeration, evaluation, and training are often coupled. Modern practices for making surrogates use an active learning procedure which alternates between training a model and using the model to identify which potential inputs to evaluate [12]. The active learning procedure may even involve enumerating new points by solving example problems using the surrogate model [5].

4.3 On-the-Fly Training of Surrogates

Early approaches to OTF learning on atomistic simulations can be found in Refs [8,10]. These approaches decompose a molecular dynamics simulation into regions where quantum-level accuracy is and is-not required, and train spline-based parametric surrogates on-the-fly in the high-fidelity regions. Instead of blending surrogate and target methods to address discontinuities, these works make use of made use of a rollback and interpolation procedures where timesteps around model updates are repeated and rerun with the new model, or in the case of Refs [8,9] rerun with a version of their surrogate interpolated between their new and old one. The target method was never used to run dynamics directly.

Similar spatially decomposed + OTF surrogate construction approaches have been employed more recently, including in the massively parallel approach described by Ref [7].

Recently an application called FLARE has been developed for OTF training of Gaussian-process based atomistic surrogates [27,29]. These model forms have the advantages of needing relatively small training datasets and separating data and parameter uncertainty, allowing the parameter uncertainty to be used to trigger surrogate updates when necessary.

5 Conclusion

We have demonstrated that control methods can ensure accurate outcomes while introducing machine-learned surrogates into dynamics simulations, provided that the transition to using the faster surrogate is smooth. We modify an existing algorithm PROXIMA [30], that enforces that surrogates are used only when appropriate, by adding a technique to adjust the blend between original and surrogate subroutines rather than toggling abruptly. This new approach, PROXIMA+BLEND, eliminates unphysical oscillations of volume in constant pressure molecular dynamics simulations and corrects the timescales of state transitions in constant-volume dynamics. Importantly, PROXIMA+BLEND required no changes in the dynamical equations. We envision that providing such autonomous control systems will enable the reduction in computational cost by providing a reliable cost/accuracy tradeoff in more software across computational sciences.

Acknowledgements. LW was supported by the Office of Defense Nuclear Nonproliferation Research and Development within the U.S. Department of Energy's National Nuclear Security Administration. MT was supported by U.S. Department of Energy, Office of Science, Office of Advanced Scientific Computing Research, Computational Science Graduate Fellowship under Award Number(s) DE-SC0023112. This report was prepared as an account of work sponsored by an agency of the United States Government. Neither the United States Government nor any agency thereof, nor any of their employees, makes any warranty, express or implied, or assumes any legal liability or responsibility for the accuracy, completeness, or usefulness of any information, apparatus, product, or process disclosed, or represents that its use would not infringe privately owned rights. Reference herein to any specific commercial product, process, or service by trade name, trademark, manufacturer, or otherwise does not necessarily constitute or imply its endorsement, recommendation, or favoring by the United States Government or any agency thereof. The views and opinions of authors expressed herein do not necessarily state or reflect those of the United States Government or any agency thereof.

References

1. Anderson, G.J., Gaffney, J.A., Spears, B.K., Bremer, P.T., Anirudh, R., Thiagarajan, J.J.: Meaningful uncertainties from deep neural network surrogates of large-scale numerical simulations (2020). https://doi.org/10.48550/arXiv.2010.13749

2. Batatia, I., et al.: A foundation model for atomistic materials chemistry (2024). https://doi.org/10.48550/arXiv.2401.00096. arXiv:2401.00096

3. Behler, J.: Perspective: machine learning potentials for atomistic simulations. J. Chem. Phys. **145**(17) (2016). https://doi.org/10.1063/1.4966192

4. Berendsen, H., Postma, J., van Gunsteren, W.F., DiNola, A., Haak, J.R.: Molecular dynamics with coupling to an external bath. J. Chem. Phys. **81**(8), 3684–3690 (1984). https://doi.org/10.1063/1.4966192

5. Bernstein, N., Csányi, G., Deringer, V.L.: De novo exploration and self-guided learning of potential-energy surfaces. NPJ Comput. Mater. **5**(1), 1–9 (2019). https://doi.org/10.1038/s41524-019-0236-6. https://www.nature.com/articles/s41524-019-0236-6

6. Brunton, S.L., Proctor, J.L., Kutz, J.N.: Discovering governing equations from data by sparse identification of nonlinear dynamical systems. Proc. Natl. Acad. Sci. **113**(15), 3932–3937 (2016). https://doi.org/10.1073/pnas.1517384113

7. Caccin, M., Li, Z., Kermode, J.R., De Vita, A.: A framework for machine-learning-augmented multiscale atomistic simulations on parallel supercomputers. Int. J. Quantum Chem. **115**(16), 1129–1139 (2015). https://doi.org/10.1002/qua.24952

8. Csányi, G., Albaret, T., Moras, G., Payne, M.C., Vita, A.D.: Multiscale hybrid simulation methods for material systems. J. Phys. Condens. Matter **17**(27), R691 (2005). https://doi.org/10.1088/0953-8984/17/27/R02

9. Csányi, G., Albaret, T., Payne, M.C., De Vita, A.: "Learn on the Fly": a hybrid classical and quantum-mechanical molecular dynamics simulation. Phys. Rev. Lett. **93**(17), 175503 (2004). https://doi.org/10.1103/PhysRevLett.93.175503

10. De Vita, A., Car, R.: A novel scheme for accurate md simulations of large systems. MRS Online Proc. Libr. **491**(1), 473–480 (1997). https://doi.org/10.1557/PROC-491-473

11. Gao, X., Ramezanghorbani, F., Isayev, O., Smith, J.S., Roitberg, A.E.: TorchANI: a free and open source PyTorch-based deep learning implementation of the ANI neural network potentials. J. Chem. Inf. Model. **60**(7), 3408–3415 (2020). https://doi.org/10.1021/acs.jcim.0c00451

12. Guo, Y., Nath, P., Mahadevan, S., Witherell, P.: Active learning for adaptive surrogate model improvement in high-dimensional problems. Struct. Multidisc. Optim. **67**(7) (2024). https://doi.org/10.1007/s00158-024-03816-9

13. Kühne, T.D., Iannuzzi, M.., Hutter, J.: Cp2k: an electronic structure and molecular dynamics software package-quickstep: efficient and accurate electronic structure calculations. J. Chem. Phys. **152**(19) (2020). https://doi.org/10.1063/5.0007045

14. Kuleshov, V., Fenner, N., Ermon, S.: Accurate uncertainties for deep learning using calibrated regression (2018). https://doi.org/10.48550/ARXIV.1807.00263. https://arxiv.org/abs/1807.00263

15. Kurth, T., et al.: Fourcastnet: accelerating global high-resolution weather forecasting using adaptive fourier neural operators. In: Proceedings of the Platform for Advanced Scientific Computing Conference, PASC 2023, pp. 1–11. ACM (2023). https://doi.org/10.1145/3592979.3593412

16. Larsen, A.H., Mortensen, J.J., Jacobsen, K.W.: The atomic simulation environment–a Python library for working with atoms. J. Phys. Condens. Matter **29**(27), 273002 (2017). https://doi.org/10.1088/1361-648X/aa680e

17. Lennard-Jones, J.E.: Cohesion. Proc. Phys. Soc. **43**(5), 461–482 (1931). https://doi.org/10.1088/0959-5309/43/5/301

18. Lubbers, N., Smith, J.S., Barros, K.: Hierarchical modeling of molecular energies using a deep neural network. J. Chem. Phys. **148**(24), 241715 (2018). https://doi.org/10.1063/1.5011181. http://arxiv.org/abs/1710.00017, arXiv:1710.00017

19. Melchionna, S., Ciccotti, G., Lee Holian, B.: Hoover NPT dynamics for systems varying in shape and size. Mol. Phys. **78**(3), 533–544 (1993). https://doi.org/10.1080/00268979300100371

20. Partee, S., et al.: Using machine learning at scale in numerical simulations with smartsim: an application to ocean climate modeling. J. Comput. Sci. **62**, 101707 (2022). https://doi.org/10.1016/j.jocs.2022.101707

21. Peterson, A.A., Christensen, R., Khorshidi, A.: Addressing uncertainty in atomistic machine learning. Phys. Chem. Chem. Phys. **19**(18), 10978–10985 (2017). https://doi.org/10.1039/c7cp00375g

22. Rasp, S., Pritchard, M.S., Gentine, P.: Deep learning to represent subgrid processes in climate models. Proc. Natl. Acad. Sci. **115**(39), 9684–9689 (2018). https://doi.org/10.1073/pnas.1810286115

23. Sastry, V.K., Maulik, R., Rao, V., Lusch, B., Renganathan, S.A., Kotamarthi, R.: Data-driven deep learning emulators for geophysical forecasting. In: Paszynski, M., Kranzlmüller, D., Krzhizhanovskaya, V.V., Dongarra, J.J., Sloot, P. (eds.) ICCS 2021. LNCS, vol. 12746, pp. 433–446. Springer, Cham (2021). https://doi.org/10.1007/978-3-030-77977-1_35

24. Smith, J.S., Isayev, O., Roitberg, A.E.: ANI-1: an extensible neural network potential with DFT accuracy at force field computational cost. Chem. Sci. **8**(4), 3192–3203 (2017). https://doi.org/10.1039/C6SC05720A. https://pubs.rsc.org/en/content/articlelanding/2017/sc/c6sc05720a

25. Smith, J.S., Nebgen, B., Lubbers, N., Isayev, O., Roitberg, A.E.: Less is more: sampling chemical space with active learning. J. Chem. Phys. **148**(24), 241733 (2018). https://doi.org/10.1063/1.5023802

26. Tran, K., Neiswanger, W., Yoon, J., Zhang, Q., Xing, E., Ulissi, Z.W.: Methods for comparing uncertainty quantifications for material property predictions. Mach. Learn. Sci. Technol. **1**(2), 025006 (2020). https://doi.org/10.1088/2632-2153/ab7e1a

27. Vandermause, J., et al.: On-the-fly active learning of interpretable Bayesian force fields for atomistic rare events. NPJ Comput. Mater. **6**(1), 1–11 (2020). https://doi.org/10.1038/s41524-020-0283-z. https://www.nature.com/articles/s41524-020-0283-z

28. Warshel, A., Levitt, M.: Theoretical studies of enzymic reactions: dielectric, electrostatic and steric stabilization of the carbonium ion in the reaction of lysozyme. J. Mol. Biol. **103**(2), 227–249 (1976). https://doi.org/10.1016/0022-2836(76)90311-9

29. Xie, Y., Vandermause, J., Ramakers, S., Protik, N.H., Johansson, A., Kozinsky, B.: Uncertainty-aware molecular dynamics from Bayesian active learning for phase transformations and thermal transport in SiC. NPJ Comput. Mater. **9**(1), 1–8 (2023). https://doi.org/10.1038/s41524-023-00988-8. https://www.nature.com/articles/s41524-023-00988-8

30. Zamora, Y., Ward, L., Sivaraman, G., Foster, I., Hoffmann, H.: Proxima: accelerating the integration of machine learning in atomistic simulations. In: Proceedings of the ACM International Conference on Supercomputing, ICS 2021, pp. 242–253. Association for Computing Machinery, New York (2021). https://doi.org/10.1145/3447818.3460370. https://dl.acm.org/doi/10.1145/3447818.3460370

Biological Community Detection with Graph Neural Network and Network Curvature Analysis on Gene Co-expression Networks

Marianna Milano[1,4(✉)], Pietro Cinaglia[2,4], Mario Cannataro[3,4], and Pietro Hiram Guzzi[3,4]

[1] Department of Experimental and Clinical Medicine, Magna Graecia University of Catanzaro, Catanzaro, Italy
[2] Department of Health Sciences, Magna Graecia University of Catanzaro, Catanzaro, Italy
[3] Department of Medical and Surgical Sciences, Magna Graecia University of Catanzaro, Catanzaro, Italy
[4] Data Analytics Research Center, Department of Medical and Surgical Sciences, University Magna Græcia of Catanzaro, Catanzaro, Italy
{m.milano,cinaglia,cannataro,hguzzi}@unicz.it

Abstract. Biomedical networks are critical for representing complex biological systems, and network curvature is a key structural property that captures topological features not highlighted by traditional graph metrics. This study introduces a Graph Neural Network (GNN)-based approach for detecting communities in cancer-specific Gene Co-expression Networks (GCNs), using *Ollivier-Ricci curvature* as an integral feature. The inclusion of curvature has shown to enhance the detection of biologically significant communities, improve network modularity, and enable finer partitioning. These preliminary results indicate that curvature-based analyses can offer new insights into the organization of gene co-expression networks, aiding in the understanding of biological modularity, disease mechanisms, and functional interactions.

Keywords: GNN · Network Curvature · Biological System

1 Introduction

The study of biological networks provides fundamental insights into the complex relationships that regulate cellular processes and disease mechanisms. Advances in bioinformatics and network science have enabled the construction of gene co-expression networks, which represent functional associations between genes based on their expression patterns across different biological conditions [12,15,24]. Identifying modular structures within these networks is a crucial task, as communities often correspond to biologically relevant functional groups, signaling pathways, or disease-related clusters. However, traditional community detection methods, such as Louvain and Greedy modularity optimization [20], may fail to capture subtle hierarchical relationships within biological networks. Recently, geometric and topological approaches, such as network curvature, have emerged as powerful tools to uncover hidden structural properties in complex networks [4,13].

Curvature measures, inspired by differential geometry, quantify local and global deviations from an idealized network structure, offering new perspectives on network robustness, modularity, and functional organization.

Recent advances in artificial intelligence, particularly in deep learning applied to graph models, have further improved our ability to analyze and interpret complex networked data. Graph Neural Networks (GNNs) have gained increasing attention due to their ability to learn node representations while preserving the underlying graph structure [3]. Unlike traditional machine learning models, which treat nodes as independent entities, GNNs incorporate relational information, allowing them to capture higher-order dependencies within a network. These models have been successfully applied in biomedical informatics, including drug discovery, protein-protein interaction prediction, and gene function annotation [5–7]. One of the key tasks that GNNs can address is community detection, which involves partitioning a graph into well-defined clusters by leveraging both topological and feature-based information. In biological networks, this approach can reveal functional modules of genes, regulatory pathways, and clusters associated with disease mechanisms [9]. However, existing GNN-based methods often rely solely on adjacency-based representations without leveraging additional structural descriptors. The incorporation of network curvature into a GNN framework could provide a richer and more informative representation of graph topology, potentially leading to improved community detection.

In this study, we propose a GNN-based framework for community detection in cancer-specific gene co-expression networks, by integrating *Ollivier-Ricci curvature* as a key feature. The primary goal of this work is to assess whether curvature enhances the detection of biologically relevant communities compared to classical clustering algorithms. To achieve this, we construct gene-disease interaction networks from the iNetModels 2.0 database, selecting five cancer-specific gene co-expression networks corresponding to breast, colon, stomach, thyroid, and pancreas tissues. Each network is analyzed by computing *Ollivier-Ricci curvature* for its edges, providing a curvature-enhanced representation that is then processed by a GNN-based community detection model. The framework is designed to extract meaningful gene clusters while capturing the underlying structural properties of the network, ensuring a more refined partitioning of biological modules.

The results demonstrate that the integration of *Ollivier-Ricci curvature* leads to an increased number of detected communities, accompanied by an improvement in modularity scores, indicating a stronger intra-community structure.

2   Related Work

The introduction of novel applications in Graph Neural Networks (GNNs) has become an essential study area [25], particularly with the incorporation of network curvature as a significant factor for improving node classification capabilities. GNNs, which have gained immense traction in recent years, exploit the topological structures of graphs to perform tasks such as node and graph classification effectively [11,14]. The foundations of GNNs were laid by early contributions such as those by Kipf and Welling, who proposed a layer-wise propagation framework for semi-supervised node classification based on Graph Convolutional Networks (GCNs) [17]. Graph Neural Networks (GNNs) operate on data structured as graphs, leveraging the graph's inherent features and the relationships between nodes. The theoretical foundation of GNNs is based on the concept of message passing, where node states are updated by recursively aggregating and transforming feature information from neighboring nodes. This iterative process can be formalized as follows:

Initially, each node v is assigned a feature vector $h_v^{(0)} = x_v$. The feature vectors are updated through layers or iterations using the rule:

$$h_v^{(k)} = \sigma \left(W^{(k)} \sum_{u \in \mathcal{N}(v)} \frac{1}{c_{vu}} h_u^{(k-1)} + B^{(k)} h_v^{(k-1)} \right),$$

where $h_v^{(k)}$ is the feature vector of node v at iteration k, $\mathcal{N}(v)$ denotes the set of neighbors of v, $W^{(k)}$ and $B^{(k)}$ are trainable parameters specific to layer k, σ is a non-linear activation function, and c_{vu} is a normalization constant often set as the cardinality of $\mathcal{N}(v)$ to average the contributions.

The objective of these iterations is to reach a stable state where the representations h_v converge to a fixed point. This approach aligns with the theoretical perspective of GNNs functioning as a form of contraction mapping in a complete metric space, wherein each iteration brings the representations closer to a point that is invariant under the mapping defined by the update rule.

In practice, this fixed point provides a powerful embedding for each node that captures both local structure—through the aggregation from immediate neighbors—and more global graph properties, as the effects of more distant nodes are indirectly incorporated through multiple iterations. The stability and convergence of this process are crucial for the practical effectiveness of GNNs and are often guaranteed under conditions like bounded weights or specially designed normalization schemes.

Moreover, these node embeddings derived from GNNs are versatile and can be used for a variety of tasks, including node classification, link prediction, and graph classification. Their performance on these tasks demonstrates the capability of GNNs to capture and utilize the complex and rich information contained within graph-structured data, adhering to both the graph's topology and the features of individual nodes.

In summary, the theoretical underpinnings of GNNs contribute significantly to their ability to generalize well across different types of graph data, making them an invaluable tool in the machine learning toolkit for handling data with intricate relational structures.

2.1 Advanced Measures of Curvature in Network Analysis

The analysis of network curvature offers insightful metrics that help understand the geometric and topological properties of networks. These measures provide crucial information on how networks deviate from being flat, affecting the dynamics within the network.

2.2 Forman-Ricci Curvature

The Forman-Ricci curvature, developed by Robin Forman, is an adaptation of *Ollivier-Ricci curvature* for discrete networks. For an edge e connecting vertices u and v with weights $w(u)$, $w(v)$, and $w(e)$, the curvature is given by:

$$F(e) = w(e) \left(\frac{1}{w(u)} + \frac{1}{w(v)} \right) - \sum_{\substack{e' \sim e \\ e' \neq e}} \frac{w(e)}{\sqrt{w(u')w(v')}}$$

This expression considers adjacent edges e' with vertices u' and v', assessing how edge weights contribute to local curvature. This curvature measure is particularly effective in identifying densely interconnected regions within a network, indicating areas of high robustness or potential fragility.

2.3 Ollivier-Ricci Curvature

The *Ollivier-Ricci curvature* provides a probabilistic measure of curvature based on optimal transport. For an edge $e = (u, v)$, it is defined as:

$$\kappa(e) = 1 - W_1(\mu_u, \mu_v)$$

Here, μ_u and μ_v are probability measures concentrated at vertices u and v respectively, and W_1 is the 1-Wasserstein distance, reflecting the cost of redistributing mass from u to v. This curvature measure is insightful for evaluating how well connected a network is, with lower curvature indicating better connectivity.

2.4 Bakry-Émery Curvature

Another important measure is the Bakry-Émery curvature, which extends the notion of *Ollivier-Ricci curvature* to graphs. It considers the behavior of the Laplacian operator on the graph and is indicative of the diffusive properties of the network. High Bakry-Émery curvature implies that the graph has good expansion properties and is well-connected.

2.5   Haantjes Curvature

The Haantjes curvature takes into account higher-dimensional structures within the graph. It evaluates the curvature formed by considering paths rather than edges alone, providing a more holistic view of the curvature within the network. This measure is particularly useful in networks where the relationships between nodes involve complex interactions, such as in biological networks or intricate social networks.

These curvature measures play vital roles in various fields. For instance, in data communication, they help in designing more efficient routing algorithms by understanding paths that minimize latency. In epidemiology, they can predict how diseases might spread through different clusters within a network, identifying potential hotspots for more focused interventions. Understanding these curvature metrics allows network engineers and data scientists to design more robust, efficient, and resilient networks, tailored to the specific dynamics and requirements of their respective fields.

3   Methods

In this study, we employ a **GNN** to perform community detection on a gene-disease interaction network, leveraging *Ollivier-Ricci curvature* as an edge feature and using the transcription factor (TF) status as a node feature. The **GNN architecture** is based on a **Graph Autoencoder (GAE)**, which learns low-dimensional node representations through an **encoder-decoder structure**. The primary goal is to evaluate whether *Ollivier-Ricci curvature* provides meaningful structural information in biological networks and improves the detection of biologically relevant communities.

3.1   Dataset Preprocessing

The dataset consists of a **gene-disease interaction network**, where nodes represent genes, and edges indicate interactions between them. Each edge is assigned a curvature value obtained from *Ollivier-Ricci curvature* **computation**. Additionally, nodes are annotated with a binary feature indicating whether the gene is a **transcription factor (TF)** ({1 if TF, 0 otherwise}). The dataset is processed as follows:

1. **Loading the network**: The edge list contains gene pairs with their corresponding *Ollivier-Ricci curvature* values.
2. **Node encoding**: Gene names are mapped to unique integer IDs, as required by PyTorch Geometric.
3. **Graph construction**: A **NetworkX graph (G)** is created, and the adjacency structure is converted into a **PyTorch Geometric Data object**.

3.2   Graph Neural Network Architecture

The GNN used for community detection was based on a **Graph Autoencoder (GAE)**, consisting of:

- **Encoder**: Two layers of **GCN** that transform the input feature matrix into a **low-dimensional latent space**.
- **Decoder**: A **single GCN layer** that attempts to reconstruct the original feature space.

Mathematical Representation. Given a graph $G = (V, E)$, let X be the feature matrix, where each node $v \in V$ has a feature vector x_v consisting of its **transcription factor status**. The encoding process is as follows:

$$H = \text{ReLU}(W_1 \cdot \text{GCNConv}(X, A)) \tag{1}$$

$$Z = \text{ReLU}(W_2 \cdot \text{GCNConv}(H, A)) \tag{2}$$

where A is the adjacency matrix, W_1, W_2 are trainable weight matrices, H is the hidden representation, and Z is the final embedding. The decoder attempts to reconstruct X from Z using another **GCN layer**.

3.3   Model Training

We used **Mean Squared Error (MSE) loss** for quantifying prediction accuracy in model training, optimizing reconstruction quality:

$$\text{MSE} = \frac{1}{N} \sum_{i=1}^{N} (Y_i - \hat{Y}_i)^2 \tag{3}$$

where Y_v, $\hat{Y}_v$ and N are the observed and predicted values, and the number of data points of the sample.

The training process follows these steps:

1. **Initialize model weights** with an Adam optimizer (learning rate $= 0.01$, weight decay $= 5 \times 10^{-4}$).
2. **Train for 200 epochs**, applying **ReLU activation** and updating weights via backpropagation.
3. **Extract node embeddings** after training.

4   A Case Study on Trascriptomic Data

Our model have been trained on real-world dataset constructed from iNetModels 2.0 database [1] that includes normal tissue and cancer-specific Gene Coexpression Networks networks. iNetModels provides 108 biological networks of different tissues. In this study we selected 5 tissues (breast, colon, stomac, tyroid, pancreas and cancer-specific networks. Table 1 summarizes the main characteristics of the networks. Also, we used Molecular signatures database (MSigDB) [18] to retrieve the information oncogenic signatures. MSigDB contain gene sets representing potential targets of regulation by transcription factors or microRNAs. For each network, we calculated the *Ollivier-Ricci curvature* a metric derived from Riemannian geometry. This measure assesses the shape of the network at each node and edge, providing insights into the overall topological structure.

Table 1. Characteristics of the GCN networks.

Network	Nodes	Edges
Breast	463	20000
Colon	950	1000
Stomac	933	1000
Tyroid	1648	2000
Pancreas	475	1000

We applied our GNN-based model to each GCN to extract communities, then, to assess the impact of *Ollivier-Ricci curvature*, we recomputed communities without curvature. Table 2 reports the number of detected communities, the modularity and ARI for each GNC networks, by considering the curvature and without curvature. We compared the results using:

– Modularity is a widely used metric for evaluating the quality of community structure in a network. It measures the strength of division of a network into modules (communities) by comparing the actual density of edges within communities to the expected density if edges were distributed at random [21]. The modularity score Q is computed as:

$$Q = \frac{1}{2m} \sum_{ij} \left[A_{ij} - \frac{k_i k_j}{2m} \right] \delta(c_i, c_j) \tag{4}$$

where A_{ij} represents the adjacency matrix, k_i and k_j are the degrees of nodes i and j, m is the total number of edges, and $\delta(c_i, c_j)$ is 1 if nodes i and j belong to the same community and 0 otherwise.
Modularity values typically range from -1 to 1, where higher values indicate a stronger community structure. Generally, values above 0.3 are considered indicative of significant community structure [8,9]. A high modularity score suggests that the network exhibits a well-defined clustering pattern, while lower values indicate weak or no community structure. However, modularity has certain limitations, including a well-known resolution limit, which can prevent the detection of small communities in large networks [10]. Alternative methods, such as modularity optimization using the Louvain algorithm [2], have been proposed to efficiently identify communities in large-scale networks.

- **Adjusted Rand Index (ARI)**: Evaluates similarity between clustering results **with and without curvature**. The Adjusted Rand Index (ARI) is a measure used to evaluate the similarity between two clustering results by adjusting for chance [16]. Given two partitions of a set, the ARI is defined as:

$$ARI = \frac{RI - E[RI]}{\max(RI) - E[RI]} \tag{5}$$

where RI (Rand Index) accounts for the number of pairs correctly classified in the same or different clusters, and $E[RI]$ is its expected value under random assignment.

The ARI ranges from -1 to 1, where 1 indicates perfect agreement, 0 corresponds to random clustering, and negative values suggest worse-than-random agreement [22]. Unlike other clustering evaluation measures, ARI is particularly robust against random assignments and provides a more reliable way to assess clustering performance. It is commonly used in applications such as gene expression analysis.

One of the advantages of ARI is its ability to handle different numbers of clusters in the compared partitions. This makes it especially useful in cases where the true number of clusters is unknown or when evaluating clustering algorithms with varying parameter settings. However, ARI is sensitive to the size of clusters; for highly imbalanced cluster distributions, alternative measures such as the Normalized Mutual Information (NMI) [23] or the Variation of Information (VI) [19] may provide complementary insights. In practical applications, ARI is often used in conjunction with other clustering metrics to obtain a more comprehensive evaluation of clustering quality.

Table 2 reports the number of detected communities, the modularity and ARI for each GNC networks, by considering the curvature and without curvature.

Table 2. Effect of *Ollivier-Ricci curvature* on Community Detection: Comparison of Number of Communities, Modularity, and Adjusted Rand Index (ARI). Results report the number of communities computed by considering the curvature (*Communities w-C*) and without considering this one (*Communities w/o-C*). Similarly, it also report the information concerning modularity (*Modularity w-C* and *Modularity w/o-C*, respectively).

Network	Communities w-C	Communities w/o-C	Modularity w-C	Modularity w/o-C	ARI
Breast	33	18	0.3731	0.2529	0.1
Colon	47	32	0.878	0.7	0.12
Stomac	51	23	0.754	0.555	0.1021
Tyroid	85	72	0.9976	0.763	0.1
Pancreas	31	26	0.721	0.584	0.153

The results demonstrate that the predicted number of communities is consistently higher when *Ollivier-Ricci curvature* is used as a feature compared to when it is not. Additionally, modularity, which quantifies the strength of community structures in a network, shows an increase when curvature is included, indicating that the detected communities are more internally cohesive and better separated from each other. A key metric we analyzed is the Adjusted Rand Index (ARI), which quantifies the similarity between two clustering results while correcting for chance. An ARI close to 1 indicates that the two clustering solutions are nearly identical, meaning that the inclusion of *Ollivier-Ricci curvature* has little to no impact on the community structure. Conversely, a low ARI (typically < 0.5) suggests that the curvature significantly alters the community structure, leading to a different partitioning of the network.

In our results, ARI values remain relatively low (ranging between 0.1 and 0.15), which might initially seem to indicate weak clustering. However, in this context, the low ARI is actually evidence that *Ollivier-Ricci curvature* has a strong influence on community formation. The fact that clustering solutions with and without curvature are markedly different suggests that curvature is reshaping the network structure in a meaningful way, leading to a new partitioning of gene interactions. This highlights that curvature is capturing additional topological information that traditional feature representations do not, thereby impacting the way communities are detected.

From a biological perspective, this means that the introduction of *Ollivier-Ricci curvature* allows the model to identify alternative, potentially more biologically relevant community structures. Traditional clustering solutions might merge functionally distinct gene groups into fewer clusters, whereas curvature-enhanced GNNs may better reflect the modular organization of biological processes, capturing subtle interactions that classical methods overlook.

For example, in the breast cancer network, the number of communities increases from 18 (without curvature) to 33 (with curvature), and modularity rises from 0.2529 to 0.3731, indicating stronger intra-community connectivity. Similarly, in the colon network, the number of communities increases from 32 to 47, and modularity improves significantly from 0.7 to 0.878, suggesting a more defined community structure. The thyroid network also exhibits a notable increase in modularity when curvature is included (from 0.763 to 0.9976), reinforcing the idea that *Ollivier-Ricci curvature* enhances community separation.

To further assess the impact of *Ollivier-Ricci curvature* on community structure, we compared the GNN-based results with classical community detection algorithms, i.e. Louvain and Greedy.

Table 3 reports the number of detected communities and the modularity values obtained with our model and with classical Louvain and Greedy algorithms. The number of detected communities in the GNN model is consistently higher, often by a significant margin. For instance, in the pancreas network, our GNN identified 248 communities, whereas Louvain and Greedy detected only 84. This discrepancy suggests that curvature-based representations may reveal finer-scale biological structures that traditional methods tend to merge into larger, less specific communities.

Moreover, modularity values in our GNN model remain superior compared to those obtained with Louvain and Greedy. Notably, in the colon network, the modularity score for our GNN is 0.878, whereas Louvain and Greedy produce very low scores (0.002), indicating that traditional algorithms struggle to detect well-separated clusters in this case. Similarly, in the stomach network, our GNN model achieves a modularity of 0.754, significantly outperforming Louvain (0.117) and Greedy (0.101), reinforcing the idea that curvature-enhanced GNNs offer a more refined and biologically relevant partitioning of the gene interaction networks. These results highlight the potential biological significance of using *Ollivier-Ricci curvature* in gene interaction networks. The increased number of communities detected when curvature is included suggests that this feature helps to identify finer subdivisions within the biological network, which could correspond to distinct functional modules, signaling pathways, or disease-related gene clusters. Traditional community detection methods, such as Louvain and Greedy, may be too coarse to capture these subtle subdivisions, leading to overly merged clusters that obscure underlying biological structures.

Table 3. Comparison of GNN-based results with Louvain and Greedy algorithms

	Breast	Colon	Stomac	Thyroid	Pancreas
N. community GNN	33	47	51	85	248
N. community Louvain	32	2	7	64	84
N. community Greedy	31	2	3	64	84
Modularity GNN	0.3731	0.878	0.754	0.9976	0.721
Modularity Louvain	0.193	0.002	0.117	0.89	0.994
Molularity Greedy	0.1882	0.002	0.101	0.89	0.994

5   Conclusion

In this study, we introduced a framework for community detection in cancer-specific Gene Co-expression Networks by leveraging Graph Neural Networks (GNNs) and network curvature measures. Our results demonstrate that the incorporation of *Ollivier-Ricci curvature* as a node feature significantly influences the structure of detected communities. Specifically, we observed that when curvature is included, the number of predicted communities increases, and the modularity values improve, indicating stronger intra-community connectivity and clearer separation of biological clusters.

Furthermore, when comparing the GNN-based community detection results with classical algorithms such as Louvain and Greedy, we found that the GNN model detects a significantly higher number of communities, often with improved modularity. This implies that curvature-based graph representations enhance the resolution of community detection, potentially uncovering subtle biological relationships that conventional methods overlook.

From a biological standpoint, these findings suggest that *Ollivier-Ricci curvature* could serve as a valuable tool for improving the functional interpretation of gene co-expression networks, aiding in the identification of biological modules, signaling pathways, and disease-related clusters. By refining the way we detect and interpret network communities, curvature-enhanced GNNs could contribute to a deeper understanding of cancer-related gene interactions, ultimately supporting advancements in biomedical research and precision medicine.

Future work will focus on further validating these findings through biological enrichment analyses, integrating additional multi-omic datasets, and exploring other curvature measures to assess their impact on community detection. Additionally, extending this approach to larger and more diverse cancer datasets could provide further insights into the biological relevance of curvature-based network partitioning.

Acknowledgments. This work was funded by the Next Generation EU - Italian NRRP, Mission 4, Component 2, Investment 1.5, call for the creation and strengthening of 'Innovation Ecosystems', building 'Territorial R&D Leaders' (Directorial Decree n. 2021/3277) - project Tech4You - Technologies for climate change adaptation and quality of life improvement, n. ECS0000009. This work reflects only the authors' views and opinions, neither the Ministry for University and Research nor the European Commission can be considered responsible for them.

We acknowledge the support of the PNRR project FAIR - Future AI Research (PE00000013), Spoke 9 - Green-aware AI, under the NRRP MUR program funded by the NextGenerationEU.

The scientific activities of the CINI InfoLife Laboratory supported this research.

Disclosure of Interests. The authors have no competing interests to declare that are relevant to the content of this article.

References

1. Arif, M., et al.: inetmodels 2.0: an interactive visualization and database of multi-omics data. Nucleic Acids Res. **49**(W1), W271–W276 (2021)
2. Blondel, V.D., Guillaume, J.L., Lambiotte, R., Lefebvre, E.: Fast unfolding of communities in large networks. J. Stat. Mech: Theory Exp. **2008**(10), P10008 (2008)
3. Bronstein, M.M., Bruna, J., LeCun, Y., Szlam, A., Vandergheynst, P.: Geometric deep learning: going beyond Euclidean data. IEEE Signal Process. Mag. **34**(4), 18–42 (2017)
4. Chatterjee, T., DasGupta, B., Albert, R.: A review of two network curvature measures. Nonlinear Anal. Global Optim., 51–69 (2021)
5. Cinaglia, P.: Multilayer biological network alignment based on similarity computation via graph neural networks. J. Comput. Sci. **78**, 102259 (2024). https://doi.org/10.1016/j.jocs.2024.102259
6. Cinaglia, P.: PyMulSim: a method for computing node similarities between multilayer networks via graph isomorphism networks. BMC Bioinf. **25**(1), 211 (2024)

7. Cinaglia, P., Cannataro, M.: Identifying candidate gene-disease associations via graph neural networks. Entropy (Basel) **25**(6) (2023)
8. Clauset, A., Newman, M., Moore, C.: Finding community structure in very large networks. Phys. Rev. E **70**(6), 066111 (2004)
9. Fortunato, S.: Community detection in graphs. Phys. Rep. **486**(3–5), 75–174 (2010)
10. Fortunato, S., Barthélemy, M.: Resolution limit in community detection. Proc. Nat. Acad. Sci. USA **104**(1), 36–41 (2007)
11. Gu, S., Jiang, M., Guzzi, P.H., Milenković, T.: Modeling multi-scale data via a network of networks. Bioinformatics **38**(9), 2544–2553 (2022)
12. Guzzi, P.H., Tradigo, G.: Biological network analysis: trends, approaches, and challenges. Brief. Bioinform. **21**(6), 1935–1953 (2020)
13. Guzzi, P.H., Milano, M.: Exploring network curvature differences in gene expression networks. In: 2024 IEEE International Conference on Bioinformatics and Biomedicine (BIBM), pp. 4352–4354. IEEE (2024)
14. Guzzi, P.H., Milenković, T.: Survey of local and global biological network alignment: the need to reconcile the two sides of the same coin. Brief. Bioinform. **19**(3), 472–481 (2018)
15. Guzzi, P.H., Roy, S.: Biological Network Analysis: Trends, Approaches, Graph Theory, and Algorithms. Elsevier (2020)
16. Hubert, L., Arabie, P.: Comparing partitions. J. Classif. **2**(1), 193–218 (1985)
17. Kipf, T., Welling, M.: Semi-supervised classification with graph convolutional networks (2016). https://doi.org/10.48550/arxiv.1609.02907
18. Liberzon, A., Subramanian, A., Pinchback, R., Thorvaldsdóttir, H., Tamayo, P., Mesirov, J.P.: Molecular signatures database (msigdb) 3.0. Bioinformatics **27**(12), 1739–1740 (2011)
19. Meila, M.: Comparing clusterings–an information based distance. J. Multivar. Anal. **98**(5), 873–895 (2007)
20. Milano, M., Cinaglia, P., Guzzi, P.H., Cannataro, M.: A novel local alignment algorithm for multilayer networks. Inf. Med. Unlocked **44**, 101425 (2024)
21. Newman, M.: Modularity and community structure in networks. Proc. Nat. Acad. Sci. USA **103**(23), 8577–8582 (2006)
22. Steinley, D.: Properties of the Hubert-Arabie adjusted rand index. Psychol. Methods **9**(3), 386–396 (2004)
23. Vinh, N.X., Epps, J., Bailey, J.: Information theoretic measures for clusterings comparison: variants, properties, normalization and correction for chance. J. Mach. Learn. Res. **11**, 2837–2854 (2010)
24. Zitnik, M., et al.: Current and future directions in network biology. arXiv preprint: arXiv:2309.08478 (2023)
25. Zitnik, M., et al.: Current and future directions in network biology (2024)

Incorporating Performance Ordering in MCDA: A Study of the Frobenius SPOTIS Method

Andrii Shekhovtsov[1] , Jean Dezert[2] , and Wojciech Sałabun[1,3(✉)]

[1] National Institute of Telecommunications, Szachowa 1, 04-894 Warsaw, Poland
{a.shekhovtsov,w.salabun}@il-pib.pl
[2] The French Aerospace Lab - ONERA, 91120 Palaiseau, France
jean.dezert@onera.fr
[3] West Pomeranian University of Technology, Żołnierska 49, 71-210 Szczecin, Poland

Abstract. Most Multi-Criteria Decision Analysis (MCDA) methods encode a decision-maker's preferences through criterion weights, yet even a well-weighted model can yield ties or virtually indistinguishable scores. We propose Frobenius SPOTIS (Fro-SPOTIS), which is a generalization of the classical SPOTIS method that incorporates ordering information to resolve such ambiguities. Each alternative's attribute-based ranking of criteria is compared with a reference ranking derived from the criteria weights. This comparison is performed by converting both rankings into pairwise preference-score matrices and computing the Frobenius distance between them. This distance, modulated by a tolerance parameter $\tau \in [0, 1]$, is used to modify to the native SPOTIS score: $\tau = 0$ recovers the original SPOTIS results, while higher values increasingly favor alternatives whose performance ordering aligns with the reference. A three-alternative example shows how Fro-SPOTIS untangles an otherwise unresolved tie, and two sensitivity analysis studies trace how rankings shift with (i) changes in the underlying data and (ii) variations in τ. The results confirm that Fro-SPOTIS retains the simplicity of SPOTIS while offering a more flexible and expressive approach to tie-breaking in MCDA.

Keywords: SPOTIS · Frobenis Distance · MCDA

1 Introduction

Multi-Criteria Decision Analysis (MCDA) is a subfield of operational research focused on providing methodologies and algorithms to support decision-makers in complex decision scenarios involving multiple, often conflicting criteria. MCDA methods assist in structuring decision problems, incorporating both qualitative and quantitative data, and deriving well-informed, rational choices. Many MCDA methods rely on subjective expert knowledge to determine the relative importance of criteria, whereas others employ mathematical formulations

© The Author(s), under exclusive license to Springer Nature Switzerland AG 2025
M. H. Lees et al. (Eds.): ICCS 2025, LNCS 15905, pp. 297–309, 2025.
https://doi.org/10.1007/978-3-031-97632-2_21

to objectively analyze alternatives and establish rankings or recommendations [15]. Consequently, these techniques offer a comprehensive framework, balancing subjective judgment with rigorous quantitative analysis, thereby enhancing the reliability and transparency of decision-making processes.

Most MCDA methods incorporate importance weights to capture the knowledge and preferences of the decision maker or expert [19]. These weights reflect the relative significance of each criterion and play a key role in shaping the final ranking of alternatives. However, in practice, weighting alone may be insufficient—particularly when it leads to ties or nearly indistinguishable rankings [17]. To overcome this limitation, some approaches incorporate additional information that assesses how well alternatives align with the desired order of performance across criteria [3].

While traditional MCDA techniques effectively encode expert input via criterion weights, they often fall short when it comes to distinguishing between alternatives with similar aggregated scores. Moreover, they typically lack the means to evaluate the degree to which alternatives adhere to the intended prioritization of criteria. This limitation highlights the need for more advanced methods that not only consider the relative importance of criteria but also quantify the alignment of alternatives with these priorities in a meaningful and discriminative way.

In this paper, we propose the Frobenius Stable Preference Ordering Toward Ideal Solution (Fro-SPOTIS) method, a generalization of the classical SPOTIS method, which utilizes the Frobenius distance [3] to measure discrepancies between the desirable ordering of criteria derived from their assigned weights and their actual performance ordering across alternatives. By introducing this ordering-based distance measure, criteria that deviate significantly from the preferred ordering are naturally penalized, resulting in their lower positioning within the final ranking of alternatives. Consequently, Fro-SPOTIS effectively addresses and resolves ranking ties and closely positioned alternatives, providing clearer discrimination among them.

Furthermore, we introduce a tolerance parameter $\tau \in [0, 1]$, enabling analysts to regulate the strictness of incorporating ordering discrepancies into the final assessment. When $\tau = 0$, the method reduces to the original SPOTIS formulation, ignoring ordering information, whereas increasing τ progressively amplifies the influence of ordering consistency. To illustrate and validate the capabilities of Fro-SPOTIS, we present an illustrative three-alternative example and perform two sensitivity analyses that explore the impact of varying the tolerance parameter and examine the robustness of rankings under data perturbations. Our results confirm that Fro-SPOTIS maintains the intuitive simplicity of the SPOTIS approach while enhancing its flexibility, interpretability, and practical utility in handling tie-breaking scenarios within multi-criteria decision analysis.

The remainder of the paper is structured as follows. In Sect. 2, we provide context for our study, iterating on recent work in the domain. In Sect. 3 we describe the algorithms and methods used in the study, the Frobenius distance algorithm and the SPOTIS method. In Sect. 3.3, we describe the proposed Frobenius SPO-

TIS method. In Sect. 4, we provide experiments showing the features and limitations of the proposed method and discuss its possible extensions. Finally, in Sect. 5, we conclude our work and provide some ideas for future research directions.

2 Related Works

To compare different alternatives in the decision-making process, many MCDA methods rely on distance metrics to assess the proximity of each alternative to an ideal solution. One of the most commonly used metrics for this purpose is the Euclidean distance. It plays a central role not only in classical MCDA approaches such as the Technique for Order Preference by Similarity to Ideal Solution (TOPSIS) [1], but also in more advanced modern methods, including COmbinative Distance-based ASsessment (CODAS) [13] and Preference Ranking On the Basis of Ideal-average Distance (PROBID) [18]. The Euclidean distance is also employed in the Stable Preference Ordering Towards Ideal Solution (SPOTIS) method [5] to evaluate the performance of alternatives.

In certain cases, researchers and decision-makers utilize even more sophisticated methodologies based on generalized fuzzy sets, necessitating adaptations or variations of classical distance metrics tailored specifically to the chosen fuzzy generalization [14,16]. Such generalizations often involve intuitionistic, hesitant, neutrosophic, or type-2 fuzzy sets, each of which requires unique adaptations of standard metrics to appropriately handle increased uncertainty and imprecision. Consequently, this underscores the significance of developing diverse distance measures, as well as thoroughly investigating their suitability, effectiveness, and robustness across different MCDA contexts. A comprehensive exploration and comparative analysis of these adapted metrics can significantly enhance the accuracy and interpretability of multi-criteria evaluations in complex decision environments.

The Frobenius distance is a recently introduced metric designed specifically to measure distances between rankings. It is constructed as a genuine distance metric, explicitly satisfying Kemeny's axiomatic principles for ranking comparisons [3,10]. Furthermore, similar metrics, notably the Kemeny distance itself, have found extensive applications beyond MCDA, such as in voting theory, preference aggregation, and social choice, highlighting their broader relevance and applicability across various decision-making domains.

Several recent studies have demonstrated the versatility of the Kemeny distance in various analytical contexts. For instance, in [6], the authors proposed robust fuzzy clustering methods combining Kemeny distance with medoid-based clustering algorithms. Their results indicated that the proposed approach effectively mitigates the impact of noise and outliers in datasets, improving the robustness of the clustering outcomes. Another noteworthy application of the Kemeny distance involves constructing median rankings from multiple individual rankings. This problem was addressed by Emond and Mason [7], who introduced a weighted version of the Kemeny-Snell distance for consensus ranking problems. Their comparative analysis highlighted the advantages of their method over Kendall's Tau measure, demonstrating superior performance in generating consensus rankings. Additionally, Kemeny distance has been successfully applied within Multi-Criteria Decision Analysis, notably in the KEmeny Median Indicator Ranks Accordance (KEMIRA) method, to determine criteria weights when handling two distinct groups of criteria [12].

Although the Frobenius distance was originally introduced to demonstrate that metrics other than the Kemeny distance could also satisfy Kemeny's axiomatic principles, its applicability and properties have been further investigated in contexts involving consensus among multiple rankings. For instance, Dezert et al. [4] proposed utilizing both Kemeny and Frobenius distances to find optimal solutions in compromise ranking problems. However, their analysis revealed that not all scenarios yielded results aligning intuitively with 'common sense' expectations. Additionally, in [2], the authors explored the potential of the Frobenius distance by proposing a methodology specifically designed to quantify differences between partial orderings, further expanding its applicability in ranking-related problems.

This highlights the relevance of both Kemeny and Frobenius distances within the MCDA domain, underlining the need for continued exploration of their theoretical properties, comparative performance, and practical applicability across diverse decision-making contexts. Further research in this direction may lead to the development of more robust, interpretable, and context-sensitive MCDA tools that can better accommodate complex preference structures and ranking-based evaluations.

3   Methodology

In this section, we briefly introduce the methods used in this study: the Frobenius distance and the SPOTIS method. The former is used to measure deviations between performance matrices, while the latter supports ranking alternatives based on their distance to ideal solutions.

3.1   Frobenius Distance

Consider a set X consisting of $n \geq 2$ objects, each ranked by two information sources. We denote the total preference orderings (TPOs) provided by these sources as Pref_1 and Pref_2. For example, consider a set of three objects $X = \{x_1 = A, x_2 = B, x_3 = C\}$. Source 1 might provide the preference Pref_1 and source 2 might provide Pref_2, with the following TPOs: $\text{Pref}_1 \triangleq A \succ B \succ C$ and $\text{Pref}_2 \triangleq B \succ C \succ A$.

The Frobenius distance between two TPOs (orderings) of N objects is computed by first constructing an $N \times N$ pairwise Preference-Score Matrix (PSM) based on the ordering given by each source. By convention, the row index i of the PSM corresponds to the index of elements x_i on the left side of the preference ordering $x_i \succ x_j$, and the column index j corresponds to the index of the element x_j on the right side of the preference ordering $x_i \succ x_j$. Thus, we denote a pairwise Preference-Score Matrix $\boldsymbol{M}(X) = [M(i,j)]$ where its components $M(i,j)$ for $i,j = 1,2,\ldots,N$ are defined as

$$\boldsymbol{M}(i,j) = \begin{cases} 1, & \text{if } x_i \succ x_j, \\ -1, & \text{if } x_i \prec x_j, \\ 0, & \text{if } x_i = x_j. \end{cases} \qquad (1)$$

Note, that all diagonal elements $M(i,i)$ ($i = 1,2,\ldots,N$) of the matrix $\boldsymbol{M}$ are always zero. Additionally, the PSM is inherently anti-symmetric because the preference $x_i \succ x_j$ implies $x_j \prec x_i$. Therefore, if $x_i \succ x_j$ holds, meaning $M(i,j) = 1$, then necessarily $x_j \succ x_i$ is false, implying $x_j \prec x_i$ is true, and thus $M(j,i) = -1$, and vice versa. As a result, $\boldsymbol{M}(X)^T = -\boldsymbol{M}(X)$, and $\text{Tr}(\boldsymbol{M}(X)) = 0$.

The distance between two TPOs, Pref_1 and Pref_2, is defined using the Frobenius distance as follows [3]:

$$d_F(\boldsymbol{M}_1, \boldsymbol{M}_2) = ||\boldsymbol{M}_1 - \boldsymbol{M}_2||_F, \qquad (2)$$

where $||\boldsymbol{M}||_F$ is the Frobenius norm of a square matrix $\boldsymbol{M} = [M(i,j), i,j = 1,\ldots,N]$, defined by [8,9]

$$||\boldsymbol{M}||_F = \sqrt{\sum_{i=1}^{n} \sum_{j=1}^{n} |M(i,j)|^2} = \sqrt{\text{Tr}(\boldsymbol{M}^T \boldsymbol{M})}, \qquad (3)$$

and where $\boldsymbol{M}^T$ is the transpose of the matrix $\boldsymbol{M}$, and $\text{Tr}(.)$ is the trace operator for matrix.

Frobenius distance can be normalized to $[0,1]$ range by dividing the value $d_F(\boldsymbol{M}_1, \boldsymbol{M}_2)$ by the maximum distance value $d_F^{\max}$ computed by considering two TPOs in full contradiction (i.e. a preference and its opposite defined by reversing the preference order). For instance, if a preference ordering is $\text{Pref} = A \succ B \succ C \succ D$, its opposite is $\neg\text{Pref} = A \prec B \prec C \prec D = D \succ C \succ B \succ A$.

3.2 Stable Preference Ordering Towards Ideal Solution (SPOTIS)

The Stable Preference Ordering Toward Ideal Solution (SPOTIS) method, introduced by [5], is a MCDA technique that employs reference objects to assess decision alternatives. Unlike other methods that typically derive reference objects from the decision matrix data, SPOTIS requires the decision-maker to explicitly define these reference objects by defining criteria bounds.

To apply the SPOTIS method, the decision-maker must first establish the criteria bounds that will serve as reference objects for evaluating alternatives. For each criterion C_j ($j \in \{1,2,\ldots,N\}$), the maximum S_j^{max} and minimum S_j^{min} bounds must be specified. Subsequently, the Ideal Solution Point (ISP) $\boldsymbol{S}^* = \{S_1^*, \ldots, S_j^*, \ldots, S_m^*\}$ is determined such that $S_j^* = S_j^{max}$ is for the profit criteria and $S_j^* = S_j^{min}$ for cost criteria. The decision matrix is represented as $S = (S_{ij})_{M \times N}$, where S_{ij} denotes the attribute value of the i-th alternative A_i for the j-th criterion C_j.

The full algorithm of the SPOTIS method presented in [5] is as follows:

Step 1. Calculation of the normalized distances to ISP (4).

$$d_{ij}(A_i, S_j^*) = \frac{|S_{ij} - S_j^*|}{|S_j^{max} - S_j^{min}|} \tag{4}$$

Step 2. Calculation of the weighted normalized distances from ISP $d(A_i, \mathbf{S}^*) \in [0, 1]$, according to (5).

$$d(A_i, \mathbf{S}^*) = \sum_{j=1}^{N} w_j d_{ij}(A_i, S_j^*) \tag{5}$$

Step 3. Determine the final ranking by ordering the alternatives by the values $d(A_i, \mathbf{S}^*)$. The better alternatives have smaller values of $d(A_i, S^*)$.

The important features of this method include its simplicity, robustness against the rank reversal paradox, and the ability to utilize the Expected Solution Point (ESP). The ESP allows the decision-maker to define an expected outcome and construct the ranking based on this point rather than the Ideal Solution Point (ISP). To apply the SPOTIS method with a selected ESP $\mathbf{S}^+$, one should follow the standard SPOTIS procedure but replace the values of the Ideal Solution Point S_j^* with the values of the ESP S_j^+. The decision-maker should select the values of $\mathbf{S}^+$ to align with the specific decision problem.

However, it is important to ensure that the selected ESP falls within the problem's scope, meaning S_j^+ should satisfy $S_j^{min} \leq S_j^+ \leq S_j^{max}$ for every value of j.

3.3 Proposed Fro-SPOTIS Method

We propose using Frobenius distance to incorporate information about performance order in an alternative to the final result. In order to do that, we need to extract information about the preferred performance order from the criteria weights. The intuition behind this is as follows: if criterion C_i is most important to us according to the weights, then we want it to have the best performance in the top alternatives. To create this order, we need to create a ranking of $R(C)$ for the criteria in the problem.

For example, the importance weight vector $\mathbf{w} = [0.3, 0.5, 0.2]$ indicates that C_2 is the most important criterion, C_1 is the second most important criterion and C_3 is the least important criterion in the given problem. This means that the importance-based (decreasing) order of the criteria is $C_2 \succ C_1 \succ C_3$, which is characterized by the ranking vector $\mathbf{r}(C) = [r(C_i), i = 1, 2, 3]$ where $r(C_i)$ is the rank of the criterion C_i in the importance-based decreasing ordering. In this example, we have $\mathbf{r}(C) = [2, 1, 3]$.

Step 1. Calculation of the normalized distances to the Ideal Solution Point as was shown in (4).

Step 2. Calculation of the normalized weighted distances $d_i = d(A_i, S^*) \in [0, 1]$, according to (5).

Step 3. The ranking matrix $\mathbf{R}$ as defined (6) is created by the ranked distances normalized from the ISP $(d_{ij}(A_i, S_j^*))$ for each alternative. The ranking should be done from the lowest to the highest values, as the values $d_{ij}(A_i, S_j^*)$ are lower if the alternative performs better (that is, it is closer to the ideal solution).

$$\mathbf{R} = \begin{bmatrix} r(A_1)_1 & r(A_1)_2 & \cdots & r(A_1)_m \\ r(A_2)_1 & r(A_2)_2 & \cdots & r(A_2)_m \\ \vdots & \vdots & \ddots & \vdots \\ r(A_n)_1 & r(A_n)_2 & \cdots & r(A_n)_m \end{bmatrix} \tag{6}$$

Step 4. Calculate the vector of normalized Frobenius distances between each row of the ranking matrix and the ideal order of the criteria $d_F(\mathbf{r}(C), \mathbf{r}(A_i))$ according to (7).

$$f_i = \{d_F(\mathbf{r}(C), \mathbf{r}(A_i))\}, \quad i \in \{1, 2, \ldots, n\} \tag{7}$$

Step 5. Define the parameter $\tau \in [0, 1]$, which defines the tolerance within which two alternatives are considered to be in a tie.

Next, for each two alternatives A_i and A_j apply the Eq. (8), which modifies their preferences d_i based on the τ. Note that Eq. (8) should be applied to both A_i and A_j for each pair. This means that if d_i is modified and increased, d_j should be decreased accordingly.

$$d_i' = \begin{cases} d_i - \frac{\tau}{2} & \text{if} \quad |d_i - d_j| \leq \tau \text{ and } f_i < f_j \\ d_i + \frac{\tau}{2} & \text{if} \quad |d_i - d_j| \leq \tau \text{ and } f_i > f_j \\ d_i & \text{otherwise} \end{cases} \tag{8}$$

Finally, when there is no pairs of alternatives for which a change in d_i is required, the final ranking should be determined based on d_i' values, where smaller values suggest that the alternative is better. Flowchart presented in Fig. 1 summarize Frobenius-SPOTIS method algorithm by visually representing it.

For a better understanding of the proposed approach, see a simple example in the following section.

4   Experiments and Results

In this section, we present the results of our experiments using the proposed Fro-SPOTIS method. We begin with a simple example that demonstrates how the Fro-SPOTIS method can resolve tie situations in preferences. Following this, we conduct two sensitivity analysis experiments to further explore the features and limitations of the proposed method. Lastly, we discuss the application of the Fro-SPOTIS method with the Expected Solution Point (ESP).

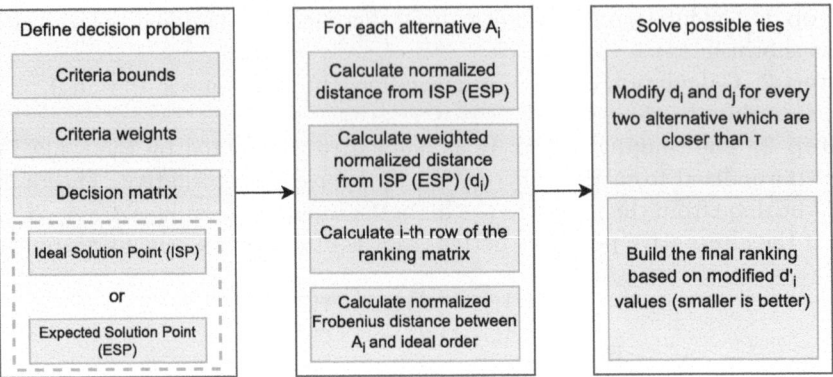

Fig. 1. Flowchart of the Fro-SPOTIS method.

4.1 Simple Example

Assume that the following decision problem is consistent with the three alternatives presented in Table 1, as well as with the criteria weights and the criteria bounds. To keep the example simple, we consider using three criteria with values in the range $[0, 10]$ and criteria weights $\mathbf{w} = [0.1, 0.3, 0.6]$. All criteria are considered profit; therefore, a value of 10 is the most desirable value in all criteria.

Table 1. Decision matrix, criteria weights and criteria bounds for the simple example.

A_i	C_1	C_2	C_3
A_1	8	3	3
A_2	5	8	2
A_3	2	3	5
w_j	0.1	0.3	0.6
S_j^{min}	0	0	0
S_j^{max}	10	10	10

If we proceed with this example with the original SPOTIS algorithm, the alternative A_1 will be evaluated as 0.65, while the other two alternatives will be in a tie, with the weighted normalized distance from the ISP equal to 0.59 (for each of them). This makes these two alternatives incomparable, making it impossible to prioritize one or another in the decision-making process. However, analyzing the performance of the alternatives, we can see that A_3 performs 2.5 times better in C_3 (which is most important to us) than in A_2. Of course, A_2 performs better in C_1 and C_2; however, the importance of these criteria is low for us, making them less desirable.

However, if we use the proposed Fro-SPOTIS algorithm that uses additional information on the performance order in the alternatives, we will get the calculation shown in Table 2. First, for each alternative, the normalized distance from ISP $d_{ij}(A_i, S_j^*)$ is calculated; then it is multiplied by weights. Furthermore, based on $d_{ij}(A_i, S_j^*)$ performance rankings $\mathbf{r}(A_i)$ are determined for each alternative.

Table 2. Calculation process for Frobenius SPOTIS.

	C_1	C_2	C_3
$d_{1j}(A_1, S_j^*)$	0.2000	0.7000	0.7000
$d_{2j}(A_1, S_j^*)$	0.5000	0.2000	0.8000
$d_{3j}(A_2, S_j^*)$	0.8000	0.7000	0.5000
$w_j d_{1j}(A_1, S_j^*)$	0.0200	0.2100	0.4200
$w_j d_{2j}(A_1, S_j^*)$	0.0500	0.0600	0.4800
$w_j d_{3j}(A_2, S_j^*)$	0.0800	0.2100	0.3000
$\mathbf{r}(A_1)$	1	2	2
$\mathbf{r}(A_2)$	2	1	3
$\mathbf{r}(A_3)$	3	2	1

Next, if we apply Eqs. (7)–(8), we obtain the results shown in Table 3. The value d_i is the original SPOTIS output, and f_i is the Frobenius distance between $\mathbf{r}(A_i)$ and $\mathbf{r}(C)$. Finally, the values P_i show the final evaluation, and R_i determines the rank of the alternatives; these values were obtained with $\tau = 0.05$. The addition of the Frobenius distance makes all three alternatives comparable, and according to the results, A_3 is the best.

Table 3. Final calculations for the Fro-SPOTIS method.

A_i	d_i	f_i	P_i	R_i
A_1	0.6500	0.8660	0.6500	3
A_2	0.5900	0.8165	0.6150	2
A_3	0.5900	0.0000	0.5650	1

This small experiment shows how the proposed Fro-SPOTIS method solves ties in the rankings and preferences obtained with the original SPOTIS method. The tolerance parameter τ also introduces some level of flexibility in terms of adapting the method to the specific decision problem.

4.2   Sensitivity Analysis

In this section, we want to demonstrate the limitation of the proposed Frobenius SPOTIS method, which appears due to the nature of the Frobenius SPOTIS in some cases. In specific cases, the proposed method can violate the dominance principle [11]. The dominance principle states that if one option is better than another in at least one aspect and not worse in all other aspects, then the former option dominates the latter, meaning that dominant one should be chosen over the dominated one. Respectively, a situation in which a dominated alternative is preferred violates the dominance principle. To demonstrate how Fro-SPOTIS can violate the dominance principle, we designed a sensitivity analysis experiment presented further in this Section.

Consider a four-criterion decision problem, with criteria bounds $[0, 10]$ for each of the four criteria. Criteria weights are selected as $\mathbf{w} = [0.4, 0.3, 0.2, 0.1]$, so we can show how the preference value changes with the change of the performance order in the alternative. Each criterion is considered profit, therefore, making 10 the most desirable value, and in all calculations, $\tau = 0.01$ was used.

Suppose that we have the alternative $A = [4, 3, 2, 1]$, which is quite far from the ISP. Figure 2 is divided into four parts, and each of them demonstrates the change in preference values for the Frobenius SPOTIS and the classic SPOTIS depending on the value of C_j in the alternative. We can see that in three of the four cases, Frobenius SPOTIS provides the same results as the original SPOTIS because there are no ties or close values. However, in the case of criterion C_4, the gradual change of its value causes a violation of the dominance principle. The cause of this behavior is that with a small weight $w_4 = 0.1$, a change of the C_4 criterion value can trigger correction of the preference according to (8).

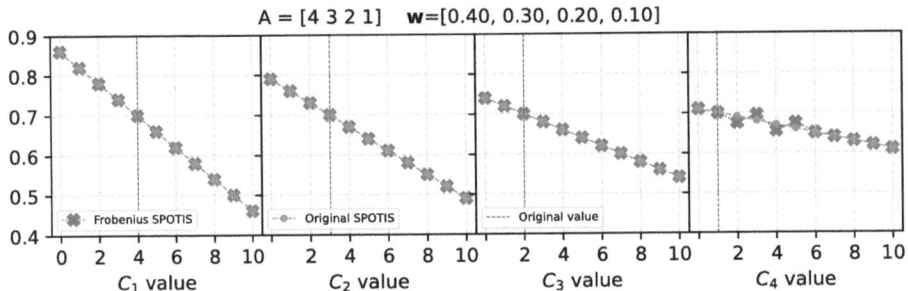

Fig. 2. Change in the preference value depending on C_j value for alternative $A = [4, 3, 2, 1]$ for SPOTIS and Fro-SPOTIS methods.

This case, however, is synthetically created, and normally such situations should not occur. Additionally, if one wants to avoid such situations, Frobenius distance can be used with a small value of τ or only in the case of ties in the ranking.

4.3   Influence of τ Value on Results

In this section, we describe the experiment, which demonstrates how different values of τ influence the final results after modification of the original SPOTIS preferences. To address this, we design a simple simulation study on random decision matrices. Each decision matrix consists of four criteria and ten alternatives and is filled with integers drawn from the uniform probability distribution with a range $[0, 100]$. The first two criteria were profit and the other two costs. The weights of the criteria were drawn from a uniform probability distribution $[0, 1]$ and normalized to sum up to 1. Next, alternatives were evaluated with both original SPOTIS and proposed Fro-SPOTIS (with $\tau \in \{0.01, 0.05, 0.10, 0.15\}$), providing the same or different results, depending on the data. The correlation between two corresponding rankings was calculated using Weighted Spearman's correlation r_w. The described experiment was repeated 10,000 times, and results were presented as boxplots in Fig. 3 grouped by different values of τ.

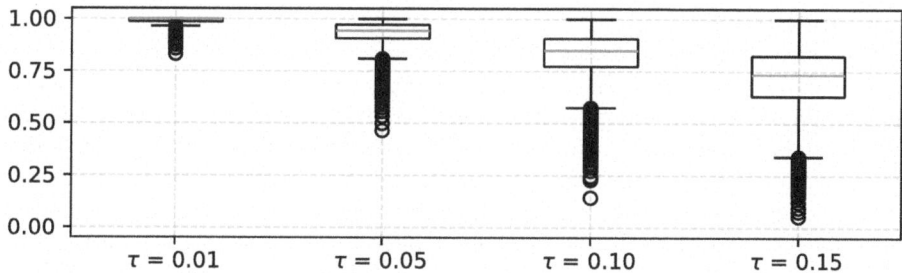

Fig. 3. Distibution of r_w correlation values calculated between original and Frobenius SPOTIS rankings for different values of τ.

In Fig. 3, we can see how much Fro-SPOTIS results are correlated with original SPOTIS results. It can be seen that for $\tau = 0.01$, values of the r_w correlation are pretty close to 1, which means that there were little changes in the rankings, but some outliers are present. For this value of τ, we expect that only ties on very similar alternatives will be affected. For $\tau = 0.05$, the correlation falls, with worst outliers below $r_w = 0.5$, while the median is still quite close to 1.0. In the other two cases, the median falls further, but we do not have any outliers lower than zero, which means that even despite the low correlation, the results are still correlated.

However, even if it is possible to use high values of τ, we advise against it. The reason is that with the larger value of τ, it is possible for the Fro-SPOTIS to violate the dominance principle. Therefore, we suggest using only small values of τ, mainly to help bring additional information to tie-solving or to strongly differentiate close alternatives.

4.4   Usage of the Expected Solution Point

The SPOTIS method stands out among most MCDA methods due to its flexibility in allowing the use of the Expected Solution Point (ESP) in place of the

Ideal Solution Point (ISP) during calculations. This feature enables the decision-maker to guide the ranking process toward a solution that better reflects specific goals, expectations, or preferences relevant to the decision problem.

Similarly, in the Fro-SPOTIS variant, the ESP can be used instead of the ISP, maintaining compatibility with the core SPOTIS mechanism. To apply this, the ISP values are simply replaced with ESP values. This substitution is valid because the Frobenius distance values f_i, which capture deviation across the entire performance matrix, are calculated independently of the reference point (ISP or ESP). Meanwhile, the distance values d_i retain their role by representing the weighted normalized distance from each alternative to the selected reference point, whether it be the ISP or ESP.

5   Conclusions

In this paper, we propose a novel MCDA method named Fro-SPOTIS, which explicitly incorporates information regarding preference ordering into the decision-making framework. The proposed approach leverages the Frobenius distance to quantify deviations between the desired ordering of alternatives, defined by the decision-maker's preferences, and the actual observed performance order of alternatives. Additionally, Fro-SPOTIS introduces a tolerance parameter τ, enabling the decision-maker to control the conditions under which certain alternatives are classified as incomparable, thereby providing greater flexibility in handling decision uncertainty or preference ambiguity.

We illustrate the practical application of Fro-SPOTIS through a numerical example, complemented by sensitivity analyses examining the robustness of the method to variations in both the attribute values of alternatives and the tolerance parameter τ. These analyses demonstrate how the ranking outcomes evolve in response to changes in input data and preference structures, providing insight into the method's adaptability to diverse decision-making scenarios. Furthermore, we present an experimental investigation highlighting potential limitations of Fro-SPOTIS, notably cases where the method may violate the dominance principle. Identifying these limitations offers valuable direction for future research, in which we aim to address these issues and subsequently apply Fro-SPOTIS to more complex, real-world decision problems to further validate and enhance its practical utility.

Disclosure of Interests. The authors have no competing interests to declare that are relevant to the content of this article.

References

1. Çelikbilek, Y., Tüysüz, F.: An in-depth review of theory of the TOPSIS method: an experimental analysis. J. Manage. Anal. **7**(2), 281–300 (2020)
2. Dezert, J., Shekhovtsov, A., Salabun, W.: Distances between partial preference orderings. arXiv preprint: arXiv:2407.19869 (2024)

3. Dezert, J., Shekhovtsov, A., Sałabun, W.: A new distance between rankings. Heliyon **10**(7) (2024)
4. Dezert, J., Shekhovtsov, A., Sałabun, W., Tchamova, A.: On optimal solution of the compromise ranking problem. In: 2024 27th International Conference on Information Fusion (FUSION), pp. 1–8. IEEE (2024)
5. Dezert, J., Tchamova, A., Han, D., Tacnet, J.M.: The SPOTIS rank reversal free method for multi-criteria decision-making support. In: 2020 IEEE 23rd International Conference on Information Fusion (FUSION), pp. 1–8. IEEE (2020)
6. D'Urso, P., Vitale, V.: A Kemeny distance-based robust fuzzy clustering for preference data. J. Classif. **39**(3), 600–647 (2022)
7. Emond, E.J., Mason, D.W.: A new rank correlation coefficient with application to the consensus ranking problem. J. Multi-Criteria Dec. Anal. **11**(1), 17–28 (2002)
8. Golub, G.H., Van Loan, C.F.: Matrix Computations. JHU Press (2013)
9. Horn, R.A., Johnson, C.R.: Matrix Analysis. Cambridge University Press (2012)
10. Kemeny, J.G.: Mathematics without numbers. Daedalus **88**(4), 577–591 (1959)
11. Kourouxous, T., Bauer, T.: Violations of dominance in decision-making. Bus. Res. **12**(1), 209–239 (2019). https://doi.org/10.1007/s40685-019-0093-7
12. Krylovas, A., Zavadskas, E.K., Kosareva, N., Dadelo, S.: New KEMIRA method for determining criteria priority and weights in solving MCDM problem. Int. J. Inf. Technol. Dec. Mak. **13**(06), 1119–1133 (2014)
13. Kumari, A., Acherjee, B.: Selection of non-conventional machining process using CRITIC-CODAS method. Mater, Today: Proc. **56**, 66–71 (2022)
14. Rani, P., Mishra, A.R.: Single-valued Neutrosophic SWARA-TOPSIS-based group decision-making for prioritizing renewable energy systems. Comput. Dec. Mak. Int. J. **2**, 425–439 (2025)
15. Sałabun, W., Wątróbski, J., Shekhovtsov, A.: Are MCDA methods benchmarkable? A comparative study of TOPSIS, VIKOR, COPRAS, and PROMETHEE II methods. Symmetry **12**(9), 1549 (2020)
16. Suvitha, K., Saraswathy, R., Thilagasree, C.S., Kalaiselvan, S., Narayanamoorthy, S.: Designing and analyzing Fermatean fuzzy decision models: a comprehensive decision making approach. J. Intell. Dec. Mak. Inf. Sci. **2**, 272–288 (2025)
17. Tammi, I., Kalliola, R.: Spatial MCDA in marine planning: experiences from the Mediterranean and Baltic Seas. Mar. Policy **48**, 73–83 (2014)
18. Wang, Z., Rangaiah, G.P., Wang, X.: Preference ranking on the basis of ideal-average distance method for multi-criteria decision-making. Ind. Eng. Chem. Res. **60**(30), 11216–11230 (2021)
19. Więckowski, J., Kizielewicz, B., Shekhovtsov, A., Sałabun, W.: RANCOM: a novel approach to identifying criteria relevance based on inaccuracy expert judgments. Eng. Appl. Artif. Intell. **122**, 106114 (2023)

Author Index

A
Anitha, J. 228

B
Bahsoon, Rami 60
Bielecka, Marzena 48
Bielecki, Andrzej 48
Borisova, Julia 104

C
Cannataro, Mario 285
Cardoso, Pedro J. S. 119
Chard, Kyle 270
Chen, Huangxin 198, 213
Chen, Yanbin 135
Chen, Yuxiang 213
Cinaglia, Pietro 285

D
Das Sharma, Amartya 241
Deelman, Ewa 75
Dezert, Jean 297
Dhanyalakshmi, R. 228
Duraj, Agnieszka 167

F
Feng, Xiaoyu 198
Firkowski, Mateusz 3
Foster, Ian 270
Fujita, Kohei 183

G
Geng, Liru 17
Gepner, Pawel 75
Guzzi, Pietro Hiram 285

H
Hemanth, Jude 228
Hori, Muneo 183
Hu, Andong 33

I
Ichimura, Tsuyoshi 183
Ito, Hideaki 183

J
Jarmatz, Piet 241
Jiang, Zhengwei 256

K
Korzeń, Marcin 3
Kou, Jisheng 213
Koziel, Slawomir 90
Krawiec, Jerzy 75
Kuznetsov, Andrey 104

L
Lapegna, Marco 151
Leifsson, Leifur 90
Lemos, Marco 119
Li, Hao 256
Liu, Baoxu 256
Liu, Fengqi 17
Liu, Yinlong 17

M
Ma, Wei 17
Markidis, Stefano 33
Mendl, Christian B. 135
Milano, Marianna 285
Moroz, Leonid 75

N
Nakao, Kai 183
Neumann, Philipp 241
Nikitin, Nikolay O. 104

O
Olas, Mateusz 151
Olas, Tomasz 151

M. H. Lees et al. (Eds.): ICCS 2025, LNCS 15905, pp. 311–312, 2025.
https://doi.org/10.1007/978-3-031-97632-2

P
Peng, Ivy   33
Pennati, Luca   33
Pietrenko-Dabrowska, Anna   90
Poczekajlo, Pawel   75
Preuß, Hauke   241

R
Ren, Fangli   256
Rodrigues, João M. F.   119
Romanchuk, Natalia   228

S
Sałabun, Wojciech   297
Seidl, Helmut   135
Shang, Shang   256
Shekhovtsov, Andrii   297
Sklyar, Grigory   3
Solovev, Gleb   104
Suchorab, Aleksander   48
Sun, Jiyan   17
Sun, Shuyu   198, 213
Szczepaniak, Piotr S.   167
Szustak, Lukasz   151

T
Taufer, Michela   75
Theodoropoulos, Georgios   60
Tynes, Michael   270
Tziritas, Nikos   60

V
Vergara-Marcillo, Christian   60
Viot, Louis   241

W
Wang, Qiuyun   256
Wang, Zehui   256
Ward, Logan   270
Wijerathne, Lalith   183
Witkowski, Konrad   167
Wojnicki, Igor   48
Woźniak, Jarosław   3
Wyrzykowski, Roman   151

X
Xiao, Dunhui   213
Xu, Xuejun   213

Y
Yu, Bo   198
Yu, Haitao   213

Z
Zakharov, Alexander   228
Zhang, Tao   213
Zhang, Wei   17
Zhuang, Shangyuan   17
Zhuang, Xiaoying   213

The manufacturer's authorised representative in the EU is Springer
Nature Customer Service Centre GmbH, Europaplatz 3, 69115 Heidelberg,
Germany. If you have any concerns regarding our products, please
contact ProductSafety@springernature.com

Printed and bound by CPI Group (UK) Ltd, Croydon, CR0 4YY
29/04/2026
02099458-0008